LEAST ACTION PRINCIPLE OF CRYSTAL FORMATION OF DENSE PACKING TYPE

AND

KEPLER'S CONJECTURE

NANKAI TRACTS IN MATHEMATICS

Published

Vol. 1 Scissors Congruences, Group Homology and Characteristic Classes
by J. L. Dupont

Vol. 2 The Index Theorem and the Heat Equation Method
by Y. L. Yu

Vol. 3 Least Action Principle of Crystal Formation of Dense Packing Type and Kepler's Conjecture
by W. Y. Hsiang

Vol. 4 Lectures on Chern–Weil Theory and Witten Deformations
by W. P. Zhang

Nankai Tracts in Mathematics – Vol. 3

LEAST ACTION PRINCIPLE OF CRYSTAL FORMATION OF DENSE PACKING TYPE AND KEPLER'S CONJECTURE

Wu-Yi Hsiang

University of California, Berkeley, USA
Hong Kong University of Science & Technology, Hong Kong

World Scientific
New Jersey • London • Singapore • Hong Kong

Published by

World Scientific Publishing Co. Pte. Ltd.
P O Box 128, Farrer Road, Singapore 912805
USA office: Suite 1B, 1060 Main Street, River Edge, NJ 07661
UK office: 57 Shelton Street, Covent Garden, London WC2H 9HE

Library of Congress Cataloging-in-Publication Data
Hsiang, Wu Yi, 1937–
Least action principle of crystal formation of dense packing type and Kepler's conjecture / Hsiang, Wu-Yi. (Nankai Tracts in Mathematics – Vol. 3)
p. cm.
Includes bibliographical references.
ISBN 9810246706 (alk. paper)
1. Kepler's conjecture. 2. Sphere packings. 3. Crystallography, Mathematical. I. Title.

QA166.7 .H85 2001
511'.6--dc21 2001045504

British Library Cataloguing-in-Publication Data
A catalogue record for this book is available from the British Library.

Printed in Singapore by World Scientific Printers

To my wife

Myrtle Wan-Zen

Foreword

The geometry of spheres in three–dimensional Euclidean space has always been of central importance in the study of solid geometry. Its fundamental problem is the sphere–packing problem, which seeks to determine the *optimal density* of packings of spheres of identical size. This was first studied by Kepler in 1611 and he made a statement which can be interpreted as a conjecture that $\pi/\sqrt{18}$ *should be the optimal density of sphere packings.*

The physical shape of atoms can be regarded as microscopic spheres, while a piece of crystal of a monoatomic element often consists of trillions of such microscopic spheres aggregated into a specific type of regular arrangement. The crystal structures of forty-eight monoatomic elements are *hexagonal close packings* which achieve the highest known density of $\pi/\sqrt{18}$. Therefore, the general philosophical belief of the least action principle naturally leads to the study of whether hexagonal close packings are the *geometric consequence of the optimization of density.* Thus, the Kepler conjecture and the least action principle of crystal formation are a pair of closely related central issues of the sphere packing problem.

Hsiang's book approaches both problems from the geometrical viewpoint and contains a wealth of interesting, original results. One of the major steps of his approach is the reduction of both problems to the *optimal estimation* of a local invariant which only depends on the double-layer local surrounding of a given sphere, together with the determination of such local structures that achieve the optimality of that local invariant (i.e. the *locally averaged density*, cf. Chap. 1).

While the expansion of mathematics has extended into abstract domains, this book shows the great vitality of the tradition of classical geometry. I cannot help feeling that the subject, while basically elementary, will see a great number of developments.

S. S. Chern

Acknowledgment

This book is the revised version of the lecture notes for the graduate courses on sphere packings that I gave at the Hong Kong University of Science and Technology in the Fall of 1999 and again in the Spring of 2001. Most of the results presented here were obtained in the period of 1990–1994, supported by grants from the National Science Foundation of the United States. The support and hospitality of IHES and the University of Geneva for my visit in 1990 are gratefully acknowledged.

I would like to thank many people for their enthusiasm, criticism and comments on the sphere packing problem. Specifically, I would like to express my gratitude to J. Reed and C. M. Zong. The daily discussions on sphere packings with J. Reed in the Fall of 1999 and with C. M. Zong in the Spring of 2001 contributed to many of the improvements, clarifications and simplifications presented in this book. Special thanks are due to W. L. Lee for his great help in the preparation of the manuscript and computer graphics.

Contents

List of Symbols

$\mathcal{P}$ packing of spheres
$\mathcal{P}^*$ single-layer extension of $\mathcal{P}$
$\mathcal{P}^{**}$ double-layer extension of $\mathcal{P}$
$\mathcal{C}$ (finite) cluster of spheres
$\mathcal{C}^*$ single-layer extension of $\mathcal{C}$
$\mathcal{C}^{**}$ double-layer extension of $\mathcal{C}$
$\rho(S_j, \mathcal{P})$ local density of $\mathcal{P}$ at S_j
$\rho(\mathcal{P}, \mathcal{P}^*)$ relative global density of the first kind of $\mathcal{P}$ in $\mathcal{P}^*$
$\rho(\mathcal{P})$ intrinsic density of $\mathcal{P}$ of the first kind
$\overline{\rho}(S_j, \mathcal{P})$ locally averaged density of $\mathcal{P}$ at S_j
$\overline{\rho}(\mathcal{P}, \mathcal{P}^{**})$ relative global density of the second kind of $\mathcal{P}$ in $\mathcal{P}^{**}$
$\overline{\rho}(\mathcal{P})$ intrinsic global density of $\mathcal{P}$ of the second kind
$\rho(\{S\})$ optimal local density
$\overline{\rho}(\{S\})$ optimal locally averaged density
ρ_N optimal intrinsic density of the first kind for N-cluster of spheres
$\mathcal{L}(S, \mathcal{P})$ local packing of S in $\mathcal{P}$
$\mathcal{L}^2(S, \mathcal{P})$ double-layer local packing of S in $\mathcal{P}$
$\mathcal{L}'(S, \mathcal{P})$ reduced-packing of S in $\mathcal{P}$
$\widehat{\mathcal{L}}(S, \mathcal{P})$ core packing of S in $\mathcal{P}$
$C(S, \mathcal{P})$ local cell (Voronoi cell) of S in $\mathcal{P}$
$C'(S, \mathcal{P})$ reduced local cell of S in $\mathcal{P}$
$\Sigma(\mathcal{L}(S))$ directional profile of $\mathcal{L}(S)$

$\delta(\Sigma)$ minimal (spherical) separation of points in Σ
$I(\Sigma)$ inscribing (convex) polyhedron with Σ as the set of vertices
$T(\Sigma)$ tangent polyhedron with Σ as the set of touching points
$\mathcal{S}(\Sigma)$ spherical configuration of Σ
$\mathcal{S}'(\Sigma)$ triangulated spherical configuration of Σ
$\mathcal{S}^*(\Sigma)$ dual spherical configuration of Σ
$T(\Sigma, \{h_j\})$ generalized tangent polyhedron with buckling heights $\{h_j\}$
$\mathcal{S}^*(\Sigma, \{h_j\})$ spherical configuration obtained by radial projection of $T(\Sigma, \{h_j\})$
$T(\sigma)$ tangent subpolyhedron of $T(\Sigma)$ inside of the solid angle cone spanned by $\sigma \in \mathcal{S}(\Sigma)$
$f(\sigma)$ volume function of $T(\sigma)$ (when σ contains its own circumcenter)
$\overline{\sigma}_\triangle$ equilateral triangle with area $\triangle$
L_n lower bound volume estimate of $T(\Sigma)$ with n faces
$L(h)$ lower bound volume estimate of rectilinear slab with thickness $h \leq 0.3$
$St(A)$ star configuration at A in $\mathcal{S}(\Sigma)$
$St'(A)$ induced star configuration by $St(A)$ at A' on a neighboring sphere S'
$\Omega(St(A))$ complementary region of $St(A)$ in the central sphere S
$\Omega'(St'(A))$ complementary region of $St'(A)$ in the neighboring sphere S'
F_3, F_4, F_5 number of triangular, quadrilateral, pentagonal faces in a spherical configuration
$\triangle_a$ area of equilateral triangle with side-lengths a
$\square_a$ area of cocircular equilateral quadrilateral with side-lengths a
$\pentagon_a$ area of cocircular equilateral pentagon with side-lengths a
$\ell(h, h')$ lower bound value of edge-length with buckling heights h, h' at end points
$\triangle(h_1, h_2, h_3)$ area of triangle with side-lengths equal to $\ell(h_1, h_2)$, $\ell(h_2, h_3)$ and $\ell(h_3, h_1)$

$\square(h_1, h_2, h_3, h_4)$ area of (cocircular) quadrilateral with side-lengths equal to $\ell(h_1, h_2)$, $\ell(h_2, h_3)$, $\ell(h_3, h_4)$ and $\ell(h_4, h_1)$

$\sigma(A)$ spread of $St(A)$

Chapter 1
Introduction

1.1 Sphere Packings and the Sphere Packing Problem

Among all the kinds of geometric shapes, the sphere is clearly the most beautiful and useful. It is the only perfectly symmetric solid body and it is the unique solution of the isoperimetric problem in space. Most atoms and celestial bodies are spherical in shape. For example, the sun, the moon and the planets are visibly round spheres and the modern satellite-photo of mother earth is a beautiful spherical globe, while a piece of gold crystal, i.e. of microscopic spherical gold atoms is a natural example of sphere packing. In daily life, the packing of soybeans into a silo is an example of packing a large number of spheres of identical size into a huge container.

Basically, there are three different kinds of sphere packings, namely

1. sphere packings with containers (cf. [CFG]),
2. finite sphere packings without containers (cf. [BHW]),
3. infinite sphere packings (cf. [Fej-2], [Rog-2]).

For example, putting marbles into a jar, oranges into a box or soybeans into a silo are daily life examples of sphere packings *with containers*, while pieces of crystal of gold, silver, lead etc. are *naturally-created* examples of finite sphere packings *without* a container. On the other hand, *infinite sphere packings* (i.e. with the whole space as the container) are just *mathematical models*, the conceptual limiting situations of the other two kinds of finite sphere packings.

The central problem in the study of sphere packings is the problem of *optimal density*. Of course, it is necessary to first give a *precise definition* of the *concept of density* before studying the problem of optimal density. In the case of a sphere packing with a container, it is natural to define the dentisy of such a packing, $\mathcal{P} \subset \Omega$ (container), to be the ratio between the volumes of $\mathcal{P}$ and Ω, namely

$$\rho(\mathcal{P}, \Omega) = \frac{\mathrm{Vol}(\mathcal{P})}{\mathrm{Vol}(\Omega)}\,. \tag{1}$$

However, in the cases of the other two kinds of sphere packings, the precise, proper definition of the concept of density already becomes a rather complex and subtle problem in itself. Therefore, one of the foremost topics in the study of sphere packing problem is, in fact, the proper definition of the concept of densities for finite sphere packings *without a container* (resp. infinite sphere packings). Let us first discuss the case of infinite packings.

1.1.1 *Density of an infinite packing*

Intuitively, the whole space $\mathbf{E}^3$ can be regarded as the limit of a sequence of infinitely expanding regions. To be more precise, a sequence of compact regions with piecewise smooth boundary $\{\Omega_n\}$ is said to be an *admissible exhaustion sequence* of $\mathbf{E}^3$ if

$$\Omega_n \subset \Omega^\circ_{n+1}, \quad \mathbf{E}^3 = \bigcup\{\Omega_n\}, \quad \operatorname*{Lim}_{n\to\infty} \frac{\mathrm{Area}(\partial\Omega_n)}{\mathrm{Vol}(\Omega_n)} = 0\,. \tag{2}$$

Let $\mathcal{P}$ be a given infinite packing of unit spheres in $\mathbf{E}^3$ and $\{\Omega_n\}$ be such an exhaustion sequence of $\mathbf{E}^3$. Set $\mathcal{P}_n$ to be the maximal subpacking of $\mathcal{P}$ contained in Ω_n. If $\mathrm{Lim}_{n\to\infty}\,\rho(\mathcal{P}_n,\Omega_n)$ always exists and moreover, is independent of the choice of the exhaustion sequences, then it would be quite natural to *define* the density of such an infinite packing $\mathcal{P}$ to be that common limiting value. However, such a nice situation only occurs for very special packings such as the face-centered cubic lattice packing and the hexagonal close packings to be defined below. In general, firstly, the limit of $\{\rho(\mathcal{P}_n,\Omega_n)\}$ may not exist, and secondly, even for those admissible exhaustion sequences with converging $\{\rho(\mathcal{P}_n,\Omega_n)\}$, their limiting values will depend on the choice of $\{\Omega_n\}$, cf. examples of such packings in Chap. 3. Since we are mainly interested in the problem of "optimal density," we can define the concept of density of a given infinite packing $\mathcal{P}$ in the following simple-minded way by taking the *least upper bounds* at every stage, namely

$$\rho(\mathcal{P}) = l.u.b.\left\{ \overline{\lim_{n\to\infty}}\, \rho(\mathcal{P}_n,\Omega_n) \right\} \tag{3}$$

where $\{\Omega_n\}$ run through all possible admissible exhaustion sequences of $\mathbf{E}^3$. The definition of (3) is, of course, a precise mathematical definition of the concept of density for infinite packings in $\mathbf{E}^3$ (cf. [Fej-2], [Hla], [Rog-2]). However, the density $\rho(\mathcal{P})$ of (3) is, in general, far from representative of the geometric intuition of the averaged density over the whole space. Anyhow, we shall use the above definition to provide a precise mathematical formulation of both the sphere packing problem and Kepler's Conjecture.

The sphere packing problem *What is the least upper bound of the densities of all possible infinite sphere packings in* $\mathbf{E}^3$, *namely*

$$l.u.b.\{\rho(\mathcal{P}), \mathcal{P} \subset \mathbf{E}^3\} = \,? \tag{4}$$

1.2 Kepler's Conjecture on Sphere Packings

One of the earliest publications on sphere packings occurred in a Latin booklet in 1611 by J. Kepler, entitled "A new year's gift — on six cornered snow-flakes" (see [Kep] for an English translation). He began with some discussions on the hexagonal structure of snowflakes and honeycombs, and then moved on to study the geometry of sphere packings, such as the number of touching neighbors and the shape of local cells for various examples of regular packings. Among them, he found that a special example of such regular arrangements achieved the maximal number of *twelve touching neighbors*, which is nowadays called *face-centered cubic lattice* packing (or simply f.c.c. packing). Then, he boldly asserted that this should be the one which would achieve the "optimal density," namely, in his own words:

"The (f.c.c.) packing will be the tightest possible, so that in no other arrangement could more pellets be stuffed into the same container."

Notice that he did *not* specify the *shape* of the container, nor the relative size between the container and the pellets. In fact, he *neither* gave a definitive formulation, *nor* any intuitive reasoning for the above bold assertion. He simply made such a very far reaching statement on sphere packings and moved on to discuss other issues.

In terms of the modern setting of coordinate geometry, the packing that Kepler "conjectured" to be the tightest possible can be represented as the packing consisting of the spheres of radii $1/\sqrt{2}$, centered at points with *integral coordinates of even sum*, namely, the f.c.c. lattice

$$\mathcal{L} = \{(x, y, z); x, y, z \in \mathbf{Z} \text{ and } x + y + z \text{ even}\}. \tag{5}$$

The whole space can be decomposed into the tiling of $2 \times 2 \times 2$-cubes with vertices of even coordinates. As indicated in Figure 1.1, each cube contains eight octant-spheres and six half-spheres and hence

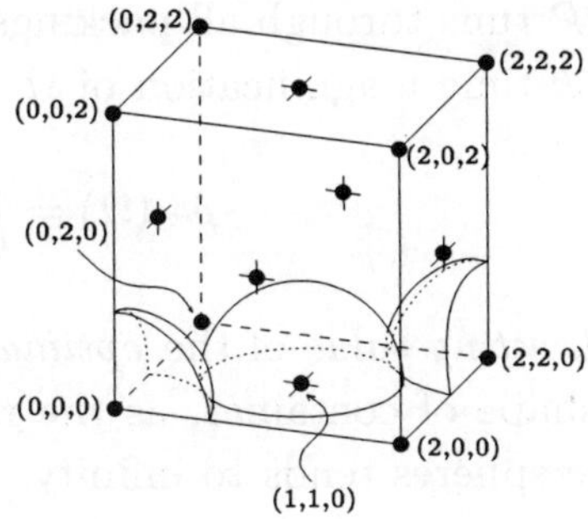

Figure 1.1

a *sum total of four whole spheres.* Thus, the density of the f.c.c. packing is equal to

$$4 \cdot \frac{4\pi}{3} \left(\frac{1}{\sqrt{2}} \right)^3 : 8 = \frac{\pi}{\sqrt{18}} = 0.740480 \cdots . \tag{6}$$

Therefore, Kepler's bold assertion of 1611 amounts to the following:

Kepler's Conjecture $\pi/\sqrt{18}$ *should be the optimal density for all possible sphere packings in Euclidean 3-space.*

1.2.1 *Another mathematical formulation of the sphere packing problem and Kepler's Conjecture*

We mention here another formulation of the sphere packing problem and Kepler's Conjecture, which provides the natural link between the problem of optimal density for finite packings *with containers* and that of the infinite packings in $\mathbf{E}^3$. For a given *shape* of container and a given *relative size*, say packings of *unit spheres* into a given container Ω, the optimal density is defined to be

$$\hat{\rho}(\Omega) = l.u.b.\{\rho(\mathcal{P}, \Omega)\} \tag{7}$$

where $\mathcal{P}$ runs through all packings of unit spheres into Ω. Set $k\Omega$ to be the k-time magnification of Ω. Then

$$\hat{\rho}_\infty(\Omega) = \overline{\lim_{k\to\infty}}\, \hat{\rho}(k\Omega) \tag{8}$$

is the *limiting value* of the *optimal density* of sphere packings into a given shape of container, as the relative size between the container and the spheres tends to infinity.

The sphere packing problem (Another formulation) *Is the limiting value, $\hat{\rho}_\infty(\Omega)$, of (8) independent of the shape of Ω? If so, what is the exact value of this universal constant of $\mathbf{E}^3$? (See* [Hla] *for the case of convex Ω.)*

Kepler's Conjecture (Another formulation) *The $\hat{\rho}_\infty(\Omega)$ is a universal constant of $\mathbf{E}^3$ (i.e. independent of the shape of Ω) and it is equal to $\pi/\sqrt{18}$.*

1.3 Density of Finite Packings without a Container and the Least Action Principle of Crystal Formation

The physical shape of atoms can be regarded as microscopic spheres while a small piece of crystal of a given mono-atomic element, such as gold, silver, lead etc., often consists of billions of trillions of such microscopic spheres (of identical sizes) aggregated into a specific type of regular arrangement. Thus, crystals are *naturally-created* examples of *finite sphere packings without containers*, which exhibit fascinating geometric regularity and remarkable precision. Upon close examination of the crystal structures of those eighty-one elements with mono-atomic stable form, one finds that forty-eight of them are of the type of *hexagonal close packings*, which are sphere packings of the highest known density of $\pi/\sqrt{18}$.

1.3.1 *Hexagonal close packings*

In his booklet of 1611, Kepler did *not* describe f.c.c. packing by the previous coordinate representation, notice that 1611 was decades before the invention of the Cartesian coordinate system. He constructed f.c.c. packing by stacking up layers of "planary" packings, just as one stacks up a large number of oranges or cannon balls. He pointed out that there are two ways of constructing f.c.c. packing, namely, either by closely stacking up layers of square arrangements (cf. Figure 1.2(i)) or by closely stacking up layers of triangular arrangements (cf. Figure 1.2(ii)). The natural coincidence of the above two different ways of stacking spheres was, in fact, admired by Kepler as a kind of natural wonder which might also play a role in convincing him that f.c.c. packing must be *nature's choice* of being the tightest packing.

However, Kepler actually overlooked something rather simple, which was pointed out by Barlow in 1883 [Bar]. Although there is a unique way of closely stacking up layers of square arrangements, there are, in fact, *infinitely many geometrically distinct* ways of closely stacking up layers of triangular arrangements.

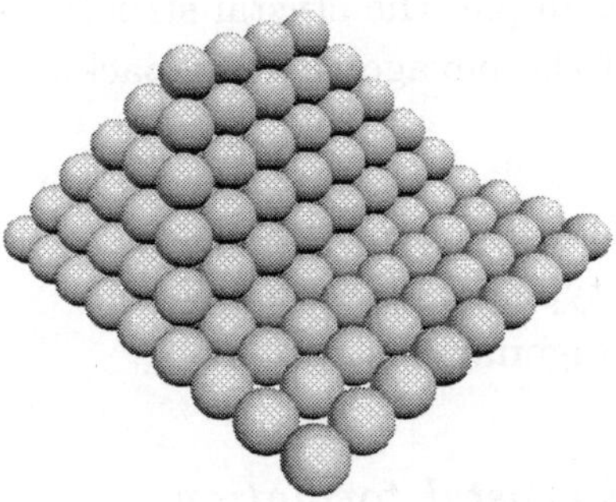

(i)

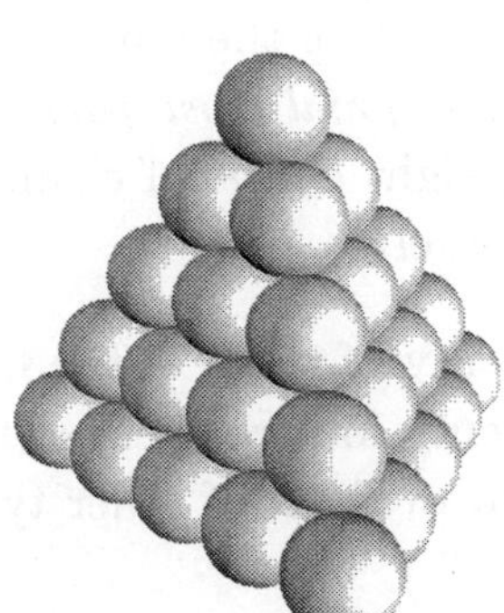

(ii)

Figure 1.2

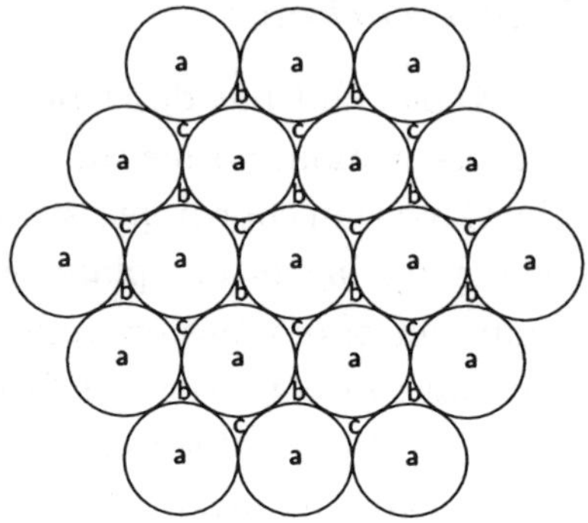

Figure 1.3

As indicated in Figure 1.3, the relative positions between layers of triangular arrangement can be specified by a, b and c while two *adjacent layers* of close stacking must occupy different ones. Therefore, such a stacking can be labeled by an infinite sequence of $\{a, b, c\}$ with distinct adjacent ones, while two such stackings are geometrically the same (resp. distinct) if and only if their labeling sequences are the same (resp. distinct). For example, only the $\{\cdots abcabcabc\cdots\}$-stacking will give f.c.c. packing, while the other infinitely many distinct stackings are, in fact, *non-lattice* packings. Of course, all of them have the same density of $\pi/\sqrt{18}$, and they are called *hexagonal close packings.* For example, the crystal structures of forty-eight chemical elements are of the hexagonal close packings type, namely

(i) twenty of them of the f.c.c. type, i.e. $\{\cdots abcabc\cdots\}$,
(ii) twenty-four of them of the h.c.p. type, i.e. $\{\cdots ababab\cdots\}$,
(iii) four of them of other types of hexagonal close packings.

1.3.2 *Least action principle and crystal formation*

The least action principle in its broad generality is actually a kind of philosophical belief that Nature always chooses the "best" among all possibilities. If one applies such a general principle to the study of the underlying mechanism of crystal formation, then the crystal structure of those forty-eight elements can be regarded as a strong assurance of Nature that $\pi/\sqrt{18}$ should be the optimal density of sphere packing. This, of course, greatly enhances our confidence in the validity of Kepler's Conjecture. However, it does *not*, in anyway, lessen the need for providing a mathematical proof of Kepler's Conjecture. In fact, in order to establish the *applicability* of the least action principle to the realm of crystal formation, one needs to prove a much stronger result which not only proves the *optimality* of $\pi/\sqrt{18}$ for the density of finite packings, but also proves that a finite cluster of spheres with the optimal density of $\pi/\sqrt{18}$ is necessarily a *hexagonal cluster* (cf. Sec. 1.5 for its definition). Only when such a strong *uniqueness theorem of finite packings with optimal density* is proven, can we say that the geometric regularity of those crystals of hexagonal close packing type is, indeed, the consequence of the *optimization of density.*

Notice that such a strong uniqueness theorem *cannot* hold in the setting of infinite packings because the omission of an arbitrary finite subset in a given infinite packing will not change its density. Therefore, it is more important to study the *problem of optimal density* in the setting of *finite packings without a container* than that of the setting of infinite packings. However, unlike the case of a finite packing with a given container, the concept of density of a finite packing *without* a container is something *yet to be defined properly*, mainly due to the lack of an obvious choice for the denominator, such as the volume of the given container in the former case. Therefore, the foremost issue in the study of finite packings without a container is the *search for a proper definition of the (intrinsic) density of a finite packing without a container.*

In the philosophical belief of the least action principle, the very existence of many crystal structures of the hexagonal close packing type can be regarded as Nature's assurance of the existence of a workable global density (for finite packings without a container) while our task is just to find such a workable one. In fact, a moment of further philosophical reflection suggests that there is another message from Nature, namely, such a workable one must be *highly localizable*, meaning that it should be the *weighted average* of a certain *local invariant* and moreover, the *optimization of the global density* should be the consequence of the *optimization* of *that local invariant everywhere.* Therefore, the *geometric localization* of sphere packings is a basic step that will provide the needed geometric structures and insights for the proper definition of such a workable global density of finite packings without a container together with the corresponding local invariant.

1.3.3 *Local cell decomposition, local density and the Dodecahedron Conjecture*

In his booklet of 1611, Kepler had already introduced the basic concept of the local cell, which is nowadays often called the Voronoi cell, thus: to each given sphere S_i in a given packing $\mathcal{P}$, the local cell of S_i in $\mathcal{P}$, denoted by $C(S_i, \mathcal{P})$, is defined to be the set of points that are as close to the center of S_i as to the centers of others. Let S_j be another sphere in $\mathcal{P}$ and O_i, O_j be the centers of S_i, S_j, respectively. Then the set of points that are as close to O_i as to O_j is a *half-space* $H_{i,j}$ with the perpendicular bisector of $\overline{O_iO_j}$ as its boundary. Therefore $C(S_i, \mathcal{P})$ is by definition the intersection of all such half spaces, namely

$$C(S_i, \mathcal{P}) = \bigcap_{j \neq i} H_{i,j}\,. \tag{9}$$

In the case of bounded $C(S_i, \mathcal{P})$, the above infinite intersection can always be reduced to that of a finite irreducible one

$$C(S_i, \mathcal{P}) = \bigcap_{j \neq i} H_{i,j} = H_{i,j_1} \cap \cdots \cap H_{i,j_a}\,, \tag{9'}$$

where $\{j_1, \ldots, j_a\}$ is such a minimal subset of indices that makes the above equality hold. Note that the index j belongs to this minimal subset if and only if $C(S_j, \mathcal{P})$ and $C(S_i, \mathcal{P})$ have a *common face.* Such a pair of spheres are *defined to be neighbors of one another.* Anyhow, local cell and neighbor are the two most basic concepts in sphere packings that all the other important geometric invariants of sphere packings are based upon. First of all, the following local invariant is clearly of fundamental importance.

Definition of local density The local density of $\mathcal{P}$ at S_i is defined to be the ratio between the volumes of the sphere and its local cell, namely

$$\rho(S_i, \mathcal{P}) = \mathrm{vol}\, S_i / \mathrm{vol}\, C(S_i, \mathcal{P}) . \tag{10}$$

For example, in the case that $\mathcal{P}$ is the f.c.c. (resp. h.c.p.) packing, all of its local cells are congruent to each other and all local densities $\{\rho(S_i, \mathcal{P})\}$ are equal to $\pi/\sqrt{18}$. In the case that $\mathcal{P}$ is another hexagonal close packing (i.e. different from f.c.c. and h.c.p.), it has two types of local cells; one of them is the same as that of f.c.c. while the other is the same as that of h.c.p., and moreover, the local cells of the spheres of the same layer are always of the same type. The sphere packings of crystals of those forty-eight elements are *sub-packings* of a hexagonal close packing.

Definition of relative density of the first kind Let $\mathcal{P}$ be a finite subpacking of $\mathcal{P}^*$. Suppose that all the local cells $\{C(S_i, \mathcal{P}^*), S_i \in \mathcal{P}\}$ are bounded. Then the relative density $\rho(\mathcal{P}, \mathcal{P}^*)$ is defined to be

$$\rho(\mathcal{P}, \mathcal{P}^*) = \frac{\mathrm{vol}(\mathcal{P})}{\sum \mathrm{vol}\, C(S_i, \mathcal{P}^*)} . \tag{11}$$

Definition of intrinsic density of the first kind For a given finite packing $\mathcal{P}$ without a container, the intrinsic density of $\mathcal{P}$ (of the first kind) is defined to be the least upper bound of all those relative densities $\{\rho(\mathcal{P}, \mathcal{P}^*)\}$ where $\mathcal{P}^*$ runs through all possible extensions of $\mathcal{P}$, namely

$$\rho(\mathcal{P}) = l.u.b.\{\rho(\mathcal{P}, \mathcal{P}^*)\}\,. \tag{12}$$

In the very special case that $\mathcal{P}$ consists of a single sphere, the above intrinsic density of the first kind is, by definition, the *optimal local density*, namely

$$\rho(\{S\}) = l.u.b.\{\rho(S, \mathcal{L}(S))\}\,, \tag{13}$$

where $\mathcal{L}(S)$ runs through all possible local packings surrounding a single sphere. Of course, the determination of the optimal local density is an important basic problem for which one has the following Dodecahedron Conjecture.

The Dodecahedron Conjecture (See [Fej-2]) *The smallest local cell is a regular circumscribing dodecahedron, or equivalently, the local packing $\mathcal{L}(S)$ with the optimal local density surrounds the central sphere by twelve symmetrically arranged touching neighbors.* (*cf. Chap.* 6 *for a proof of the above conjecture. We refer to* [Bli], [Gau], [Lin2], [Mud], [Ran] *and* [Rog-1] *for results on upper bound estimates of local density down to* 0.7731.)

Note that the volume of a regular circumscribing dodecahedron of the unit sphere is given by

$$20\left(1 - 2\cos\frac{2\pi}{5}\right)\tan\frac{\pi}{5} = 5.5502910\cdots \tag{14}$$

and hence, the local density of the above local packing is equal to

$$\frac{4\pi}{3} : 20\left(1 - 2\cos\frac{2\pi}{5}\right)\tan\frac{\pi}{5} = 0.754697\cdots\,, \tag{14$'$}$$

which is considerably larger than $\pi/\sqrt{18} = 0.740480\cdots$. Moreover, a regular dodecahedron is *not* a tiling domain of $\mathbf{E}^3$. Anyhow, the *optimization* of the above *intrinsic density of the first kind cannot* serve as the least action principle of crystal formation, while the optimization of local density at one sphere always makes the local densities of all its twelve neighbors much smaller than $\pi/\sqrt{18}$. Thus, we must search for some other local invariants that will work. This leads to the discovery of the *locally averaged density* and the definition of the *intrinsic density of the second kind.*

1.4 Locally Averaged Density and the Intrinsic Density of the Second Kind

Geometrically, the local density is the simplest local invariant which only depends on the *single-layer coating* (or surrounding) of a sphere. The next simplest types of local invariants are those which only depend on the *double-layer coating* of a sphere. Thus, it is natural to search for a workable local invariant in the realm of this type, if it exists. The locally averaged density of [Hsi-2] is exactly such a workable local invariant.

Definition A pair of spheres $\{S_i, S_j\}$ are said to be close neighbors of each other if their center-distance is at most equal to 2.18-times of the radii. The core packing of S_i in $\mathcal{P}$, $\widehat{\mathcal{L}}(S_i, \mathcal{P})$, consists of S_i and all its close neighbors in $\mathcal{P}$.

Set

$$\begin{gathered}\widehat{\mathcal{L}}(S_i, \mathcal{P}) = \{S_{ij};\ 1 \le j \le \#\widehat{\mathcal{L}}(S_i, \mathcal{P})\}\,, \\ \lambda_{ij} = 13/\#\widehat{\mathcal{L}}(S_{ij}, \mathcal{P})\,, \quad w_{ij} = \lambda_{ij} \operatorname{vol} C(S_{ij}, \mathcal{P})\,. \end{gathered} \tag{15}$$

Definition The locally averaged density of S_i in $\mathcal{P}$ is defined to be the following weighted average:

$$\overline{\rho}(S_i, \mathcal{P}) = \sum_j w_{ij} \rho(S_{ij}, \mathcal{P}) \Big/ \sum_j w_{ij}\,. \tag{15$'$}$$

A finite packing (without a container), $\mathcal{C}$, is called a *cluster* if any pair of them can be linked through neighboring pairs with center-distances less than $2\sqrt{2}$ times the radii. An extension, $\mathcal{C}^*$, of $\mathcal{C}$ is called a *saturated coating* of $\mathcal{C}$ if all spheres of $\mathcal{C}^*$ are neighbors of some sphere in $\mathcal{C}$ and it is impossible to add any more such neighbors of $\mathcal{C}$ to $\mathcal{C}^*$. $\mathcal{C}^{**}$ is a double-layer coating of $\mathcal{C}$ if $\mathcal{C}^{**}$ is a single-layer coating of $\mathcal{C}^*$, which is itself a saturated single-layer coating of $\mathcal{C}$.

Definition of relative global density of the second kind Let $\mathcal{C}$ be a cluster and $\mathcal{C}^{**}$ be a double-layer coating of $\mathcal{C}$. Then

$$\overline{\rho}(\mathcal{C},\mathcal{C}^{**}) = \sum_{S_i \in \mathcal{C}} \overline{w}_i \overline{\rho}(S_i,\mathcal{C}^{**}) \Big/ \sum_{S_i \in \mathcal{C}} \overline{w}_i\,, \tag{16}$$

where

$$\overline{w}_i = \sum_j w_{ij} = \sum_j \lambda_{ij} \operatorname{vol} C(S_{ij},\mathcal{C}^{**})\,. \tag{16'}$$

Definition of intrinsic global density of the second kind Let $\mathcal{C}$ be a cluster and $\mathcal{C}^{**}$ be a double-layer coating of $\mathcal{C}$. Then

$$\overline{\rho}(\mathcal{C}) = l.u.b.\{\overline{\rho}(\mathcal{C},\mathcal{C}^{**})\}\,,$$

where $\mathcal{C}^{**}$ runs through all possible double-layer coatings of $\mathcal{C}$.

In the very special case that $\mathcal{C} = \{S\}$ is a single sphere, $\overline{\rho}(\{S\})$ is, by definition, the *optimal locally averaged density.*

Question *What is $\overline{\rho}(\{S\})$ equal to and what are those double-layer local packing $\mathcal{L}^2(S)$ that achieves the optimal locally averaged density?*

1.5 Main Theorems on Sphere Packings

The main results of [Hsi-2] on sphere packings are the following:

Main Theorem I $\overline{\rho}(\{S\}) = \pi/\sqrt{18}$ *and* $\overline{\rho}(S,\mathcal{L}^2(S)) = \pi/\sqrt{18}$ *if and only if $\mathcal{L}^2(S)$ is isometric to the double-layer local packing of a certain hexagonal close packing.*

Definition A cluster $\mathcal{C}$ is called a *hexagonal cluster* if $\mathcal{C}$ is the *close stacking* of *planary layers* of subpackings $\{\mathcal{C}_n\}$, each of them consisting of *planary hexagonal clusters.*

Remark A hexagonal cluster C is a subcluster of a global hexagonal packing when and only when *all* the hexagonal clusters of a *given planary layer*, C_n, of C are of the same $\{a, b, c\}$-type.

Main Theorem II (Least Action Principle) *Let C be a (finite) cluster of spheres. Then*

$$\overline{\rho}(C) \leq \pi/\sqrt{18}$$

and the equality holds if and only if C has a double-layer coating C^{**} which is a hexagonal cluster.

The above *optimal estimates* on both the locally averaged density and the intrinsic global density of the second kind, together with the strong uniqueness on the geometric structures that achieve the optimal densities, provide the mathematical understanding of the least action principle of crystal formation, namely, the fact that *geometric regularity* of crystals of hexagonal close packings is, indeed, a consequence of the *optimization of density* and moreover, that the optimization of the global density is also a consequence of the optimization of the locally averaged density *everywhere*!

Let $\mathcal{P}$ be an infinite packing in $\mathbf{E}^3$ and $\mathcal{P}_n$ be a given finite subpacking of $\mathcal{P}$. A sphere $S_i \in \mathcal{P}_n$ is said to be an *interior sphere* of $\mathcal{P}_n$ if all the neighbors of S_i in $\mathcal{P}$ are already in $\mathcal{P}_n$, otherwise S_i is said to be a *boundary sphere* of $\mathcal{P}_n$. Set $N(\mathcal{P}_n)$ (resp. $N'(\mathcal{P}_n)$) to be the number of spheres (resp. boundary spheres) in $\mathcal{P}_n$.

Definition An exhaustion sequence $\{\mathcal{P}_n\}$ of $\mathcal{P}$ is called *admissible* if all spheres of $\mathcal{P}_n$ are interior sphere of $\mathcal{P}_{n+1}$ and moreover

$$\operatorname*{Lim}_{n\to\infty} N'(\mathcal{P}_n)/N(\mathcal{P}_n) = 0\,. \tag{17}$$

The global density of $\mathcal{P}$ is *redefined* to be the following least upper bound, namely

$$\rho(\mathcal{P}) = l.u.b.\left\{ \overline{\operatorname*{Lim}_{n\to\infty}}\, \rho(\mathcal{P}_n, \mathcal{P}_{n+1}) \right\}, \tag{18}$$

where $\{\mathcal{P}_n\}$ runs through all admissible exhaustion sequences of $\mathcal{P}$.

Main Theorem III (Kepler's Conjecture)

$$\rho(\mathcal{P}) \leq \frac{\pi}{\sqrt{18}} \tag{19}$$

for all possible infinite packings in $\mathbf{E}^3$.

1.5.1 *Relationship between the two kinds of relative densities and the proof of Kepler's Conjecture*

On the one hand, it follows directly from Main Theorem I that $\overline{\rho}(\mathcal{P}_n, \mathcal{P})$ is at most equal to $\pi/\sqrt{18}$, and on the other hand, the conjecture of Kepler asserts that

$$\overline{\operatorname{Lim}}_{n\to\infty} \rho(\mathcal{P}_n, \mathcal{P}) \leq \frac{\pi}{\sqrt{18}}$$

for any admissible exhaustion sequence of $\mathcal{P}$. Therefore, the following relationship between the two kinds of relative densities enable us to derive Kepler's Conjecture as a direct consequence of Main Theorem I.

Set I_n (resp. I_n°, I_n') to be the index sets of spheres (resp. interior, boundary spheres) of $\mathcal{P}_n$. Then

$$\begin{aligned}
0 &< \sum_{i\in I_n} \overline{w}_i \overline{\rho}(S_i, \mathcal{P}) - 13 \sum_{i\in I_n^\circ} \operatorname{vol}(S_i)\,, \\
&\sum_{i\in I_n} \overline{w}_i \overline{\rho}(S_i, \mathcal{P}) - 13 \sum_{i\in I_n} \operatorname{vol}(S_i) \leq \sum_{i\in I_n'} \overline{w}_i \overline{\rho}(S_i, \mathcal{P})\,, \\
0 &< \sum_{i\in I_n} \overline{w}_i - 13 \sum_{i\in I_n^\circ} \operatorname{vol} C(S_i, \mathcal{P})\,, \\
&\sum_{i\in I_n} \overline{w}_i - 13 \sum_{i\in I_n} \operatorname{vol} C(S_i, \mathcal{P}) \leq \sum_{i\in I_n'} \overline{w}_i\,.
\end{aligned} \tag{20}$$

Hence

$$-13 \sum_{i\in I_n'} \operatorname{vol}(S_i) \leq \sum_{i\in I_n} \overline{w}_i \overline{\rho}(S_i, \mathcal{P}) - 13 \sum_{i\in I_n} \operatorname{vol}(S_i) \leq \sum_{i\in I_n'} \overline{w}_i \overline{\rho}(S_i, \mathcal{P})\,,$$

$$-13 \sum_{i \in I_n'} \operatorname{vol} C(S_i, \mathcal{P}) \leq \sum_{i \in I_n} \overline{w}_i - 13 \sum_{i \in I_n} \operatorname{vol} C(S_i, \mathcal{P}) \leq \sum_{i \in I_n'} \overline{w}_i \,. \quad (21)$$

Therefore, it follows from the *admissible condition* of $\{\mathcal{P}_n\}$ and the assumption that the volumes of all local cells of a given $\mathcal{P}$ are uniformly bounded that

$$\lim_{n \to \infty} \frac{\sum_{i \in I_n} \overline{w}_i \overline{\rho}(S_i, \mathcal{P}) - 13 \sum_{i \in I_n} \operatorname{vol}(S_i)}{\sum_{i \in I_n} \overline{w}_i \overline{\rho}(S_i, \mathcal{P})} = 0 \,,$$

$$\lim_{n \to \infty} \frac{\sum_{i \in I_n} \overline{w}_i - 13 \sum_{i \in I_n} \operatorname{vol} C(S_i, \mathcal{P})}{\sum_{i \in I_n} \overline{w}_i} = 0 \,.$$

Hence

$$\overline{\lim_{n \to \infty}}\, \rho(\mathcal{P}_n, \mathcal{P}) = \overline{\lim_{n \to \infty}}\, \overline{\rho}(\mathcal{P}_n, \mathcal{P}) \leq \frac{\pi}{\sqrt{18}} \,.$$

In summary, both the least action principle of crystal formation and the proof of Kepler's Conjecture can be reduced to the proof of Main Theorem I, which amounts to the *optimal estimation* of the fundamental local invariant — the *locally averaged density* — together with determination of the *tightest double-layer local packing.* The central topic of this book is to present a complete proof of Main Theorem I, based upon the overall structure outlined in the 1993 paper [Hsi-2] and its addendum 1995 [Hsi-3]. As one might expect, the whole journey of going through such a proof, no matter how well one tries to organize it, is necessarily long and laborious! In this book, we try to make such a long journey enjoyable and such hard work worthwhile. This is what a well-organized presentation of the least action principle of crystal formation and the proof of Kepler's Conjecture can and should achieve.

Chapter 2

The Basics of Euclidean and Spherical Geometries and a New Proof of the Problem of Thirteen Spheres

Among all kinds of geometries we know so far, or whatever one may invent in the future, the geometry of Euclidean 3-space must be the most natural and the most important one. For all practical purposes, this is the space in which we live, while for all theoretical geometric studies, Euclidean geometry is and will always be the foundation to be based upon. For example, Euclidean spaces are exactly the infinitesimal model for the local structure of Riemannian manifolds.

From a modern viewpoint, all the geometric properties of Euclidean 3-space are encoded in its *structure preserving transformation group*, namely the *Euclidean group*, which is the semi-direct product of the *translation group and the orthogonal group.* The former is isomorphic to the 3-dimensional vector group over $\mathbf{R}$, and it acts simple-transitively over Euclidean 3-space, while the latter consists of the conjugacy class of *isotropy subgroups* of the transformation group, each of them acting with *orbits* of *concentric spheres* around its unique fixed point. Roughly speaking, the translation group represents the "*linearity*" of Euclidean space, while the geometry of spheres can be regarded as the manifestation of the "*orthogonality*" of

Euclidean space, namely, the geometric properties which are encoded in the orthogonal transformation group. Thus, *spherical geometry* is, actually, a part of Euclidean geometry. On the other hand, the whole Euclidean geometry can be regarded as the study of *orthogonal invariants* of the 3-dimensional vector group over $\mathbf{R}$, which clearly reveals the central importance of spherical geometry in the study of Euclidean geometry. We refer to [BöH] for the study of high dimensional volume of spherical polyhedra.

2.1 Vector Algebra and Basic Spherical Geometry

Two spheres of the same radius are obviously congruent to each other while a homothety of scaling factor k maps a sphere of radius r to a sphere of radius kr. Thus, all spheres are "*similar*" to one another (i.e. of the same shape), while their *sizes* are determined by their radii. Therefore, the geometry of spherical surfaces can easily be reduced to the normalized case of unit radius, often referred to as the *unit sphere.* The unit sphere is the natural *moduli space* of all possible *directions* in the space, and it is for this reason spherical geometry is the basic mathematical tool in the *quantitative study of astronomy.* For example, it was exactly motivated by such a need that basic spherical trigonometry had already been systematically developed by the successive efforts of Hipparchus (2nd Century B.C.), Menelous (1st Century A.D.) and Ptolemy (2nd Century A.D.). Menelous' *Sphaerica* and Ptolemy's *Almagest* represent their remarkable contributions to *spherical trigonometry* and *astronomy.* It is interesting to note that spherical trigonometry was, in fact, developed many centuries ahead of the simpler and more practical Euclidean trigonometry because the geometers of antiquity were much more enthusiastic about the study of the geometry of the heavens (i.e. astronomy) than in the practical surveying of small pieces of land.

Algebraically, the unit sphere can be systematically represented by the set of unit vectors (i.e. vectors of unit length), thus making vector algebra a powerful tool for studying spherical geometry. In this section, we shall choose such a modern approach to develop the basic theory of spherical geometry. From a global overview of geometry, spherical geometry is, of course, just a part of Euclidean geometry. In view of the fact that spheres are exactly the orbits of the orthogonal groups, spherical geometry studies the part of Euclidean geometry that is *encoded in the orthogonal* group. Somehow, the more essential and deeper aspects of solid geometry are often contained in this part.

2.1.1 *Basic properties of the unit sphere*

Let $S^2(1)$ be the sphere of unit radius centered at the origin O. The position vectors of points on $S^2(1)$ are exactly the set of unit vectors. It is an *orbit* of the orthogonal group, $O(3)$, fixing the origin.

(i) **Reflectional symmetries** Let π be a plane passing through the center O. The reflectional symmetry of the whole space with respect to π is an order 2 element of $O(3)$ fixing all points of π. Thus, its restriction to $S^2(1)$ is an order 2 *isometry* fixing all points of the intersection of π and $S^2(1)$, a *great circle* on $S^2(1)$. We shall call this isometry the *reflectional symmetry of* $S^2(1)$ *with respect to that great circle.*

(ii) **Rotational symmetries** Let l be a line passing through the center O. The intersection of l and $S^2(1)$ consists of a pair of points that are *antipodal* to each other, such as the north and the south poles on the earth. The restriction to $S^2(1)$ of the rotational symmetry of the whole space with l as the axis, is called the *rotational symmetry of* $S^2(1)$ *with respect to that pair of antipodal points.*

(iii) **Spherical polar coordinate system** Let N, S be a given pair of antipodal points and $SO(2)$ be the rotation group with the line NS as the axis. Then the orbits of $(SO(2), S^2(1))$ are the "*latitude*"

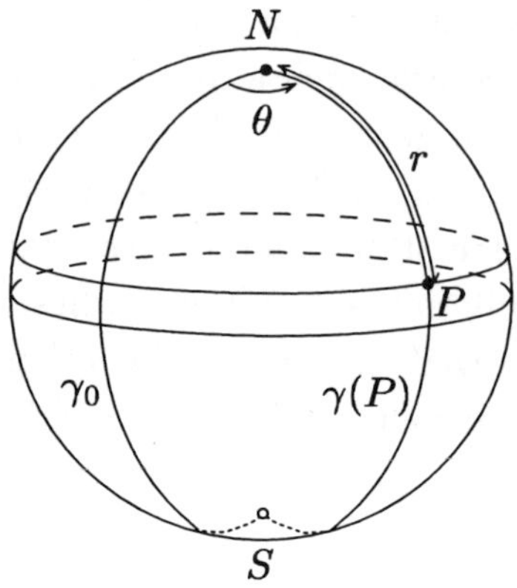

Figure 2.1

circles which are everywhere perpendicular to all the great circles passing through N and S, i.e. to the "*longitude*" *arcs* (cf. Figure 2.1).

Let P be any other point on $S^2(1)$ and $\gamma(P)$ be the unique longitude arc passing through P. Set r to be the arc length of $\overset{\frown}{NP}$ and θ to be the angle between $\gamma(P)$ and a fixed reference longitude arc γ_0. Then (r, θ) is called the *polar coordinates* of P with respect to the *spherical polar coordinate system* on $S^2(1)$ *with* N *as the base point and* γ_0 *as the reference longitude arc.* If we choose the Cartesian coordinate system of the space with $\overrightarrow{ON}$ along the z-axis and use the the longitude arc passing through (1, 0, 0) as the reference one, then the x, y, z-coordinates of the point P with (r, θ) as its spherical polar coordinates are given as

$$x = \sin r \cos \theta\,, \quad y = \sin r \sin \theta\,, \quad z = \cos r\,. \tag{1}$$

Hence,

$$ds^2 = dx^2 + dy^2 + dz^2 = dr^2 + \sin^2 r d\theta^2\,. \tag{2}$$

It follows easily from (2) that the longitude arc $\overset{\frown}{NP}$ $(P \neq S)$ is the *unique shortest path* linking N and P, thus proving the unique existence of the shortest path linking a given pair of *non-antipodal* points, which is the shorter one of the great circle arcs linking them.

(iv) **Archimedes' Theorem** *The total area of* $S^2(1)$ *is equal to* 4π.

Historical Remark Archimedes (287–212 B.C.), the greatest scientist and mathematician of antiquity, had many outstanding achievements both in science and in geometry. Among them, that which he was certainly most proud of, was his proof of the area formula of the sphere, namely

$$\begin{aligned}\text{Area}\,(S^2(R)) &= \text{Area (circular cylinder of radius } R \text{ and height } 2R)\\ &= 4\pi R^2 = 4\cdot\text{Area (circle of radius } R)\,. \qquad (3)\end{aligned}$$

This is the famous theorem whose geometric proof (cf. Figure 2.2) was engraved on his tombstone in accordance with his wish.

As indicated in Figure 2.2, Archimedes placed the sphere inside the circular cylinder and then cut both surfaces into very, very thin strips with horizontal parallel planes. He applied the Eudoxian principle of approximation to prove that the corresponding thin strips are always of the same area! In terms of modern terminology, the

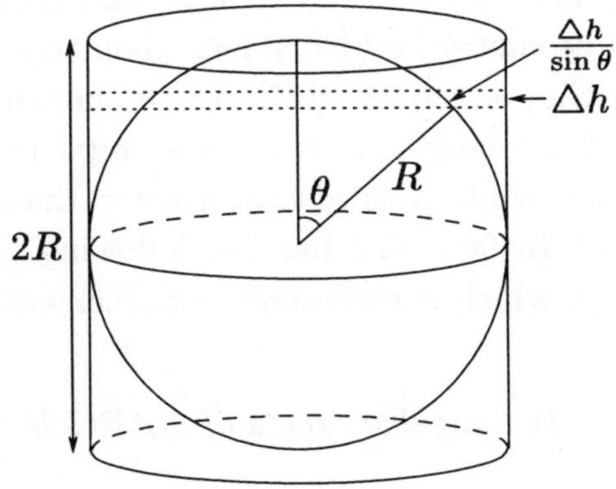

Figure 2.2

cylindrical strip is of length $2\pi R$ and of width $\triangle h$, while the corresponding spherical strip is approximately of length $2\pi R \sin\theta$ and of width $\frac{\triangle h}{\sin\theta}$.

(v) **Spherical triangles** Let $\{A, B, C\}$ be a given triple of points inside a suitable open hemi-sphere. They are pair-wise non-antipodal. Set $\overset{\frown}{AB}$, $\overset{\frown}{BC}$, $\overset{\frown}{CA}$ to be the shorter great circle arcs linking them. The two-dimensional spherical region enclosed by the above triple of arcs, is called a *spherical triangle* with $\{A, B, C\}$ as its vertices. We shall denote it by $\overset{\frown}{\triangle}ABC$ in order to distinguish it from the flat triangle $\triangle ABC$ on the plane with the same set of vertices. (The reason is that there are occasions when one needs to use both of them in the same discussion.) However, we shall use the same set of notations for the lengths, angles, area etc. of spherical triangles as that of Euclidean triangles, for example, A, B, C for the angles and a, b, c for the arc lengths of the sides opposite to A, B, C, respectively.

SAS congruence axiom and theorem of isosceles triangles It follows from the reflectional symmetry of $S^2(1)$ that spherical triangles with side-angle-side respectively equal, are congruent to each other. Moreover, one has the same set of logically equivalent characteristic properties for isosceles *spherical* triangles as that of the Euclidean ones. Therefore, the symmetry and congruence aspects of both spherical geometry and Euclidean geometry are actually the same. However, they are quite different in the aspect of the sum of inner angles of a triangle; it is always equal to a flat angle in the Euclidean case, while it is always greater than a flat angle in the spherical case! In fact, one has the following basic theorem of spherical geometry, which is essentially a refinement of Archimedes' Theorem:

Theorem 2.1.1 *The angular excess of $\overset{\frown}{\triangle}ABC$ is equal to its area, namely*

$$A + B + C - \pi = \triangle\,. \tag{4}$$

Proof The spherical region bounded by two longitude arcs is called the *spherical lune*. Its congruence class is clearly determined by the (equal) angles at its pair of antipodal vertices. Moreover, it follows from rotational symmetry and the famous theorem of Archimedes that the area of such a lune with angle θ is equal to

$$\frac{\theta}{2\pi} \cdot 4\pi = 2\theta \,. \tag{5}$$

In other words, the area of a lune is proportional to its angle θ (consequence of rotational symmetry), while the whole sphere can be regarded as a lune of angle $\theta = 2\pi$.

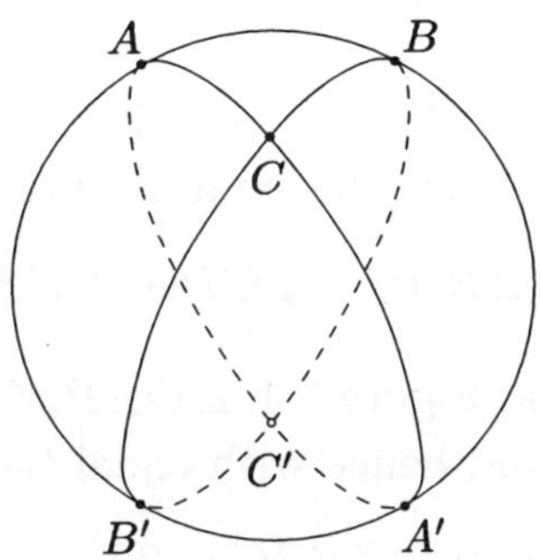

Figure 2.3

As indicated in Figure 2.3, A', B' and C' are the antipodal points of A, B and C, respectively. Applying the above area formula to the three lunes of angles A, B, C with pairs A, A', B, B', C, C' as their vertices, respectively, one has

$$\widehat{\triangle}ABC + \widehat{\triangle}A'BC = 2A \,, \quad \widehat{\triangle}ABC + \widehat{\triangle}AB'C = 2B \,,$$

$$\widehat{\triangle}ABC + \widehat{\triangle}ABC' = 2C \,,$$

$$\widehat{\triangle}ABC' = \widehat{\triangle}A'B'C \text{ (they are antipodal to one another)} \,,$$

$$\widehat{\triangle}ABC + \widehat{\triangle}A'BC + \widehat{\triangle}AB'C + \widehat{\triangle}A'B'C = 2\pi \text{ (a hemisphere)} \,. \tag{6}$$

Hence

$$2A + 2B + 2C = 2\,\widehat{\triangle}ABC + 2\pi\,, \quad A + B + C - \pi = \widehat{\triangle}ABC\,. \tag{7}$$

□

Remark Theoretically, the above set of properties (i.e. (i)–(v)) already constitutes a complete set of properties from which all the other formulas and theorems of spherical geometry (e.g. the laws of spherical trigonometry) can be deduced. However, we shall only deduce a few of them in such a classical approach in the following. In the next subsection, we shall change to the modern approach of studying spherical geometry in the framework of vector algebra.

1. Let $\widehat{\triangle}\,ABC$ and $\widehat{\triangle}\,A'BC$ be two spherical triangles of the same orientation. Then they have the same circumcircle if and only if

$$\angle CBA + \angle ACB - \angle BAC = \angle CBA' + \angle A'CB - \angle BA'C\,. \tag{8}$$

Proof As indicated in Figure 2.4, $\widehat{\triangle}OAB$, $\widehat{\triangle}OBC$ and $\widehat{\triangle}OCA$ are all isosceles triangles and hence with equal base angles. Therefore,

$$\angle CBA + \angle ACB - \angle BAC = 2\theta_1\,, \quad (\text{resp.} \ -2\theta_1) \tag{9}$$

in the case as indicated in Figure 2.4(i) (resp. Figure 2.4(ii)). □

2. **Lexell's Theorem** *Let $\widehat{\triangle}\,A_1BC$ and $\widehat{\triangle}\,A_2BC$ be two triangles with the same oriented area and B', C' be the antipodal points of B, C respectively. Then B', C', A_1 and A_2 are cocircular.*

Proof As indicated in Figure 2.5,

$$\angle 1 = \pi - B\,, \quad \angle 2 = \pi - C\,, \quad \angle 3 = A\,. \tag{10}$$

Hence,

$$\angle 1 + \angle 2 - \angle 3 = 2\pi - (A + B + C) = \pi - \widehat{\triangle}ABC\,. \tag{11}$$

□

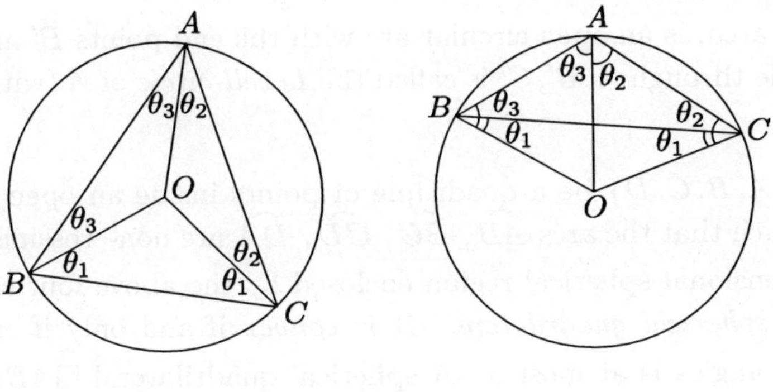

(i) (ii)

Figure 2.4

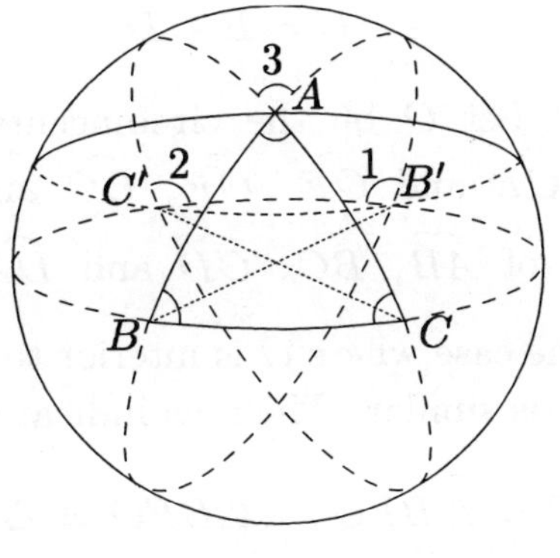

Figure 2.5

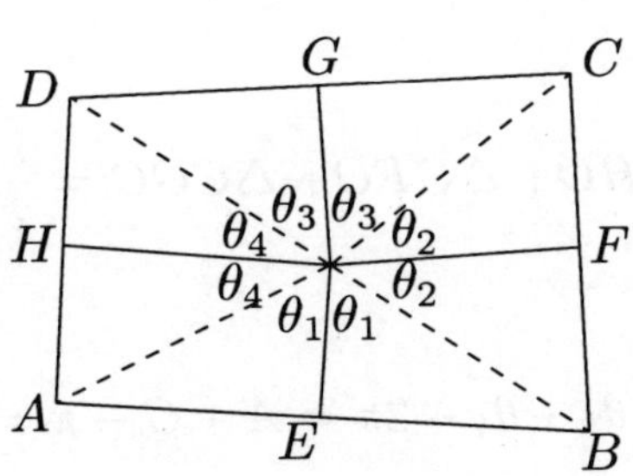

Figure 2.6

The locus of a point A such that the triangle $\widehat{\triangle}ABC$ has a constant oriented area, is an open circular arc with the end points B' and C'. The circle through A, B', C' is called the *Lexell-circle* of A (with $\widehat{BC}$ as base).

3. Let $\{A, B, C, D\}$ be a quadruple of points inside an open hemisphere such that the arcs $\widehat{AB}$, $\widehat{BC}$, $\widehat{CD}$, $\widehat{DA}$ are non-crossing. The two-dimensional spherical region enclosed by the above four arcs, is called a *spherical quadrilateral.* It is *convex* if and only if each of its inner angles is at most π. A spherical quadrilateral $\widehat{\square}ABCD$ is cocircular if and only if the sums of its two pairs of opposite angles are equal, namely

$$A + C = B + D\,. \tag{12}$$

Proof **Necessity** Let O be the circumcenter of the cocircular quadrilateral $\widehat{\square}ABCD$ and $\widehat{OE}$, $\widehat{OF}$, $\widehat{OG}$ and $\widehat{OH}$ be the perpendicular bisector of $\widehat{AB}$, $\widehat{BC}$, $\widehat{CD}$ and $\widehat{DA}$, respectively. We give the proof for the case where O is interior to $\widehat{\square}ABCD$. The case with O non-interior is similar. Then, as indicated in Figure 2.6,

$$\begin{aligned} \widehat{\triangle}AEO \cong \widehat{\triangle}BEO\,, \quad \widehat{\triangle}BFO \cong \widehat{\triangle}CFO\,,\\ \widehat{\triangle}CGO \cong \widehat{\triangle}DGO\,, \quad \widehat{\triangle}DHO \cong \widehat{\triangle}AHO\,. \end{aligned} \tag{13}$$

Therefore,

$$\widehat{\triangle}AEO + \widehat{\triangle}AHO + \widehat{\triangle}CFO + \widehat{\triangle}CGO = \frac{1}{2}\widehat{\square}ABCD\,, \tag{14}$$

and hence,

$$\begin{aligned} A + C + \theta_1 + \theta_2 + \theta_3 + \theta_4 - 2\pi = A + C - \pi = \frac{1}{2}\widehat{\square}ABCD\,,\\ B + D + \theta_1 + \theta_2 + \theta_3 + \theta_4 - 2\pi = B + D - \pi = \frac{1}{2}\widehat{\square}ABCD\,. \end{aligned} \tag{15}$$

Namely,

$$A + C = B + D = \frac{1}{2}\widehat{\square}ABCD + \pi\,. \tag{16}$$

Sufficiency Suppose $A+C = B+D$. Without loss of generality we may assume that the lune with vertex A spanned by B and D contains C, and that its angle is less that π. Suppose for a contradiction that C is *outside* the circumcircle of $\widehat{\triangle}ABD$. Let $\beta = \angle CBC'$ and $\delta = \angle CDC'$, where C' is the intersection point of the circumcircle with $\widehat{AC}$. Then $A + C' = (B - \beta) + (D - \delta)$, hence $C - C' = \beta + \delta$. It follows that

$$\widehat{\square}BCDC' = C - C' + \beta + \delta = 2\beta + 2\delta > \widehat{\triangle}BCC' + \widehat{\triangle}CDC'\,, \tag{17}$$

which is a contradiction. The assumption that C is *inside* the circumcircle of $\widehat{\square}ABD$ similarly leads to a contradiction. This shows that A, B, C, D are cocircular. $\square$

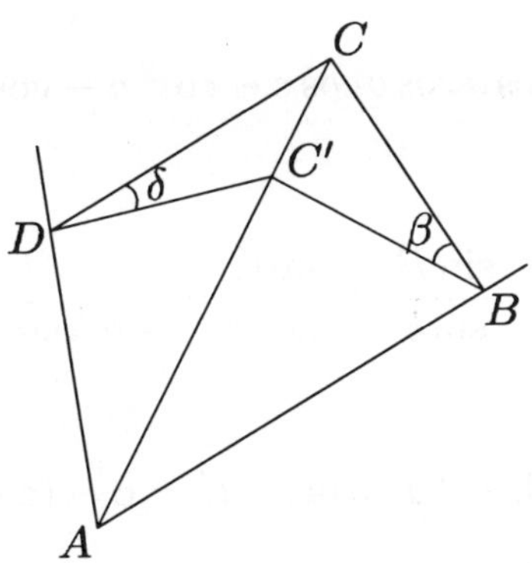

Figure 2.7

2.1.2 *Vector algebra and spherical trigonometry*

Let $\mathbf{a}, \mathbf{b}$ and $\mathbf{c}$ be the position vectors of the vertices of a positively oriented spherical triangle $\widehat{\triangle}ABC$, which are, of course, *unit vectors.*

Thus,

(i) $\cos a = \mathbf{b} \cdot \mathbf{c}, \quad \cos b = \mathbf{c} \cdot \mathbf{a}, \quad \cos c = \mathbf{a} \cdot \mathbf{b},$

(ii) the areas of $//(\mathbf{b}, \mathbf{c}), \quad //(\mathbf{c}, \mathbf{a})$ and $//(\mathbf{a}, \mathbf{b})$ are equal to $\sin a$, $\sin b$ and $\sin c$, namely

$$\sin a = |\mathbf{b} \times \mathbf{c}|, \sin b = |\mathbf{c} \times \mathbf{a}|, \sin c = |\mathbf{a} \times \mathbf{b}|, \tag{18}$$

(iii) the dihedral angle between $//(\mathbf{a}, \mathbf{b})$ and $//(\mathbf{a}, \mathbf{c})$ (resp. $//(\mathbf{b}, \mathbf{c})$ and $//(\mathbf{b}, \mathbf{a})$, $//(\mathbf{c}, \mathbf{a})$ and $//(\mathbf{c}, \mathbf{b})$) is the angle A (resp. B, C) of $\overset{\frown}{\triangle} ABC$.

Set $D = \det(\mathbf{a}, \mathbf{b}, \mathbf{c}) > 0$. Then,

$$D^2 = \begin{vmatrix} \mathbf{a} \cdot \mathbf{a} & \mathbf{b} \cdot \mathbf{a} & \mathbf{c} \cdot \mathbf{a} \\ \mathbf{a} \cdot \mathbf{b} & \mathbf{b} \cdot \mathbf{b} & \mathbf{c} \cdot \mathbf{b} \\ \mathbf{a} \cdot \mathbf{c} & \mathbf{b} \cdot \mathbf{c} & \mathbf{c} \cdot \mathbf{c} \end{vmatrix} = \begin{vmatrix} 1 & \cos c & \cos b \\ \cos c & 1 & \cos a \\ \cos b & \cos a & 1 \end{vmatrix}$$

$$= 1 + 2 \cos a \cos b \cos c - \cos^2 a - \cos^2 b - \cos^2 c. \tag{19}$$

Spherical sine law

$$\frac{\sin A}{\sin a} = \frac{\sin B}{\sin b} = \frac{\sin C}{\sin c} = \frac{D}{\sin a \sin b \sin c}. \tag{20}$$

Proof Set

$$\mathbf{b}' = \mathbf{b} - (\mathbf{b} \cdot \mathbf{a})\mathbf{a}, \quad \mathbf{c}' = \mathbf{c} - (\mathbf{c} \cdot \mathbf{a})\mathbf{a}. \tag{21}$$

Then both $\mathbf{b}'$ and $\mathbf{c}'$ are perpendicular to $\mathbf{a}$ and $|\mathbf{b}'| = \sin c, |\mathbf{c}'| = \sin b$. Hence

$$D = \mathbf{a} \cdot (\mathbf{b} \times \mathbf{c}) = \mathbf{a} \cdot (\mathbf{b}' \times \mathbf{c}') = \sin c \sin b \sin A. \tag{22}$$

This proves that

$$\frac{\sin A}{\sin a} = \frac{\sin B}{\sin b} = \frac{\sin C}{\sin c} = \frac{D}{\sin a \sin b \sin c}.$$

□

Spherical cosine law

$$\begin{cases} \sin b \sin c \cos A = \cos a - \cos b \cos c\,, \\ \sin c \sin a \cos B = \cos b - \cos c \cos a\,, \\ \sin a \sin b \cos C = \cos c - \cos a \cos b\,. \end{cases} \tag{23}$$

Proof

$$\begin{aligned} \sin b \sin c \cos A &= |\mathbf{c} \times \mathbf{a}|\,|\mathbf{b} \times \mathbf{a}| \cos A = (\mathbf{c} \times \mathbf{a}) \cdot (\mathbf{b} \times \mathbf{a}) \\ &= \begin{vmatrix} \mathbf{c} \cdot \mathbf{b} & \mathbf{a} \cdot \mathbf{b} \\ \mathbf{c} \cdot \mathbf{a} & \mathbf{a} \cdot \mathbf{a} \end{vmatrix} = \begin{vmatrix} \cos a & \cos c \\ \cos b & 1 \end{vmatrix} \\ &= \cos a - \cos b \cos c\,. \end{aligned} \tag{24}$$

This proves the first equation of (23), and the same proof will also demonstrate the other two of (23). □

Another Proof

$$\begin{aligned} \sin c \sin b \cos A &= |\mathbf{b}'| \cdot |\mathbf{c}'| \cos A = (\mathbf{b} - (\mathbf{b} \cdot \mathbf{a})\mathbf{a}) \cdot (\mathbf{c} - (\mathbf{c} \cdot \mathbf{a})\mathbf{a}) \\ &= \mathbf{b} \cdot \mathbf{c} - (\mathbf{b} \cdot \mathbf{a})(\mathbf{a} \cdot \mathbf{c}) - (\mathbf{b} \cdot \mathbf{a})(\mathbf{c} \cdot \mathbf{a}) \\ &\quad + (\mathbf{b} \cdot \mathbf{a})(\mathbf{c} \cdot \mathbf{a})(\mathbf{a} \cdot \mathbf{a}) = \mathbf{b} \cdot \mathbf{c} - (\mathbf{b} \cdot \mathbf{a})(\mathbf{a} \cdot \mathbf{c}) \\ &= \cos a - \cos c \cos b\,. \end{aligned} \tag{24$'$}$$

□

The special case of a right-angle triangle In the special case that $C = \pi/2$, the above sine law and cosine law imply the following set of simple formulas, namely

$$\cos c = \cos a \cos b\,, \quad \sin A = \frac{\sin a}{\sin c}\,, \quad \sin B = \frac{\sin b}{\sin c}\,, \tag{25}$$

$$\begin{aligned} \cos A &= \frac{\tan b}{\tan c}\,, \quad \cos B = \frac{\tan a}{\tan c}\,, \\ \tan A &= \frac{\tan a}{\sin b}\,, \quad \tan B = \frac{\tan b}{\sin a}\,. \end{aligned} \tag{26}$$

Proof (25) follows directly from (23) and (20), and the assumption that $\cos C = 0,\ \sin C = 1$. Thus, it suffices to prove the first equation of (26) as follows

$$\begin{aligned}\sin b \sin c \cos A &= \cos a - \cos b \cos c = \cos a - \cos^2 b \cos a \\ &= \sin^2 b \cos a\,. \end{aligned} \tag{27}$$

Hence,

$$\cos A = \frac{\sin b \cos a}{\sin c} = \frac{\tan b \cos b \cos a}{\sin c} = \frac{\tan b}{\tan c}\,. \tag{28}$$

□

Half-angle formulas Substituting $\cos A = 2\cos^2 \frac{A}{2} - 1$ (resp. $1 - 2\sin^2 \frac{A}{2}$) into the cosine law, one has

$$\begin{aligned}2 \sin b \sin c \cos^2 \frac{A}{2} &= \cos a - \cos b \cos c + \sin b \sin c = \cos a - \cos(b+c) \\ &= 2 \sin \frac{1}{2}(b+c-a) \sin \frac{1}{2}(a+b+c) \end{aligned} \tag{29}$$

$$\left(\begin{aligned} &\text{resp.} \\ &\qquad 2 \sin b \sin c \sin^2 \frac{A}{2} = -\cos a + \cos b \cos c + \sin b \sin c \\ &\qquad\qquad\qquad\qquad\quad = \cos(b-c) - \cos a \\ &\qquad\qquad\qquad\qquad\quad = 2 \sin \frac{1}{2}(a+b-c) \sin \frac{1}{2}(a-b+c) \end{aligned} \right). \tag{30}$$

This proves the following set of half angle formulas, namely

$$\begin{aligned} \cos \frac{A}{2} &= \sqrt{\frac{\sin(s-a) \sin s}{\sin b \sin c}}\,, \text{etc}\,, \\ \sin \frac{A}{2} &= \sqrt{\frac{\sin(s-b) \sin(s-c)}{\sin b \sin c}}\,, \ \text{etc}\,, \\ \tan \frac{A}{2} &= \sqrt{\frac{\sin(s-b) \sin(s-c)}{\sin s \sin(s-a)}}\,, \ \text{etc}\,, \end{aligned} \tag{31}$$

where $s = \frac{1}{2}(a+b+c)$.

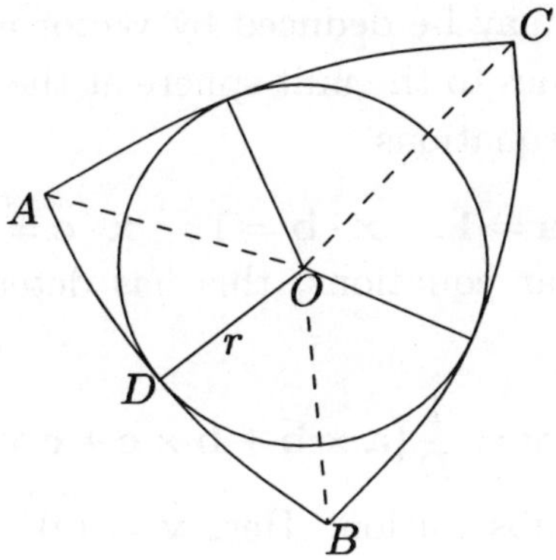

Figure 2.8

Let r be the (spherical) radius of the inscribing circle of $\widehat{\triangle}ABC$. Then, as indicated in Figure 2.8, $\widehat{\triangle}\, ADO$ is a right-angle triangle with $\angle OAD = \frac{A}{2}$ and $\widehat{AD} = s - a$. Hence, from (26) and (31)

$$\frac{\tan r}{\sin(s-a)} = \tan\frac{A}{2} = \sqrt{\frac{\sin(s-b)\sin(s-c)}{\sin s \sin(s-a)}}. \tag{32}$$

Thus, we obtain a formula for r, namely

$$\tan r = \sqrt{\frac{\sin(s-a)\sin(s-b)\sin(s-c)}{\sin s}}. \tag{33}$$

The circumradius of a spherical triangle We shall also obtain a formula for the radius R of the circumscribing circle of a spherical triangle. In the sphere packing problem, the circumcircle will play a rather important role as opposed to the inscribing circle. This is due to the fact that the circumcircles of a spherical triangle $\widehat{\triangle}\, ABC$ and the corresponding Euclidean triangle $\triangle ABC$ actually coincide, i.e. both equal to the intersection of the plane (ABC) with the unit sphere.

A formula for R may be deduced by vector algebra. We observe that the tangent planes to the unit sphere at the vertices A, B, C are given by the vector equations

$$\mathbf{x} \cdot \mathbf{a} = 1\,, \quad \mathbf{x} \cdot \mathbf{b} = 1\,, \quad \mathbf{x} \cdot \mathbf{c} = 1\,. \tag{34}$$

As a system of linear equations, this has determinant D and the unique solution

$$\mathbf{x} = \mathbf{v} = \frac{1}{D}\{\mathbf{a} \times \mathbf{b} + \mathbf{b} \times \mathbf{c} + \mathbf{c} \times \mathbf{a}\} \tag{35}$$

is seen directly by substitution. Here $\mathbf{v} = \overrightarrow{OV}$ is the position vector of the intersection point of the three tangent planes, and hence this point V is equidistant from A, B, C. The locus of all points in space which have equal distance from A, B, C, is the line OV. The circumcenter of $\widehat{\triangle}ABC$ is therefore given by the unit vector $\mathbf{v}/|\mathbf{v}|$. By the rotational symmetry of the sphere with respect to the axis OV, all points on the circumcircle of $\widehat{\triangle}ABC$ have tangent planes with the sphere that pass through V. The point V and the plane (ABC) are *dual* to each other, in the sense of projective geometry (cf. Figure 2.9).

Therefore, for the circumradius R we get

$$\begin{aligned}\sec^2 R &= |\mathbf{v}|^2 = \mathbf{v} \cdot \mathbf{v} \\ &= \frac{1}{D^2}(\mathbf{a} \times \mathbf{b} + \mathbf{b} \times \mathbf{c} + \mathbf{c} \times \mathbf{a}) \cdot (\mathbf{a} \times \mathbf{b} + \mathbf{b} \times \mathbf{c} + \mathbf{c} \times \mathbf{a}) \\ &= \frac{1}{D^2}\left\{3 - \sum \cos^2 a + 2\sum \cos a \cos b - 2\sum \cos a\right\}. \end{aligned} \tag{36}$$

By introducing the symmetric functions

$$u = 1 + \cos a + \cos b + \cos c\,, \tag{37}$$

$$w = \cos a \cos b + \cos a \cos c + \cos b \cos c\,, \tag{38}$$

we can rewrite formula (36) as

$$\sec^2 R = \frac{1}{D^2}\{4(1 + w) - u^2\}\,. \tag{39}$$

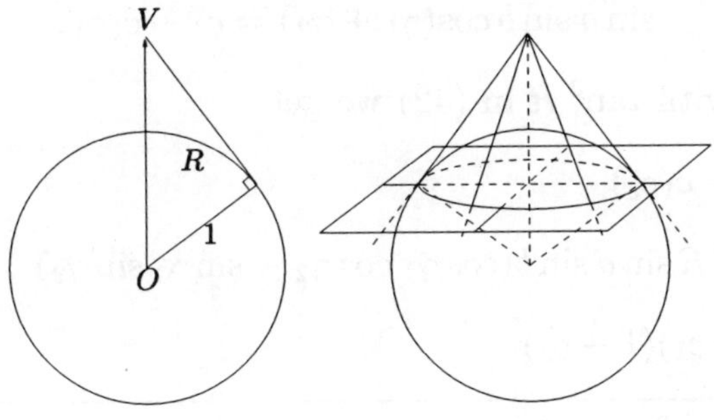

Figure 2.9

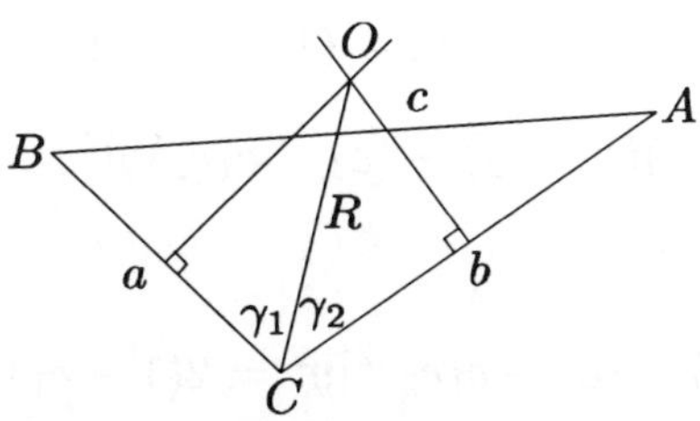

Figure 2.10

Formula (39) expresses the circumradius in terms of the cosines of the three sides. Although the proof is very simple by vector algebra, looking at the sphere from outside, it is of interest also to have an "intrinsic" proof, i.e. by spherical trigonometry. In particular, such a proof would have a parallel in hyperbolic geometry.

Let $c_1 = \cos a,\ c_2 = \cos b,\ c_3 = \cos c$. Let O be the circumcenter, and let $\gamma_1 = \angle OCB$, $\gamma_2 = \angle OCA$. Then,

$$\cos\gamma_1 = \frac{\tan\frac{a}{2}}{\tan R}, \quad \cos\gamma_2 = \frac{\tan\frac{b}{2}}{\tan R}, \tag{40}$$

$$\sin\gamma_1 = \frac{\sqrt{\tan^2 R - \tan^2\frac{a}{2}}}{\tan R}, \quad \sin\gamma_2 = \frac{\sqrt{\tan^2 R - \tan^2\frac{b}{2}}}{\tan R}, \tag{41}$$

$$\sin a \sin b \cos(\gamma_1 + \gamma_2) = c_3 - c_1 c_2 \,. \tag{42}$$

Multiplying with $\tan^2 R$ in (42) we get

$$\begin{aligned}
&\tan^2 R(c_3 - c_1 c_2) \\
&\quad = \tan^2 R \sin a \sin b(\cos\gamma_1 \cos\gamma_2 - \sin\gamma_1 \sin\gamma_2) \\
&\quad = (1 - c_1)(1 - c_2) \\
&\qquad - \sqrt{\sin^2 a \tan^2 R - (1 - c_1)^2} \times \sqrt{\sin^2 b \tan^2 R - (1 - c_2)^2} \,.
\end{aligned}$$

Let $\tan^2 R = y$. Then we get

$$\begin{aligned}
&(1 - c_1)(1 - c_2)[(1 + c_1)y - (1 - c_1)][(1 + c_2)y - (1 - c_2)] \\
&\quad = [(1 - c_1)(1 - c_2) - (c_3 - c_1 c_2)y]^2 \,,
\end{aligned}$$

which gives

$$[(1 - c_1^2)(1 - c_2^2) - (c_3 - c_1 c_2)^2]y^2 = 2(1 - c_1)(1 - c_2)(1 - c_3)y$$

or

$$D^2 y = 2(1 - c_1)(1 - c_2)(1 - c_3) \,,$$

which means

$$\tan^2 R = \frac{2(1 - \cos a)(1 - \cos b)(1 - \cos c)}{D^2} \,. \tag{43}$$

Compared with the formula (39), it is easy to see that we get the same result.

Remark This brief discussion not only provides a simple systematic proof of the classical results on spherical trigonometry as a direct application of the basic formulas of vector algebra, it also clarifies the interlocking relationship between vector algebra and spherical trigonometry. Such a clarification will, of course, enrich the understanding of both. The following results on the areas of spherical triangles and quadrilaterals are just some of its immediate effects.

2.1.3 *Some further results on areas of spherical triangles and quadrilaterals*

From a purely geometric point of view, the area formula of Theorem 2.1.1, namely

$$\triangle = A + B + C - \pi$$

is so neat and beautiful that it must already be *the optimal* one! Why bother to deduce any other area formula for spherical triangles? But, from the viewpoint of orthogonal invariant theory and for the purpose of applying spherical geometry to the study of solid geometry, it is useful to have a neat formula expressing the area *directly* in terms of the triple of inner products, namely, $\{\cos a = \mathbf{b} \cdot \mathbf{c},\ \cos b = \mathbf{a} \cdot \mathbf{c},\ \cos c = \mathbf{a} \cdot \mathbf{b}\}$.

Lemma 2.1.1 *Set*

$$\begin{aligned} u &= 1 + \cos a + \cos b + \cos c\,, \\ D &= [1 + 2\cos a \cos b \cos c - \cos^2 a - \cos^2 b - \cos^2 c]^{1/2}\,. \end{aligned} \tag{44}$$

Then

$$\tan \frac{\triangle}{2} = \frac{D}{u}\,. \tag{45}$$

Proof The sine law and cosine law assert that

$$\sin A = \frac{D}{\sin b \sin c}\,, \qquad \cos A = \frac{\cos a - \cos b \cos c}{\sin b \sin c}\,, \quad \text{etc}\,. \tag{46}$$

Therefore, straightforward substitution and simplification shows that

$$\begin{aligned} \sin \triangle &= \sin(A + B + C - \pi) \\ &= -\sin(A + B + C) = \frac{D}{\Pi \sin^2 a} \\ &\quad \times \left\{D^2 + \sum \cos a \cos^2 b - \sum \cos a \cos b - (u-1)\Pi \cos a\right\}\,. \end{aligned} \tag{47}$$

The denominator of the above algebraic expression can be rewritten as

$$\begin{aligned}\Pi(1-\cos^2 a) &= \Pi(1+\cos a)\Pi(1-\cos a)\\ &= \frac{1}{2}(u^2+D^2)\Pi(1-\cos a)\,. \qquad (48)\end{aligned}$$

It is a remarkable fact that the polynomial of $\{\cos a, \cos b, \cos c\}$ inside the $\{\cdots\}$ of (47) also contains a factor of $\Pi(1-\cos a)$, namely

$$\{\cdots\} = u\cdot\Pi(1-\cos a)\,. \qquad (49)$$

Therefore, the formula of (47) greatly simplifies to

$$\sin\triangle = \frac{2uD}{u^2+D^2}\,. \qquad (50)$$

In a similar way one gets

$$\cos\triangle = \frac{u^2-D^2}{u^2+D^2}\,. \qquad (51)$$

Hence,

$$\tan\frac{\triangle}{2} = \frac{\sin\triangle}{1+\cos\triangle} = \frac{D}{u}\,. \qquad \square$$

Remarks (i) The simple algebraic form of $\tan\frac{\triangle}{2}$ above will be very useful in the study of the sphere packing problem, in which one often needs to study the interlocking relationship between spherical geometry and Euclidean geometry. Thus, vector algebra is naturally the most advantageous setting for such problems.

(ii) We note that as a consequence of Lemma 2.1.1 we have $\triangle < \pi$ for $u > 0$, $\triangle = \pi$ for $u = 0$, and $\triangle > \pi$ for $u < 0$, assuming $D \neq 0$.

(iii) In the special cases of right-angle triangles, one has even simpler area formulas than the above general one.

Lemma 2.1.1′ *Let* $\widehat{\triangle}ABC$ *be a right-angle triangle and* $\triangle$ *be its area. Then*

$$\begin{aligned}\cos\triangle &= \frac{\cos a + \cos b}{1+\cos c},\\ \sin\triangle &= \frac{\sin a \sin b}{1+\cos c} = \frac{(1-\cos c)\sin A \cos A}{\sqrt{\cos^2 A + \cos^2 c \sin^2 A}},\\ \frac{\partial\triangle}{\partial a} &= \frac{\sin b}{1+\cos c}.\end{aligned} \tag{52}$$

Proof From the assumption that $C = \frac{\pi}{2}$, one has

$$\begin{aligned}&D = \sin a \sin b, \quad \cos c = \cos a \cos b,\\ &u = 1 + \cos a + \cos b + \cos c = (1+\cos a)(1+\cos b).\end{aligned} \tag{53}$$

Therefore,

$$\cos\triangle = \frac{u^2 - D^2}{u^2 + D^2} = \frac{\cos a + \cos b}{1+\cos c}, \tag{54}$$

$$\begin{aligned}\sin\triangle &= \frac{2uD}{u^2+D^2} = \frac{\sin a \sin b}{1+\cos c} = \frac{\sin^2 c}{1+\cos c}\frac{\sin A \cos A}{\sqrt{1-\sin^2 A \sin^2 c}}\\ &= \frac{(1-\cos c)\sin A \cos A}{\sqrt{\cos^2 A + \cos^2 c \sin^2 A}}.\end{aligned} \tag{55}$$

Hence, from the differentiation of (54), one has

$$\begin{aligned}\sin\triangle\frac{\partial\triangle}{\partial a} &= \frac{\sin a \sin^2 b}{(1+\cos a \cos b)^2}\\ \Rightarrow \frac{\partial\triangle}{\partial a} &= \frac{\sin b}{1+\cos c}.\end{aligned} \tag{56}$$

□

As an application of the area formula of Lemma 2.1.1, let us study the area of a spherical triangle with two sides of fixed lengths and the third varying.

Corollary 2.1.1 *For such a spherical triangle* $\widehat{\triangle} ABC$*, set* $c_1 = \cos a$*,* $c_2 = \cos b$*,* $x = 1 + \cos c$*, and let* φ *be the angle at the vertex* C*. Then*

$$\frac{d\triangle}{dx} = \frac{c_1 + c_2 - x}{xD}, \quad \frac{d\triangle}{d\varphi} = \frac{x - c_1 - c_2}{x}. \tag{57}$$

Proof With the notation above we have

$$\begin{aligned} u &= c_1 + c_2 + x\,, \\ D^2 &= -(c_1 + c_2)^2 + 2(c_1c_2 + 1)x - x^2\,. \end{aligned} \tag{58}$$

By differentiation of the formula $\triangle = 2\tan^{-1}\frac{D}{u}$, one gets

$$\begin{aligned} \frac{d\triangle}{dx} &= \frac{uD' - D}{\frac{1}{2}(u^2 + D^2)} = \frac{uDD' - D^2}{(1 + c_1)(1 + c_2)xD} \\ &= \frac{c_1 + c_2 - x}{xD}. \end{aligned} \tag{59}$$

On the other hand, by differentiation of the cosine law, namely, $\sin a \sin b \cos\varphi = x - 1 - c_1c_2$, one gets

$$-\sin a \sin b \sin\varphi \frac{d\varphi}{dx} = -D\frac{d\varphi}{dx} = 1\,, \quad \frac{dx}{d\varphi} = -D\,, \tag{60}$$

$$\frac{d\triangle}{d\varphi} = \frac{d\triangle}{dx}\frac{dx}{d\varphi} = \frac{x - c_1 - c_2}{x}. \tag{61}$$

□

If we let the angle φ vary from 0 to π and study the area function $\triangle = \triangle(\varphi)$, we get three different cases:

Case 1: $a + b < \pi$. Then $c_1 + c_2 > 0$. The side length $c = c(\varphi)$ increases from $|a - b|$ to $a + b$, while the area $\triangle = \triangle(\varphi)$ increases from 0 to a maximum, which is obtained for the unique solution

$$x = c_1 + c_2 \tag{62}$$

of $d\triangle/d\varphi = 0$, and then decreases to 0. The existence of a maximum for the area function at an interior point is clear from the geometric

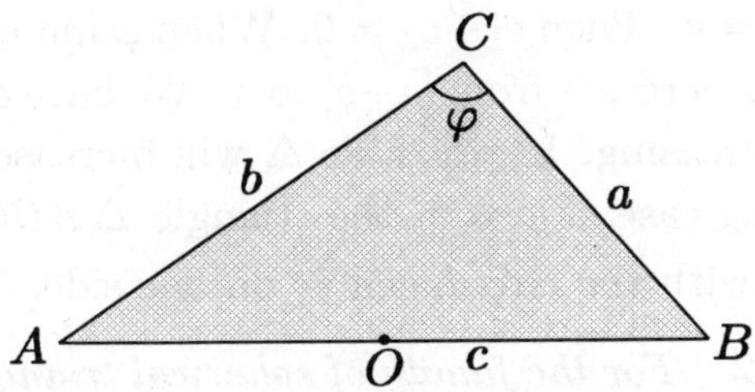

Figure 2.11

interpretation in this case, so the value (62) must be the maximum point. But it is also easy to verify this analytically, because we get from (57) that $d\triangle/d\varphi(0) > 0$ and $d\triangle/d\varphi(\pi) < 0$.

We shall shortly deduce the geometric meaning of (62), namely, the circumcenter of the triangle $\widehat{\triangle}ABC$ is *on* the side $\widehat{AB}$ (cf. Figure 2.11).

The condition $x > c_1 + c_2$ (resp. $x < c_1 + c_2$) corresponds to the circumcenter being on the inside (resp. outside) of $\widehat{AB}$. Thus as the triangle opens up with increasing angle φ, the circumcenter will move from inside the triangle to outside, across the edge $\widehat{AB}$, and the maximal area is obtained exactly when the circumcenter is on the edge.

It is worth observing that for the Euclidean case, with given a and b, the maximal area is attained by the right angle triangle, $\varphi = \pi/2$, which can also be characterized by the condition that the circumcenter is on the opposite side. However, in the spherical case, it is the circumcenter condition which generalizes, not the right angle condition.

Case 2: $a + b > \pi$. Then $c_1 + c_2 < 0$. The side length $c = c(\varphi)$ now increases from $|a - b|$ to $2\pi - (a + b)$. We have $d\triangle/d\varphi > 0$, and hence $\triangle = \triangle(\varphi)$ is increasing. The area increases from 0 towards 2π. In the limiting case that A, B, C are on a great circle, then $\widehat{\triangle}ABC$ becomes a whole hemisphere.

Case 3: $a+b=\pi$. Then $c_1+c_2=0$. When φ increases from 0 to π, the side length c increases from $|a-b|$ to π. We have $d\triangle/d\varphi=1$, thus $\triangle=\triangle(\varphi)$ is increasing. In this case $\triangle$ will increase from 0 towards π. In the limiting case of $\varphi=\pi$, the triangle $\widehat{\triangle}ABC$ converges to a quarter sphere, with the circumcenter on one side.

Corollary 2.1.2 *For the family of spherical triangles $\widehat{\triangle}ABC$ with both c and $a+b$ fixed at a given pair of constants, the area of $\widehat{\triangle}ABC$ is a monotonically decreasing function of $|a-b|$.*

Proof For $0<a,b,c<\pi$ to be the side lengths of a proper spherical triangle, we must have $0<c<a+b$ and $a+b+c<2\pi$ as conditions on the given constants, and for each triangle the condition $|a-b|<c$. Set

$$A=\cos\frac{a+b}{2}\,,\qquad C=1+\cos c\,,\qquad t=\cos\frac{a-b}{2}\,. \tag{63}$$

Then,

$$\begin{aligned} u &= \cos a+\cos b+\cos c+1 = 2At+C\,,\\ D^2+u^2 &= 2(1+\cos a)(1+\cos b)(1+\cos c)=2C(A+t)^2\,,\\ D^2 &= (C-2A^2)(2t^2-C)\,. \end{aligned} \tag{64}$$

By differentiation of the formula (45) we get

$$\frac{d\triangle}{dt}=\frac{D'u-Du'}{\frac{1}{2}(u^2+D^2)}=\frac{DD'u-D^2u'}{\frac{1}{2}(u^2+D^2)D}=\frac{2(C-2A^2)}{D(A+t)}>0\,, \tag{65}$$

since

$$\begin{aligned} C-2A^2 &= 1+\cos c-2\cos^2\frac{a+b}{2}=\cos c-\cos(a+b)\,,\\ &= 2\sin\frac{a+b+c}{2}\sin\frac{a+b-c}{2}>0\,, \end{aligned} \tag{66}$$

and

$$A+t=\cos\frac{a+b}{2}+\cos\frac{a-b}{2}$$

$$= 2\cos\frac{a}{2}\cos\frac{b}{2} > 0 \tag{67}$$

by the assumptions. Therefore, $\triangle$ is a monotonically increasing function of t, and hence a monotonically decreasing function of $|a-b|$. □

Remark If the above family of triangles contains a triangle with area $\triangle = \pi$, we have a corresponding value $u = 0$, which from (64) occurs for $t = t_0 = -C/2A$. Although the right-hand side of the formula (45) is undefined for $u = 0$, the expression for $d\triangle/dt$ we obtained in (65), is still well defined, showing that the area function $\triangle = \triangle(t)$ is in fact differentiable at $t = t_0$. This case will occur if and only if the isosceles triangle of equal sides $(a+b)/2$ and base c has area $\geq \pi$.

An immediate consequence of Corollary 2.1.2 is the *isoperimetric property* of regular spherical triangles:

Among all spherical triangles with a given circumference the regular triangles, and only these, have maximal area.

2.1.3.1 *Spherical quadrilaterals*

As in the case of Euclidean plane geometry, the congruence class of a spherical triangle is uniquely determined by its three side-lengths, while that of a spherical quadrilateral is *not* determined by its four side-lengths. A convex quadrilateral in an open hemisphere has two diagonals, each of them cut the given quadrilateral into a pair of triangles. Thus, its four side-lengths together with an additional diagonal length suffice to determine the quadrilateral up to congruence, while the six arc-lengths of a quadrilateral (i.e. four side-lengths and two diagonal-lengths) are interlocked by a functional relation. Let $\{\mathbf{a}, \mathbf{b}, \mathbf{c}, \mathbf{d}\}$ be the position vectors of the four vertices of a given spherical quadrilateral $\widehat{\square}ABCD$ with $\{A, B, C, D\}$ assumed to be inside a suitable hemisphere. Then the six pairwise inner products are the cosines of the above six arc-lengths $\{\ell_1, \ell_2, \ell_3, \ell_4; d_1, d_2\}$. One of the simplest ways of obtaining such a functional relation among the above six geometric invariants of spherical quadrilaterals, is the following seemingly trivial algebraic formula, namely

$$\begin{aligned} \det(\mathbf{a},\mathbf{b},\mathbf{c})^2 \det(\mathbf{a},\mathbf{c},\mathbf{d})^2 &= [\det(\mathbf{a},\mathbf{b},\mathbf{c}) \det(\mathbf{a},\mathbf{c},\mathbf{d})]^2 \\ (\text{resp. } \det(\mathbf{a},\mathbf{b},\mathbf{d})^2 \det(\mathbf{b},\mathbf{c},\mathbf{d})^2 &= [\det(\mathbf{a},\mathbf{b},\mathbf{d}) \det(\mathbf{b},\mathbf{c},\mathbf{d})]^2)\,. \end{aligned} \tag{68}$$

Using the product formula of determinants, one obtains the following algebraic relation for the six cosines, namely

$$\begin{aligned} &\begin{vmatrix} 1 & \cos\ell_1 & \cos d_1 \\ \cos\ell_1 & 1 & \cos\ell_2 \\ \cos d_1 & \cos\ell_2 & 1 \end{vmatrix} \cdot \begin{vmatrix} 1 & \cos d_1 & \cos\ell_4 \\ \cos d_1 & 1 & \cos\ell_3 \\ \cos\ell_4 & \cos\ell_3 & 1 \end{vmatrix} \\ &\quad = \begin{vmatrix} 1 & \cos\ell_1 & \cos d_1 \\ \cos d_1 & \cos\ell_2 & 1 \\ \cos\ell_4 & \cos d_2 & \cos\ell_3 \end{vmatrix}^2 \left(\text{resp. } \begin{vmatrix} 1 & \cos\ell_1 & \cos\ell_4 \\ \cos\ell_1 & 1 & \cos d_2 \\ \cos\ell_4 & \cos d_2 & 1 \end{vmatrix} \right. \\ &\qquad \left. \cdot \begin{vmatrix} 1 & \cos\ell_2 & \cos d_2 \\ \cos\ell_2 & 1 & \cos\ell_3 \\ \cos d_2 & \cos\ell_3 & 1 \end{vmatrix} = \begin{vmatrix} \cos\ell_1 & 1 & \cos d_2 \\ \cos d_1 & \cos\ell_2 & \cos\ell_3 \\ \cos\ell_4 & \cos d_2 & 1 \end{vmatrix}^2 \right). \end{aligned} \tag{69}$$

The following are some special types of quadrilaterals for which the above general relation becomes simpler. For example, if $\ell_1 = \ell_3$ $(= a)$ and $\ell_2 = \ell_4 (= b)$, then the above general relation becomes the following simple one, namely

$$(1 + \cos d_1)(1 + \cos d_2) = (\cos a + \cos b)^2\,. \tag{70}$$

In the case of Euclidean geometry, the equalities of the two pairs of opposite sides characterize parallelograms. Thus, the above special type of quadrilaterals may be regarded as the "*spherical parallelograms.*" Although there is no global parallelism in spherical geometry, a spherical parallelogram still has the same basic geometric structure that its two diagonals bisect each other, and it is central symmetric with respect to the intersection point of its diagonals. Using such a nice geometric structure of spherical parallelograms, the relation (70) can also be deduced directly by the cosine law, namely

$$\begin{aligned} \sin\frac{d_1}{2}\sin\frac{d_2}{2}\cos\theta &= \cos a - \cos\frac{d_1}{2}\cos\frac{d_2}{2}\,, \\ \sin\frac{d_1}{2}\sin\frac{d_2}{2}\cos(\pi-\theta) &= \cos b - \cos\frac{d_1}{2}\cos\frac{d_2}{2}\,, \end{aligned} \tag{71}$$

where θ and $\pi - \theta$ are the two angles between the intersecting diagonals. Hence,

$$\cos a + \cos b = 2\cos\frac{d_1}{2}\cos\frac{d_2}{2} = \sqrt{1+\cos d_1}\sqrt{1+\cos d_2}\,,$$

which gives (70).

Note that a spherical parallelogram is *cocircular* if and only if its *two diagonals* are of *equal length.* This special type of spherical parallelograms can be regarded as *spherical "rectangles."* Therefore, for a spherical rectangle of side-lengths $\{a, b\}$, its diagonal length is given by

$$1 + \cos d = \cos a + \cos b\,. \tag{72}$$

Remarks The relation (72) is just a reformulation of the condition (62) for a maximum in our earlier discussion of the area function for triangles. This proves that the area of spherical triangles with two given side lengths has a unique maximum when its third side becomes the diameter of its circumcircle. Moreover, deformation of a triangle with a fixed pair of side lengths, further away from the above shape, is always area decreasing.

For a given pair of side lengths $\{a, b\}$ with $a + b < \pi$, there is a continuous family of spherical parallelograms of side lengths

$\{a, b, a, b\}$ whose pair of diagonal lengths $\{d_1, d_2\}$ satisfy the relation (70). Thus, the area of the parallelogram, as a function of $x = 1 + \cos d_1$, has a unique maximum at $x = \cos a + \cos b$, which is also the condition that the spherical parallelogram becomes cocircular. Geometrically speaking, the shearing deformation of a spherical parallelogram further away from the cocircular shape is always area decreasing.

In fact, such a monotonic area decreasing property holds in general for the *shearing deformation*, i.e. side-lengths preserving deformation, of *all* spherical quadrilaterals. Before proving this, let us first consider this type of deformation a little more closely.

2.1.3.2 *Shearing deformations*

For a single spherical triangle $\overset{\frown}{\triangle} ABC$, if we fix the side lengths $\ell_1 = \overset{\frown}{AB}$, $\ell_2 = \overset{\frown}{BC}$, and let the angle θ at B increase from 0 to π, the side length $d = \overset{\frown}{AC}$ will increase (by the cosine relation) from $|\ell_1 - \ell_2|$ to the smaller $\ell_1 + \ell_2$ and $2\pi - (\ell_1 + \ell_2)$. The latter of these values comes into use when $\ell_1 + \ell_2 \geq \pi$, and $\theta = \pi$ then corresponds only to the limiting case of spherical triangles, where the region is a hemisphere, or in the special case $\ell_1 + \ell_2 = \pi$, a quartersphere.

To form a quadrilateral, consider another spherical triangle $\overset{\frown}{\triangle} A'C'D$ with fixed side lengths $\ell_3 = \overset{\frown}{C'D}$, $\ell_4 = \overset{\frown}{DA'}$. When the angle θ' at D increases from 0 to π, the side length $d' = \overset{\frown}{A'C'}$ increases from $|\ell_3 - \ell_4|$ to the smaller of $\ell_3 + \ell_4$ and $2\pi - (\ell_3 + \ell_4)$.

On the overlap of the two intervals for d and d' respectively, for each value $d = d'$ we can match the two positively oriented spherical triangles $\overset{\frown}{\triangle} ABC$ and $\overset{\frown}{\triangle} A'C'D$ by the identification $A = A'$, $C = C'$, the union of the two regions of the triangles then defining a spherical quadrilateral $\overset{\frown}{\square} ABCD$. Letting d vary, we obtain a one parameter family of quadrilaterals with the fixed side lengths $\ell_i, 1 \leq i \leq 4$. Alternatively, we may use any one of the angles θ or θ' as the

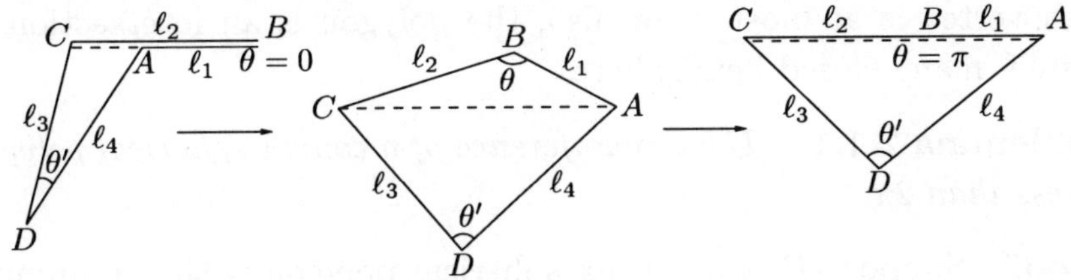

Figure 2.12

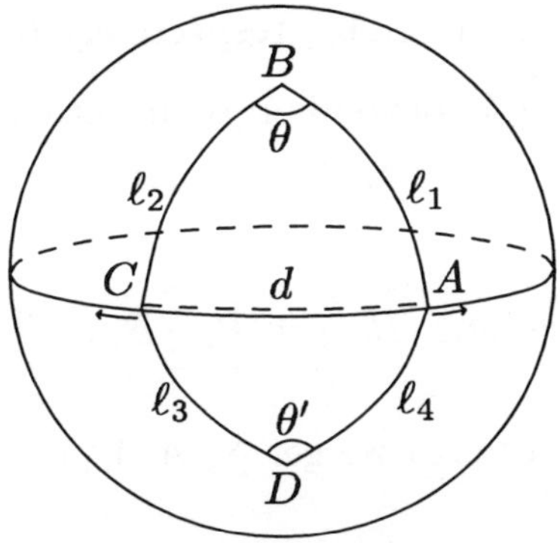

Figure 2.13

parameter in the deformation. At the beginning of the deformation one of these angles will be zero with the corresponding triangle folded, and at the end one of the angles is π with the corresponding triangle stretched, on a great circle arc (cf. Figure 2.12).

More generally we consider *spherical polygons.* They are defined as regions on the unit sphere bounded by a finite number of minimal geodesic segments, the *edges*, which are disjoint except for their end points, the *vertices.* We will assume that the region is contained in an open hemisphere, unless it is made clear from the context that this need not be the case. A spherical polygon is said to be *convex* if the minimal geodesic segment between any two of its points, is contained entirely in the polygon. Equivalently, the inner angle at each vertex is at most π, or also, the polygon is an intersection of finitely many closed hemispheres.

Sublemma 2.1.1 *The circumference of a convex spherical polygon is less than 2π.*

Proof Suppose P is a convex spherical polygon with a minimum number of vertices $A_1, \ldots, A_n$. Thus, each vertex has an inner angle $< \pi$ (otherwise it would not be counted as a vertex). Then P is contained in the lune from A_1 bounded by the longitude arcs γ_1 and γ_2 containing $\overset{\frown}{A_1A_2}$ and $\overset{\frown}{A_1A_n}$, respectively (cf. Figure 2.14). We extend each side $\overset{\frown}{A_{i-1}A_i}$ beyond A_i to an intersection point B_i with γ_2 for $2 \leq i \leq n$. By the triangle inequality for spherical distance it follows that

$$\overset{\frown}{A_{i-1}B_i} < \overset{\frown}{A_{i-1}B_{i-1}} + \overset{\frown}{B_{i-1}B_i}\,, \quad 3 \leq i \leq n \tag{73}$$

and by repeated use of (73) we get $\sum \overset{\frown}{A_iA_{i+1}} < \gamma_1 + \gamma_2 = 2\pi$. □

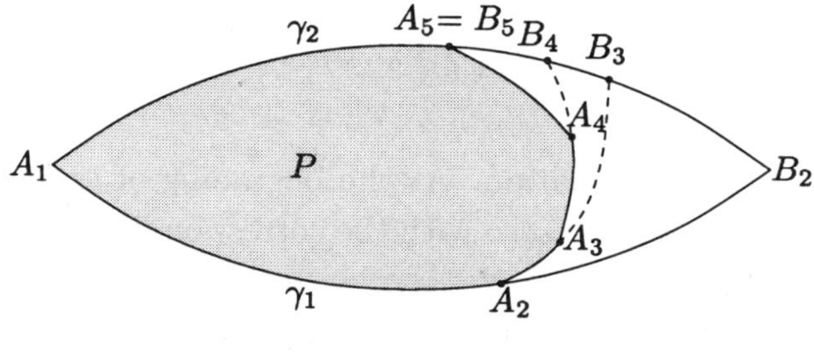

Figure 2.14

Remark One consequence of the previous result is that if a set Σ of four points is contained in a closed hemisphere, then the smallest spherical distance between pairs of the points, the *separation* of Σ, can be at most $\pi/2$. For if the separation of Σ is $> \pi/2$, and Σ is contained in a closed hemisphere, then by shrinking Σ we can assume it is in an open hemisphere, and still has a separation

$> \pi/2$. The convex hull of Σ in the open hemisphere is either a convex quadrilateral with the four points of Σ as vertices, a triangle with three of the points as vertices, or a geodesic segment. But these are all impossible when Σ has separation $> \pi/2$.

We will give a sufficient condition for a set Σ of n points on the unit sphere to be contained in an open hemisphere. For this purpose we make use of the convex polyhedron $I(\Sigma)$, which is the convex hull of Σ in the Euclidean space. As we shall see later, this polyhedron plays a basic role in the study of sphere packings (cf. Sec. 2.2). While here we only use the fact that the set Σ is contained in an open hemisphere (resp. closed hemisphere) if and only if the origin O does not belong to (resp. is not in the interior of) $I(\Sigma)$.

Sublemma 2.1.2 *Let $\Sigma = \{A_1, \ldots, A_n\}$ be a set of n points on the unit sphere and set $\ell_i = \overset{\frown}{A_i A_{i+1}}$ for $1 \leq i \leq n$ (cyclic numbering) where $n \geq 2$. If $\Sigma \ell_i < 2\pi$, then Σ is contained in an open hemisphere.*

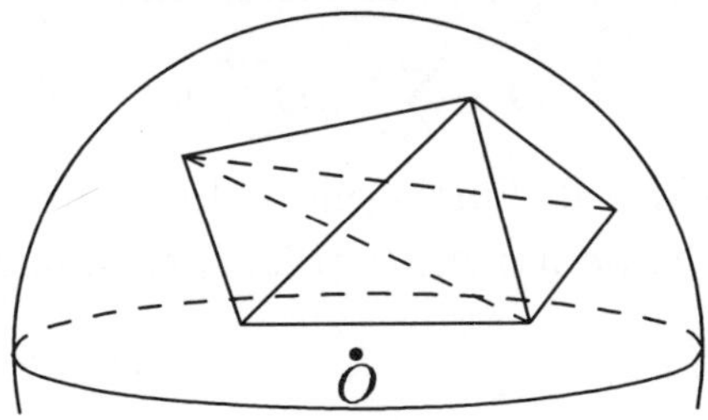

Proof The condition on Σ of having a cycle of total length less than 2π, is inherited by subsets of Σ, by the triangle inequality for spherical distance. Assume Σ is given with $\Sigma \ell_i < 2\pi$. We want to show that $O \notin I(\Sigma)$.

For $n = 3$, if $O \in I(\Sigma) = \triangle A_1 A_2 A_3$, the points A_1, A_2, A_3 must be on a great circle with $\Sigma \ell_i = 2\pi$, contrary to assumption. Hence, $O \notin I(\Sigma)$ in this case. Assume $n = 4$. Since Σ cannot contain any pair of antipodal points, its angles are well defined, namely $\angle A_iA_jA_k = \cos^{-1}(\mathbf{n}_{ji}, \mathbf{n}_{jk})$, where in general

$$\mathbf{n}_{ij} = \overrightarrow{OA_i} \times \overrightarrow{OA_j} / | \overrightarrow{OA_i} \times \overrightarrow{OA_j} |, \quad i \neq j.$$

Suppose Σ has an angle $\angle A_iA_jA_k = \pi$. Since the subset $\{A_i, A_j, A_k\}$ is contained in an open hemisphere by the previous case, it follows that $\widehat{A_iA_k} = \widehat{A_iA_j} \cup \widehat{A_jA_k}$. The subset $\Sigma \setminus \{A_j\}$ is contained in an open hemisphere D by the previous case $n = 3$. It follows that D contains the geodesic segment $\widehat{A_iA_k}$. Hence, we have $A_j \in D$ and therefore $\Sigma \subseteq D$. In this case there is nothing more to prove.

On the other hand, if all angles of Σ are less than π, then we make a shearing deformation of Σ by increasing the angle $\theta = \angle A_1A_2A_3$ (keeping all ℓ_i constant) until one of the angles of Σ has become equal to π (cf. Figure 2.12). Then, Σ will be contained in an open hemisphere at the end of the deformation.

More precisely, let $(\Sigma(\theta))_{\theta_0 \leq \theta \leq \theta_1}$ be the shearing deformation of Σ, where we have $O \notin I(\Sigma(\theta_1))$ from above. If for some parameter value θ we have $O \in I(\Sigma(\theta))$, this means that

$$O = \sum \lambda_i \overrightarrow{OA_i}(\theta)$$

with $\Sigma\lambda_i = 1$ and $\lambda_i \geq 0$. Furthermore, from the case $n = 3$ above it follows also that the points $A_i(\theta)$ of $\Sigma(\theta)$ then must be independent

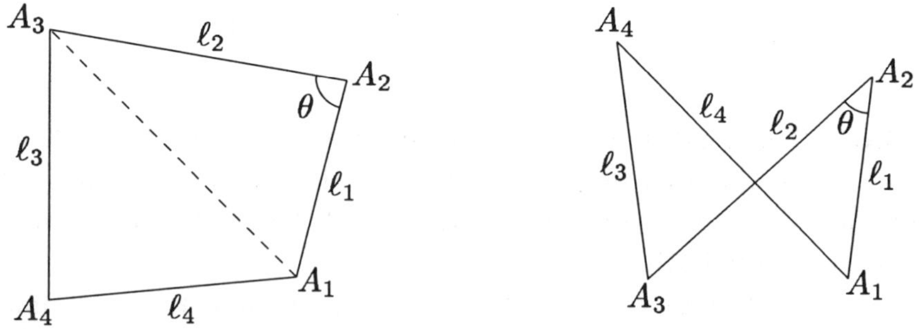

Figure 2.15

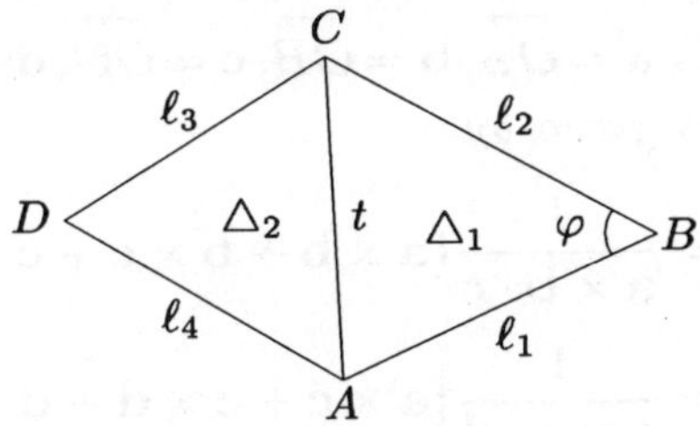

Figure 2.16

(non-coplanar) and $\lambda_i > 0$ for $1 \le i \le 4$. Therefore, the set $\{\theta : O \in I(\Sigma(\theta))\}$ must be open and closed in the interval $[\theta_0, \theta_1]$, and hence empty by connectedness of the interval. This shows that $O \notin I(\Sigma) = I(\Sigma(\theta_0))$. Thus, the case $n = 4$ is proved.

For $n > 4$, suppose we have $O \in I(\Sigma)$. Then, there exists a subset Σ' of four points in Σ such that $O \in I(\Sigma')$ (by Caratheodory's theorem for convex hulls). This gives a contradiction with the previous case $n = 4$. Therefore, the sublemma follows. □

Remark If we assume instead the condition $\Sigma\ell_i = 2\pi$, then we get that Σ is contained in a closed hemisphere. This is an easy consequence of the previous sublemma by a limit argument in the case where one of the angles $\angle A_{i-1}A_iA_{i+1}$ is less than π. On the other hand, if the angles are all equal to π, it is clear that Σ is on a great circle. Also it is easy to see that if Σ contains a pair of antipodal points, the condition $\Sigma\ell_i = 2\pi$ implies that Σ must be on the boundary of a lune.

Next, let us consider the family of spherical quadrilaterals with a given set of side-lengths $\{\ell_i, 1 \le i \le 4\}$, $\Sigma\ell_i < 2\pi$. The congruence class of such a quadrilateral is uniquely determined by an additional parameter such as the length of one diagonal or the size of one angle, namely, $t = \widehat{AC}$ or $\varphi = \angle B$ as indicated in Figure 2.16.

Therefore, the areas A of the above family of quadrilaterals can be regarded as a function of t (resp. φ).

Lemma 2.1.2 *Set* $\mathbf{a}=\overrightarrow{OA}, \mathbf{b}=\overrightarrow{OB}, \mathbf{c}=\overrightarrow{OC}, \mathbf{d}=\overrightarrow{OD}$ *and* $\{V_1, V_2\}$ *be the pair of points given by*

$$\begin{aligned}\overrightarrow{OV_1} &= \frac{1}{\mathbf{a}\times\mathbf{b}\cdot\mathbf{c}}\{\mathbf{a}\times\mathbf{b}+\mathbf{b}\times\mathbf{c}+\mathbf{c}\times\mathbf{a}\}\,,\\ \overrightarrow{OV_2} &= \frac{1}{\mathbf{a}\times\mathbf{c}\cdot\mathbf{d}}\{\mathbf{a}\times\mathbf{c}+\mathbf{c}\times\mathbf{d}+\mathbf{d}\times\mathbf{a}\}\,.\end{aligned} \tag{74}$$

Then

$$\overrightarrow{V_1V_2} = \frac{dA}{dt}\frac{\mathbf{a}\times\mathbf{c}}{|\mathbf{a}\times\mathbf{c}|}\,, \quad \frac{dA}{d\varphi} = \overrightarrow{V_2V_1}\cdot\mathbf{b}\,. \tag{75}$$

Proof Set $c_i = \cos\ell_i, 1 \le i \le 4$, and $x = 1+\cos t$. In the case of a non-convex quadrilateral, we choose $\widehat{AC}$ to be the interior diagonal. Then,

$$A = \triangle_1 + \triangle_2 = 2\tan^{-1}\frac{D_1}{u_1} + 2\tan^{-1}\frac{D_2}{u_2}\,, \tag{76}$$

where

$$\begin{aligned} u_1 &= c_1+c_2+x\,, \quad u_2 = c_3+c_4+x\,,\\ D_1^2 &= -(c_1+c_2)^2 + 2(c_1c_2+1)x - x^2\,,\\ D_2^2 &= -(c_3+c_4)^2 + 2(c_3c_4+1)x - x^2\,. \end{aligned} \tag{76'}$$

Therefore, from (57)

$$\frac{dA}{dx} = \frac{c_1+c_2-x}{xD_1} + \frac{c_3+c_4-x}{xD_2}\,. \tag{77}$$

On the other hand, both $\overrightarrow{OV_i}, i=1,2$, satisfy

$$\overrightarrow{OV_i}\cdot\mathbf{a} = 1\,, \quad \overrightarrow{OV_i}\cdot\mathbf{c} = 1\,, \tag{78}$$

and hence, there exist $k_i, i = 1,2$, such that

$$\overrightarrow{OV_i} = \frac{\mathbf{a}+\mathbf{c}}{1+\mathbf{a}\cdot\mathbf{c}} + k_i\,\mathbf{a}\times\mathbf{c}\,. \tag{79}$$

Moreover, it follows from the conditions $\overrightarrow{OV_1} \cdot \mathbf{b} = 1$ (resp. $\overrightarrow{OV_2} \cdot \mathbf{d} = 1$) that

$$k_1 = \frac{c_1 + c_2 - x}{xD_1} \quad \left(\text{resp. } k_2 = \frac{x - c_3 - c_4}{xD_2}\right). \tag{80}$$

Therefore, from (77), (79) and (80),

$$\begin{aligned}\overrightarrow{V_1V_2} &= (k_2 - k_1)\mathbf{a} \times \mathbf{c} = -\frac{dA}{dx}\mathbf{a} \times \mathbf{c} \\ &= -\frac{dA}{dt}\frac{dt}{dx}\mathbf{a} \times \mathbf{c} = \frac{dA}{dt}\frac{\mathbf{a} \times \mathbf{c}}{|\mathbf{a} \times \mathbf{c}|}, \quad \left[\frac{dx}{dt} = -|\mathbf{a} \times \mathbf{c}|\right],\end{aligned}$$

$$\begin{aligned}\overrightarrow{V_2V_1} \cdot \mathbf{b} &= (k_1 - k_2)\mathbf{a} \times \mathbf{c} \cdot \mathbf{b} = \frac{dA}{dx}(-D_1) \\ &= \frac{dA}{dx} \cdot \frac{dx}{d\varphi} = \frac{dA}{d\varphi}.\end{aligned} \tag{81}$$

□

Corollary 2.1.3 *$A(x)$ has a unique maximum when $\widehat{\square ABCD}$ is cocircular, namely, when $V_1 = V_2$. Moreover, the unique maximal point $x_0 = 1 + \cos t_0$ is given by*

$$x_0 = \frac{(c_1 + c_2)\sqrt{(1 - c_3)(1 - c_4)} + (c_3 + c_4)\sqrt{(1 - c_1)(1 - c_2)}}{\sqrt{(1 - c_3)(1 - c_4)} + \sqrt{(1 - c_1)(1 - c_2)}}. \tag{82}$$

Proof It suffices to show that the x_0 of (82) is the unique solution of

$$\left.\frac{dA}{dx}\right|_{x_0} = \frac{c_1 + c_2 - x_0}{x_0 D_1(x_0)} + \frac{c_3 + c_4 - x_0}{x_0 D_2(x_0)} = 0. \tag{83}$$

Case 1: $c_1 + c_2 = c_3 + c_4$

Then $x_0 = c_1 + c_2 = c_3 + c_4$.

Case 2: $c_1 + c_2 \neq c_3 + c_4$ (we may assume that $c_1 + c_2 > c_3 + c_4$).

In this case, (83) implies that

$$\frac{c_1 + c_2 - x_0}{x_0 - c_3 - c_4} = \frac{D_1(x_0)}{D_2(x_0)} > 0 . \tag{84}$$

Set it to be λ. Then

$$\begin{aligned}\lambda^2 &= \frac{D_1(x_0)^2}{D_2(x_0)^2} = \frac{(c_1 + c_2 - x_0)^2}{(-x_0 + c_3 + c_4)^2} \\ &= \frac{D_1(x_0)^2 + (c_1 + c_2 - x_0)^2}{D_2(x_0)^2 + (c_3 + c_4 - x_0)^2} = \frac{(1 - c_1)(1 - c_2)}{(1 - c_3)(1 - c_4)} .\end{aligned} \tag{85}$$

Therefore,

$$\lambda = \sqrt{\frac{(1 - c_1)(1 - c_2)}{(1 - c_3)(1 - c_4)}} \tag{86}$$

and hence x_0 is given by (82). □

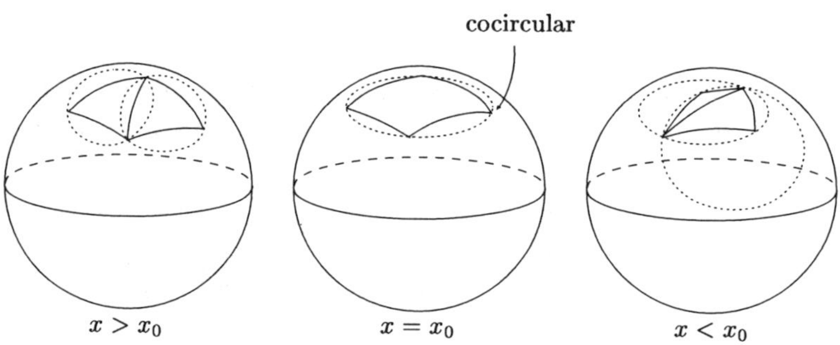

Figure 2.17

Corollary 2.1.4 *Among the spherical polygons with a set of given side lengths in cyclic order and total length less than 2π, the area maximizing one is exactly the cocircular one.*

Proof With a set of n given side lengths of sum less than 2π the n-gon is contained in an open hemisphere. We may assume the *existence* of a maximum solution (by a compactness argument).

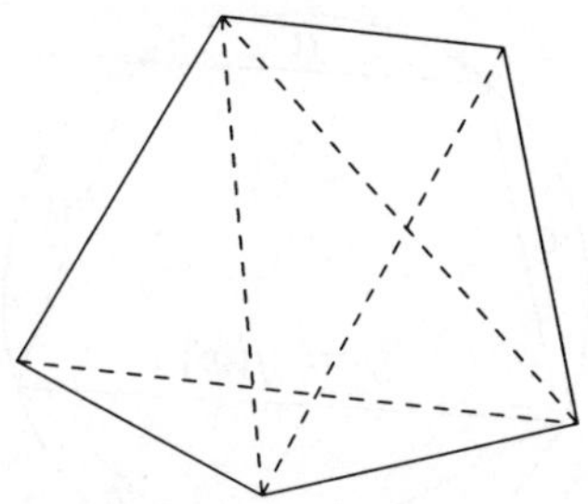

Figure 2.18

This must clearly be a convex n-gon, otherwise we could increase its area by reflecting part of it outwards. Also, each four consecutive of its vertices must be cocircular (cf. Figure 2.18), otherwise, by Lemma 2.1.2, we could increase the area by a shearing deformation, keeping all except two of the n vertices fixed. It follows that the n-gon must be cocircular. □

For $n \geq 3$ and $0 < a < 2\pi/n$ given, let $R_n(a)$ denote the circumradius of the regular spherical n-gon with side length a. Let $A_n(x)$ be the area of the cocircular spherical n-gon with $n-1$ sides of fixed length a and the remaining side of length x. Such an n-gon exists and is unique up to congruence when $0 < x < (n-1)a$ in the case that $(n-1)a < \pi$ and when $0 < x < 2\pi - (n-1)a$ in the case that $(n-1)a \geq \pi$. Furthermore, x is increasing with the circumradius.

Sublemma 2.1.3 (i) *If $(n-1)a < \pi$, then $A_n(x)$ is increasing for $0 < x < 2R_{2n-2}(a)$ and decreasing for $2R_{2n-2}(a) < x < (n-1)a$.*

(ii) *If $(n-1)a \geq \pi$, then $A_n(x)$ is increasing for $0 < x < 2\pi - (n-1)a$.*

Proof For $n = 3$ this is just the isosceles case of Corollary 2.1.1 of Lemma 2.1.1. Assume $n > 3$.

Let $\triangle(x, y)$ denote the area of the spherical triangle of side lengths $\{a, x, y\}$ (cf. Figure 2.19), and consider the function defined by

$$B_n(x, y) = \triangle(x, y) + A_{n-1}(y)\,. \tag{87}$$

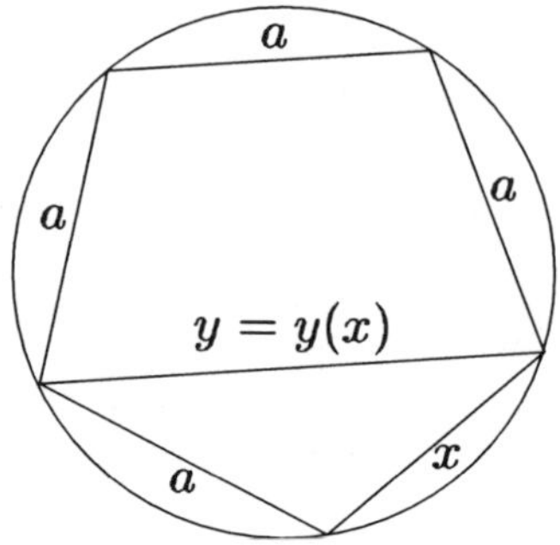

Figure 2.19

Then, we have

$$A_n(x) = B_n(x, y(x))\,, \tag{88}$$

where $y(x)$ is the length of the diagonal of the cocircular spherical n-gon with side lengths $\{a, \ldots, a, x\}$, separating the adjacent sides of lengths a and x from the rest.

Differentiation of (88) then gives

$$\begin{aligned} A_n'(x) &= \frac{\partial B_n}{\partial x}(x, y(x)) + \frac{\partial B_n}{\partial y}(x, y(x))y'(x) \\ &= \frac{\partial B_n}{\partial x}(x, y(x)) \\ &= \frac{\partial \Delta}{\partial x}(x, y(x))\,, \end{aligned} \tag{89}$$

because $\frac{\partial B_n}{\partial y}(x, y(x)) = 0$ by the previous corollary, since the spherical n-gon is cocircular. But by Corollory 2.1.1 we have $\partial\Delta/\partial x > 0$ when the circumcenter is inside the edge of length x, and $\partial\Delta/\partial x < 0$ when it is outside. This means that the function $A_n(x)$ is increasing up to the point where x becomes the diameter of the circumcircle and from there decreasing, in the case where $(n-1)a < \pi$, and that the function $A_n(x)$ is increasing for all x, in the case $(n-1)a \geq \pi$. □

Example 2.1.1 Cocircular quadrilaterals with an equal pair of opposite side lengths Let $\widehat{\Box}ABCD$ be a cocircular quadrilateral with side-lengths $\ell_2 = \ell_4 = a$, $\ell_1 = c$ and $\ell_3 = b$. Then, the pair of diagonals are equal and $\det(\mathbf{a}, \mathbf{b}, \mathbf{c}) = \det(\mathbf{a}, \mathbf{b}, \mathbf{d})$. Hence,

$$\det(\mathbf{a}, \mathbf{b}, \mathbf{c}) \cdot \det(\mathbf{a}, \mathbf{b}, \mathbf{d}) - \det(\mathbf{a}, \mathbf{b}, \mathbf{c})^2 = 0\,, \tag{90}$$

namely

$$\begin{vmatrix} 1 & \cos c & \cos d \\ \cos c & 1 & \cos a \\ \cos a & \cos d & \cos b \end{vmatrix} - \begin{vmatrix} 1 & \cos c & \cos d \\ \cos c & 1 & \cos a \\ \cos d & \cos a & 1 \end{vmatrix} = 0\,. \tag{90$'$}$$

Therefore,

$$\cos d = \cos a - \sqrt{(1 - \cos b)(1 - \cos c)} \tag{91}$$

and it is straightforward to compute $\cos B$ and $\cos D$ by the cosine law, namely

$$\begin{aligned} \cos B &= \frac{1}{\sin a}\left\{\cos a \sqrt{\frac{1 - \cos c}{1 + \cos c}} - \sqrt{\frac{1 - \cos b}{1 + \cos c}}\right\}, \\ \cos D &= \frac{1}{\sin a}\left\{\cos a \sqrt{\frac{1 - \cos b}{1 + \cos b}} - \sqrt{\frac{1 - \cos c}{1 + \cos b}}\right\}, \end{aligned} \tag{92}$$

and the area is given by

$$A(\Box) = 2(B + D - \pi)\,. \tag{93}$$

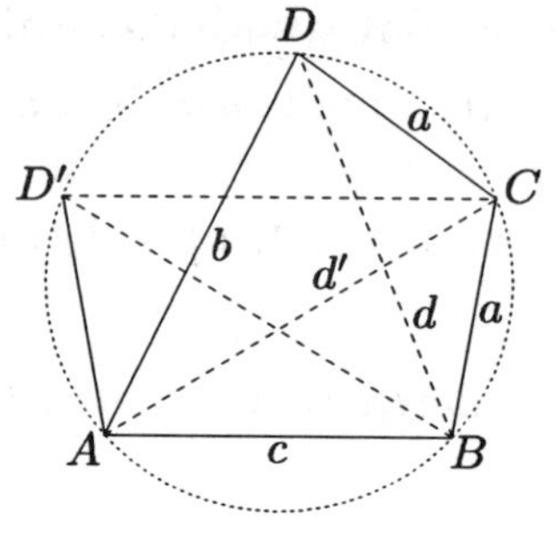

Figure 2.20

Example 2.1.2 Cocircular quadrilateral with an equal pair of adjacent side lengths Let $\widehat{\square}ABCD$ be a cocircular quadrilateral with side-lengths $\ell_1 = c$, $\ell_2 = \ell_3 = a$ and $\ell_4 = b$. As indicated in Figure 2.20, $\widehat{\triangle}ACD \cong \widehat{\triangle}CAD'$ while $\widehat{\square}ABCD'$ is just the one discussed in Example 2.1.2.

Set $d = \widehat{BD}$, $d' = \widehat{AC} = \widehat{BD'}$. Then $\cos d'$ is given by (91) while $\cos d$ is given by (82), namely

$$\cos d = x_0 - 1 = \frac{4\cos a \sin\frac{b}{2}\sin\frac{c}{2} + (\cos b + \cos c)(1 - \cos a)}{(1 - \cos a) + 2\sin\frac{b}{2}\sin\frac{c}{2}} - 1\,. \tag{94}$$

Note that $\cos B$ and $\cos D$ are again given by (92) while $A + C = B + D$. Therefore, the area is again given by (93) and the only remaining one is the following formula of $\cos A$, namely

$$\sin b \sin c \cos A = \cos d - \cos b \cos c \tag{95}$$

where $\cos d$ is given by (94).

In the special case of $a = \frac{\pi}{3}$, set $b = 2t$ and $c = 2u$. Then the formulas (94) and (95) can be simplified as follows, namely

$$\cos d = \frac{\frac{1}{2}(\cos 2t + \cos 2u - 1)}{\frac{1}{2} + 2\sin t \sin u}\,, \tag{94'}$$

$$\cos A = -\frac{\sin t \sin u + \cos 2t \cos 2u}{\cos t \cos u + \sin 2t \sin 2u}. \qquad (95')$$

Example 2.1.3 Lexell deformation and Lexell radius Lexell's Theorem asserts that the top vertices of the family of triangles, $\{\widehat{\triangle}ABC\}$, with fixed B and C and a given area $\triangle$, constitute a circular arc passing through the antipodal points of B and C, say denoted by B' and C' as indicated in Figure 2.5. We shall call the radius of the above circular arc the *Lexell radius* which is completely determined by the given length $a = \widehat{BC}$ and the given area $\triangle$, namely, it is a function $r(a, \triangle)$. Therefore, the quantitative analysis of Lexell deformation naturally begins with the explicit computation of $r(a, \triangle)$ in terms of a and $\triangle$.

For a given pair of $\{a, \triangle\}$ and fixed $\{B, C\}$, there is a special pair of triangles $\widehat{\triangle} A_i BC$, $i = 1, 2$, such that $\widehat{\square} A_1 A_2 BC$ is a *cocircular* quadrilateral with side-lengths $\{a, b, a, b\}$. As indicated in Figure 2.21, $\square A_1 A_2 C' B'$ is a cocircular quadrilateral with side-lengths $\{a, \pi - d, a, \pi - d\}$, where d is the diagonal length of $\widehat{\square} A_1 A_2 BC$.

Note that the pair of diagonals of $\widehat{\square} A_1 A_2 C' B'$ are both equal to $(\pi - b)$ and they bisect each other. Hence,

$$r(a, \triangle) = \frac{1}{2}(\pi - b), \quad \cos r(a, \triangle) = \sin\frac{b}{2}. \qquad (96)$$

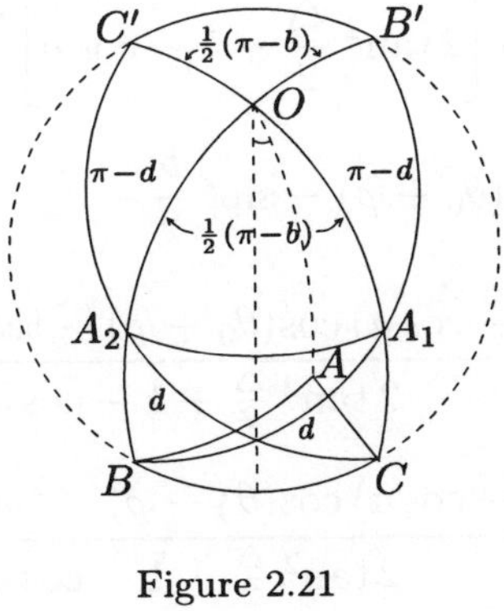

Figure 2.21

Thus, it suffices to compute $\sin\frac{b}{2}$ in terms of $\{a, \triangle\}$ as follows, namely

$$\begin{aligned} 1+\cos d &= \cos a + \cos b\,, \\ \tan\frac{\triangle}{2} &= \frac{D}{u} = \frac{1}{\sqrt{2}}\sqrt{\frac{(1-\cos a)(1-\cos b)}{\cos a + \cos b}}\,, \\ \cos b &= \frac{1-\cos a(2\tan^2\frac{\triangle}{2}+1)}{2\tan^2\dfrac{\triangle}{2}+1-\cos a}\,, \\ \cos r(a,\triangle) &= \sin\frac{b}{2} = \tan\frac{\triangle}{2}\left\{\frac{1+\cos a}{2\tan^2\frac{\triangle}{2}+1-\cos a}\right\}^{\frac{1}{2}}\,. \end{aligned} \tag{97}$$

With the above formulas of b and $r(a, \triangle)$ at hand, it is straightforward to provide an explicit parametric representation of the Lexell deformation. Set

$$\theta_0 = \frac{1}{2}\angle A_2OA_1\,, \quad \angle BOA = \theta_0 + \varphi\,, \quad \angle AOC = \theta_0 - \varphi\,. \tag{98}$$

Then,

$$\begin{aligned} \sin\theta_0 &= \frac{\sin\frac{a}{2}}{\sin r(a,\triangle)} = \left\{\frac{1-\cos a}{1+\cos b}\right\}^{\frac{1}{2}} \\ &= \frac{1}{\sqrt{2}}\cos\frac{\triangle}{2}\left[2\tan^2\frac{\triangle}{2}+1-\cos a\right]^{\frac{1}{2}}\,, \\ \cos\widehat{AB} &= \cos^2\frac{b}{2}\cos(\theta_0+\varphi) - \sin^2\frac{b}{2} \\ &= \frac{\sec^2\frac{\triangle}{2}(1-\cos a)\cos(\theta_0+\varphi) - \tan^2\frac{\triangle}{2}(1+\cos a)}{2\tan^2\frac{\triangle}{2}+1-\cos a}\,, \\ \cos\widehat{AC} &= \frac{\sec^2\frac{\triangle}{2}(1-\cos a)\cos(\theta_0-\varphi) - \tan^2\frac{\triangle}{2}(1+\cos a)}{2\tan^2\frac{\triangle}{2}+1-\cos a}\,. \end{aligned} \tag{99}$$

2.2 Spherical Configurations, Area Estimates and a New Proof of the Impossibility of Thirteen Touching Neighbors

The unit sphere is the natural parametrization space of the set of *directions* in the space, or the set of *ratios* of complex numbers (i.e. the Riemann sphere in complex analysis). Thus, many basic geometric as well as algebraic objects, such as the chart of celestial stars and the roots of a polynomial, are naturally represented by the distribution of a finite number of points on the unit sphere. Therefore, many basic geometric as well as algebraic problems can be reduced to the study of spherical geometry and the distribution of a finite number of points on $S^2(1)$.

The *global geometry of finite point sets* is a *special feature of spherical geometry* which has *no analog* in either Euclidean geometry or hyperbolic geometry. This is due to the fact that spherical spaces are *finite in expanse* while Euclidean spaces and hyperbolic spaces are both of *infinite* expanse. Thus, the distribution of just a finite number of points in a sphere already constitutes a global problem, but a set of finite points in the Euclidean or hyperbolic space is always confined to a *negligible corner* of the infinite expanse, thus having no global significance.

Many basic geometric problems can be reduced to problems on the distribution of a given number of points in $S^2(1)$ (or its higher dimensional generalizations $S^n(1)$), and such problems are often rather challenging. In this section, we shall develop some basic techniques for studying this type of problem. Let us begin with the formulation of some specific problems of this type.

2.2.1 *Examples of problems on the distribution of point on $S^2(1)$*

1. The problem of thirteen spheres If one packs three circles of identical radii close together in the plane, they form a triple touching each other, and the three intervals joining their centers form a *regular* triangle with the three touching points as their middle points. Thus, the two touching points on the same circle have an *angular separation* of $\pi/3$. Therefore, a circle can be surrounded by at most six (non-overlapping) touching circles, and such a *local circle packing* with six touching neighbors is unique up to congruence, namely, the distribution of the six touching points on a circle with angular separations of at least $\pi/3$, is unique up to a rotation. The above basic geometry on local circle packings *intuitively suggests* that the *tightest global circle packing* should be that which *replicates* the above local pattern globally. Actually, this is also the basic fact that the proof of the theorem on optimal global circle packings is based on (cf. Sec. 3.1). Therefore, as one tries to study the problem of optimal global sphere packings, it is rather natural to begin with an investigation of the *maximal number of touching neighbors* that one can arrange around a central sphere in a local sphere packing. This was most likely the reason why both I. Newton and D. Gregory were interested in such a problem.

In an unpublished notebook of Christ Church in Oxford, there is a recorded discussion of 1694 between Newton and Gregory on the above problem. Newton believed that twelve should be the maximal number of touching neighbors, while Gregory thought thirteen might be possible. This record was discovered later, and the topic of their discussion became a famous problem in solid geometry, often referred to as the *problem of thirteen spheres*, or Newton's problem. The maximal number was finally proved to be twelve in 1953 by Schütte and van der Waerden [SW-2] and in 1956 by Leech [Lee]. [In fact, there were several incorrect proofs before that.]

Let O_i, $1 \leq i \leq 3$, be the centers of three spheres of identical radii touching each other. Then the plane passing through their centers intersects the triple of spheres at a triple of great circles touching each other. This shows that the angular separation between each pair of touching points must be at least equal to $\pi/3$. Therefore, the above problem of thirteen spheres can directly be reduced to the following problem on the distribution of thirteen points on $S^2(1)$, namely

"*Is it possible to distribute thirteen points on $S^2(1)$ such that the pairwise spherical distances among them are at least equal to $\pi/3$?*"

2. Problem of spherical codes Let Σ be a finite set of points on $S^2(1)$. The *minimal* (spherical) distance between points of Σ is defined to be the *separation of* Σ.

Among all finite subsets of a *fixed cardinality* in $S^2(1)$, what is the *maximal separation*? and what are those optimal distributions of a given number of points that achieve the maximal separation? (See Tammes problem in [CFG] for references.)

Set δ_n to be the *maximal separation of n points in* $S^2(1)$. The above problem of thirteen spheres simply asks whether $\delta_{13} \geq \pi/3$? Thus, the above problem of spherical codes is a natural refinement and generalization of the former. So far, the above problem of spherical codes has only been settled for some small n such as $2 \leq n \leq 12$ and $n = 24$. For example, the beginning cases of $n = 2, 3, 4$ are quite simple, namely

(i) $\delta_2 = \pi$, the antipodal pairs are the optimal ones;
(ii) $\delta_3 = 2\pi/3$, an optimal triple is always arranged on a great circle symmetrically;
(iii) $\delta_4 = \cos^{-1}(-1/3)$, an optimal quadruple always forms the vertices of a regular tetrahedron.

The cases of $n = 5, 6, 7, 8$ and 9 were solved by Schütte and van der Waerden in a paper in 1951 [SW-1], in which they developed the techniques used in their proof of the impossibility of thirteen touching neighbors in [SW-2], (see [Fej-2], [C-S] for references of problems of this type).

3. Isoperimetric problem for polyhedra of a given number of faces The original *isoperimetric problem* seeks the solution of plane (or spherical) regions of a given *perimeter length* which enclose the *maximal possible area*, or turning the table around, of a *given area* with *minimal* possible perimeter-length. It is natural to generalize it to regions in higher dimensional Euclidean, spherical or hyperbolic spaces. This beautiful geometric problem of basic importance was solved in its full generality for the above three families of spaces in the 1940's [Tal]. The above three families of spaces are commonly characterized by the symmetry condition that the local isometry group at an arbitrary point, ISO (p, M^n), is isomorphic to $O(n)$. An isoperimetric region in such a space is *proven to be invariant* under the local isometry group fixing its center of gravity, and hence it is a spherical ball.

Although the above isoperimetric problem for regions in general has already been successfully solved, the isoperimetric problem restricted to polyhedra of a given number of faces remains unsolved for most cases (i.e. except for a few special numbers of faces such as 4, 5, 6 and 12, cf. [Gol]). We mention here the following theorem of L. Lindelöf, in a strengthened form due to Minkowski, which reduces the above isoperimetric problem of polyhedra to a problem of spherical geometry.

Theorem (Lindelöf-Minkowski) *To a given convex polyhedron K of k faces, set Σ to be the k points on the unit sphere representing the (outward) normal directions of its k faces and $T(\Sigma)$ to be the circumscribing polyhedron with Σ as the set of tangent points. Then*

$$\frac{\operatorname{Vol} T(\Sigma)^2}{\operatorname{Area} \partial T(\Sigma)^3} \geq \frac{\operatorname{Vol} K^2}{\operatorname{Area} \partial K^3} \tag{1}$$

and equality holds if and only if $T(\Sigma)$ and K are homothetic to each other.

The above theorem shows that the isoperimetric ratio, namely Vol K^2/Area ∂K^3, always improves when one replaces the given polyhedron by its associated circumscribing polyhedron with parallel corresponding faces, thus reducing the isoperimetric problem of polyhedra to the following optimization problem, namely

Problem *Among all circumscribing polyhedra of the unit sphere with k faces, which one has the minimal volume?*

We also mention here the dual of the above problem, namely

Problem *Among all inscribing polyhedra of the unit sphere with k vertices, which one has the maximal volume?*

Now, let us proceed to develop some basic techniques for studying such types of problems on the distributions of points on the unit sphere. First of all, let us introduce the concept of spherical configurations.

2.2.2 *Spherical configuration and some basic techniques of area estimation*

Intuitively, it is quite clear that the spherical trigonometry and vector algebra we discussed in Sec. 2.1 already constitute a complete set of basic tools for the study of the spherical geometry of finite points. However, we need to develop some good organization and a systematic method in order to analyze the geometry of finite sets of points by efficient applications of such basic tools.

Let Σ be a given set of n points in $S^2(1)$ whose position vectors are the n unit vectors $\{\mathbf{a}_i, 1 \le i \le n\}$. The set of $\frac{1}{2}n(n-1)$ pairwise spherical distances (or equivalently, pairwise inner products $\{\mathbf{a}_i \cdot \mathbf{a}_j = \cos d_{ij},\ 1 \le i < j \le n\}$), of course, already constitutes a *complete set of congruence invariants* of the given set Σ. However, for all cases of $n \ge 4$, the above set of inner products has many relations of the type of Eq. (1.69) in Sec. 2.1. Therefore, the moduli space of congruence classes of sets of n points in $S^2(1)$ is already quite complicated, even for rather small n such as $n = 13$. (This is why the problem of thirteen spheres remained open for more than 250 years (1694-1953).)

Thus, the study of spherical geometry of finite point sets should begin with the important first step of providing a "*good organization*" of the above large collection of inner products with lots of interlocking relations, namely, to *select* a *well-organized*, *concise subset of complete invariants* from the whole set of pairwise inner products. The following definition of the *associated spherical configuration* is exactly one of the canonical ways of providing such an efficient organization.

Definition Given a finite point set Σ in $S^2(1)$, set $I(\Sigma)$ to be the inscribing polyhedron with Σ as the set of its vertices (i.e. $I(\Sigma)$ is the convex hull of Σ in the Euclidean space). The radial projections of all the edges of $I(\Sigma)$ constitute a collection of spherical arcs which subdivide the unit sphere into spherical polygons. Such a spherical configuration is called the *associated spherical configuration of* Σ. We shall denote it by $\mathcal{S}(\Sigma)$.

(i) $\mathcal{S}(\Sigma)$ and $I(\Sigma)$, of course, have the same set of vertices and they also have one-to-one correspondences both between their edges and between their faces. Each face of $I(\Sigma)$ is, of course, planary, thus each face of $\mathcal{S}(\Sigma)$ is a *cocircular* spherical polygon. In the case of a non-triangular face of $\mathcal{S}(\Sigma)$ (such as a quadrilateral or pentagonal one), it follows from the cocircular property that all its diagonal lengths are already determined by its side-lengths. Therefore, the cosines of the set of edges of $\mathcal{S}(\Sigma)$, indeed, constitute a concise subset of pairwise inner products which determine $\mathcal{S}(\Sigma)$ up to congruence.

(ii) It follows from the convexity of $I(\Sigma)$ that the whole polyhedron is located on one side of the support plane of one of its faces. Therefore, to each given face of $\mathcal{S}(\Sigma)$ (which is always a *cocircular* polygon), there can be *no* points of Σ inside of its circumcircle. We shall call this important property of $\mathcal{S}(\Sigma)$ the *convexity property* because it originated from the convexity of $I(\Sigma)$.

Remark If the origin is in the interior of $I(\Sigma)$, then each face of $\mathcal{S}(\Sigma)$ is strictly contained in a hemisphere. Otherwise, the whole set of Σ is contained in a suitable hemisphere. Such a distribution of points will, for most geometric problems, play *no* significant role. Therefore, we shall *always assume* that $I(\Sigma)$ *contains the origin in its interior* without saying so explicitly.

Now, let us take the problem of spherical codes and the problem of thirteen spheres as the "*motivating problems*" for the development of basic techniques in the study of the geometry of spherical configurations.

Let $\delta(\Sigma)$ be the *minimal* pairwise distances of points of Σ, i.e. the *separation of* Σ. Then, it is not difficult to see that $\delta(\Sigma)$ is also equal to the *minimal edge-length* of $\mathcal{S}(\Sigma)$, namely, it follows from the above *convexity property of* $\mathcal{S}(\Sigma)$ that the distances between pairs of vertices *not* linked by edges, are always larger than $\delta(\Sigma)$. Thus, the proof of *non-existence* of a set of thirteen points in $S^2(1)$ with a *separation of at least* $\pi/3$ amounts to showing the *non-existence* of a spherical configuration with thirteen vertices and with edge-lengths of at least $\pi/3$. Intuitively speaking, what one needs to prove is that $S^2(1)$ does not have enough "room" to accommodate thirteen points with $\pi/3$-separation. The total area of $S^2(1)$ is equal to 4π (Archimedes' Theorem), and this can be regarded as a quantitative measurement of the total amount of "room" in $S^2(1)$. Such an intuition suggests that the following *method of area estimation* will be a natural approach to prove such a non-existence result, namely, to show such a hypothetical spherical configuration must have a total area *exceeding* 4π. Such a contradiction will, certainly, prove the non-existence of such a hypothetical spherical configuration. Of course, the effectiveness of the above method of area estimation will mainly depend on the "efficiency" of the techniques of area estimates that one possesses.

2.2.3 *Techniques of area estimates*

Definition A finite set of points Σ in $S^2(1)$ is said to be *a-saturated* if the adding of any other point to Σ will *change the separation from at least equal to* a to *less than* a, namely, $\delta(\Sigma) \geq a$ but $\delta(\Sigma + pt) < a$.

Lemma 2.2.1 *If Σ is an a-saturated set with at least four points, then all faces in $\mathcal{S}(\Sigma)$ are either triangles, quadrilaterals or pentagons, and the area of any face is at least equal to the area of the cocircular spherical polygon with the same number of sides of lengths equal to a.*

Proof The condition that the origin is an interior point of $I(\Sigma)$ is here always satisfied. For if $a \leq \pi/2$, this follows since Σ is a-saturated. On the other hand, suppose $a > \pi/2$. Then, with at least four points and separation more than $\pi/2$, the set Σ cannot be contained in a closed hemisphere, see remark to Sublemma 2.1.1. Hence, it follows again that the origin is in the interior of $I(\Sigma)$.

The assumption of a-saturacy of Σ implies that the circumradii of all faces of $\mathcal{S}(\Sigma)$ are less than a. For otherwise, the existence of a face with its circumradius at least equal to a will enable us to extend Σ by adding its circumcenter and still have separation at least equal to a, by the convexity of $\mathcal{S}(\Sigma)$.

Let r_1 (resp. r_2, r_3) be the circumradius of the regular spherical triangle (resp. quadrilateral, pentagon) of side length equal to a, defined when $a < 2\pi/3$ (resp. $a < \pi/2$, $a < 2\pi/5$). Then, the top angle of the isosceles triangle with side lengths $\{r_1, r_1, a\}$ (resp. $\{r_2, r_2, a\}$, $\{r_3, r_3, a\}$) is equal to $2\pi/3$ (resp. $\pi/2, 2\pi/5$). Hence,

$$\sin r_1 = \frac{2}{\sqrt{3}} \sin\frac{a}{2} \left(\text{resp. } \sin r_2 = \sqrt{2}\sin\frac{a}{2},\ \sin r_3 = \frac{\sin\frac{a}{2}}{\sin\frac{\pi}{5}}\right). \tag{2}$$

Let $\triangle(r,a)$ (resp. $\square(r,a)$, $\pentagon(r,a)$) be the area of the spherical triangle (resp. cocircular quadrilateral, pentagon) with circumradius equal to r and two (resp. three, four) of its side lengths equal to a.

From Sublemma 2.1.3 it follows that $\triangle(r,a)$ as a function of r is either ($a < \pi/2$) increasing for $r_1 < r < r_2$ and decreasing for

$r_2 < r < a$, or $(\pi/2 \leq a < 2\pi/3)$ increasing for $r_1 < r < a$. The minimal value of $\triangle(r,a)$ on the interval $r_1 \leq r < a$ is therefore $\triangle(r_1,a) = \triangle(a,a) = \triangle_a$ in the case that $a < \pi/2$, and $\triangle(r_1,a) = \triangle_a$ in the case that $\pi/2 \leq a < 2\pi/3$.

From Sublemma 2.1.3 it follows also that $\square(r,a)$ is increasing for $r_2 < r < a$, and $\pentagon(r,a)$ is increasing for $r_3 < r < a$, because the regular spherical hexagon and octagon of side length a have circumradii always exceeding a. The minimal values of $\square(r,a)$ on the interval $r_2 \leq r < a$ and $\pentagon(r,a)$ on $r_3 \leq r < a$ are therefore $\square(r_2,a) = \square_a$ and $\pentagon(r_3,a) = \pentagon_a$, respectively.

On the other hand, it follows from the Lexell theorem (cf. Figure 2.22) that the area of a spherical triangle (resp. cocircular quadrilateral, pentagon) with circumradius r and all its side lengths at least equal to a, is at least equal to $\triangle(r,a)$ (resp. $\square(r,a)$, $\pentagon(r,a)$). Therefore, the area of each triangular (resp. quadrilateral, pentagonal) face of $\mathcal{S}(\Sigma)$ is at least equal to $\triangle_a$ (resp. $\square_a$, $\pentagon_a$). Moreover, $\mathcal{S}(\Sigma)$ cannot have any face with six or more sides, because the circumradius of such a cocircular polygon always exceeds a. $\square$

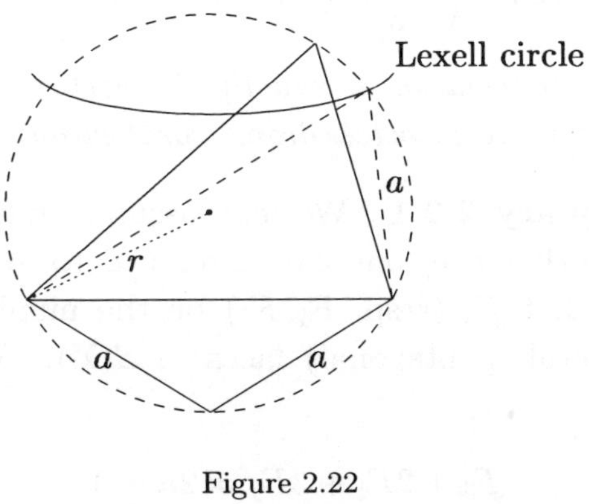

Figure 2.22

For the areas of the regular triangle, quadrilateral and pentagon with side lengths a we may use the expressions (2) for the circumradii to get

$$
\begin{aligned}
\triangle_a &= 3\cos^{-1}\frac{\cos a}{1+\cos a} - \pi\,, & a \le 2\pi/3\,;\\
\square_a &= 4\cos^{-1}\frac{\cos a - 1}{\cos a + 1} - 2\pi\,, & a \le \pi/2\,;\\
\pentagon_a &= 10\cos^{-1}\sqrt{\frac{\cos a - \cos\frac{2\pi}{5}}{1+\cos a}} - 3\pi\,, & a \le 2\pi/5\,,
\end{aligned}
\tag{3}
$$

(including the limit cases of closed hemispheres).

Corollary 2.2.1 *If Σ is a set of n points ($n \ge 4$) in $S^2(1)$ with its separation equal to a, then*

$$(2n-4)\triangle_a \le 4\pi \tag{4}$$

and the above equality holds when and only when $\mathcal{S}(\Sigma)$ consists of $(2n-4)$ regular triangles of side-length equal to a.

Corollary 2.2.2 *Among all finite subsets of $S^2(1)$ with 4 (resp. 6 or 12) points, the maximal separation is equal to*

$$\cos^{-1}\left(-\frac{1}{3}\right) \quad (\textit{resp. } \pi/2, \tan^{-1} 2) \tag{5}$$

and the optimal distribution is exactly the vertices of the inscribing regular tetrahedron (resp. octahedron, icosahedron).

Proof of Corollary 2.2.1 We may assume that Σ is already *a-saturated.* For otherwise, one can extend Σ to an a-saturated set with bigger n. Let F_3 (resp. F_4, F_5) be the number of triangular (resp. quadrilateral, pentagonal) faces in $\mathcal{S}(\Sigma)$. Then, by Euler's formula

$$F_3 + 2F_4 + 3F_5 = 2n - 4\,. \tag{6}$$

On the other hand, it is easy to check directly that

$$\square_a > 2\triangle_a\,, \quad \pentagon_a > 3\triangle_a\,. \tag{7}$$

Therefore, it follows from (6) and Lemma 2.2.1 that

$$(2n-4)\triangle_a \le F_3\triangle_a + F_4\square_a + F_5 \pentagon_a \le 4\pi \tag{8}$$

and the equality holds only when $\mathcal{S}(\Sigma)$ consists of $(2n-4)$ regular triangles of side-length equal to a. □

Proof of Corollary 2.2.2 What one needs to prove is that the equality of (4) holds when and only when $n = 4$, 6 or 12 and a is given by (5). Notice that the angles of such regular triangular faces of $\mathcal{S}(\Sigma)$ can only be equal to $\frac{2\pi}{3}$, $\frac{\pi}{2}$ or $\frac{2\pi}{5}$ while the corresponding number of vertices must be equal to 4, 6 or 12. Hence, Corollary 2.2.2 follows directly from Corollary 2.2.1. □

Corollary 2.2.3 *The separation of an arbitrary set of n points in $S^2(1)$ is bounded above by*

$$\cos^{-1}\left(\frac{\cos\frac{n\pi}{3n-6}}{1-\cos\frac{n\pi}{3n-6}}\right)(n \ge 4)\,. \tag{9}$$

Proof Let a be the separation of a set Σ of n points in $S^2(1)$. Then, by Corollary 2.2.1,

$$\triangle_a = \left(3\cos^{-1}\frac{\cos a}{1+\cos a} - \pi\right) \le \frac{2\pi}{n-2}\,. \tag{10}$$

Therefore,

$$\cos^{-1}\frac{\cos a}{1+\cos a} \le \frac{n\pi}{3n-6} \tag{11}$$

and hence,

$$a \le \cos^{-1}\left(\frac{\cos\frac{n\pi}{3n-6}}{1-\cos\frac{n\pi}{3n-6}}\right)\,. \tag{12}$$

Remarks (i) The above general upper bound (i.e. for all n) becomes the *optimal* one *only* for the cases $n = 4, 6$ or 12.

(ii) For $n = 14$, the above upper bound is smaller than $\pi/3$, thus providing a proof that it is *impossible* to pack fourteen touching neighbors around a central sphere. However, the above upper bound is still larger than $\pi/3$ for $n = 13$. This means the area estimate of Lemma 2.2.1 is still *not good enough* to provide a proof of the impossibility of thirteen touching neighbors! Therefore, one needs to develop some additional area estimation techniques in order to prove that "thirteen is impossible."

Let us begin with the following analysis of the geometric situation:

The geometry of arranging points on $S^2(1)$ with *at least* $\pi/3$ *separation* corresponds to that of *packing* $\pi/6$*-circles* on $S^2(1)$. In the *tight local pattern* of three spherical $\pi/6$-circles touching each other, the triple of their centers span a spherical regular triangle of side-length equal to $\pi/3$. Its three angles are equal to $\cos^{-1}\frac{1}{3}$, which is slightly less than $\frac{2\pi}{5}$ but considerably larger than $2\pi/6$. Therefore, in the geometry of local packing of $\pi/6$-circles on $S^2(1)$, there is more than enough room to surround a circle by five *touching* neighbors, but there is far from adequate room to accommodate six touching neighbors. Hence, in the case of a local packing of $\pi/6$-circles on $S^2(1)$ with six (or more) neighbors, some of them must be considerably further away from touching. Translating the above geometric fact on local packings of $\pi/6$-circles back to the geometry of spherical configurations with edge-lengths of at least $\pi/3$, the total area of the star-configuration of a vertex in $\mathcal{S}(\Sigma)$ consisting of six triangles, must be considerably larger than $6\triangle_{\pi/3}$.

Now, in a hypothetical spherical configuration, $\mathcal{S}(\Sigma)$, with thirteen vertices and with edge-lengths at least equal to $\pi/3$, one may add diagonals to those non-triangular faces of $\mathcal{S}(\Sigma)$, if any, so as to cut $\mathcal{S}(\Sigma)$ into a triangulation of $S^2(1)$. It will have 22 triangles and 33 edges. Each edge is, of course, shared by a pair of vertices and

$$2 \times 33 = 66 = 5 \times 13 + 1\,. \tag{13}$$

The above simple arithemetic shows that the star-configuration of at least one vertex must have *more* than five triangles. Thus, if one can prove a sufficiently strong area estimation lemma on the total area of a $6\triangle$-star configuration, then such an additional area estimate may already be good enough to show that the total area of the hypothetical spherical configuration $\mathcal{S}(\Sigma)$ must exceed 4π, thus succeeding in proving the impossibility of thirteen touching neighbors by the method of area estimation.

2.2.4 *Star configurations*

The *star configuration* of a vertex A in a given spherical configuration $\mathcal{S}(\Sigma)$ is, by definition, the subconfiguration consisting of those faces of $\mathcal{S}(\Sigma)$ with A as their common vertex. Therefore, the geometric structure of $St(A)$ is exactly the localization of $\mathcal{S}(\Sigma)$ to the surrounding of A, i.e. a kind of *local spherical configuration.* Notice that a star configuration also satisfies the following *convexity property*, which is simply the restriction of the convexity property of spherical configurations, namely, every face of $St(A)$ is *cocircular* and the *interior* of each circumcircle contains *no* vertices of $St(A)$. Set $\{B_i, 1 \le i \le k\}$ to be the vertices on the boundary edges of $St(A)$ in cyclic order. We shall always cut those non-triangular faces of $St(A)$, if any, into triangles by adding their radial diagonals. Thus, a star configuration $St(A)$ with k boundary edges (resp. vertices) has exactly k triangular faces, and will be referred as a $k\triangle$-star. For example, the pair of $6\triangle$-stars indicated in Figure 2.23 both consist of two $\frac{\pi}{3}$-equilateral triangles and two $\frac{\pi}{3}$-equilateral quadrilaterals. Their total areas are both equal to $2\cos^{-1}(-\frac{1}{3})$, namely

$$\begin{aligned} 2\square_{\frac{\pi}{3}} + 2\triangle_{\frac{\pi}{3}} &= 8\cos^{-1}\left(-\frac{1}{3}\right) + 6\cos^{-1}\frac{1}{3} - 6\pi \\ &= 2\cos^{-1}\left(-\frac{1}{3}\right) \\ &= 6\triangle_{\frac{\pi}{3}} + 0.51355288\ldots . \end{aligned} \tag{14}$$

Lemma 2.2.2 *Let $St(A)$ be a $6\triangle$-star with all edge-lengths at least equal to $\frac{\pi}{3}$. Then, its total area is at least equal to $2\cos^{-1}(-\frac{1}{3})$ and it*

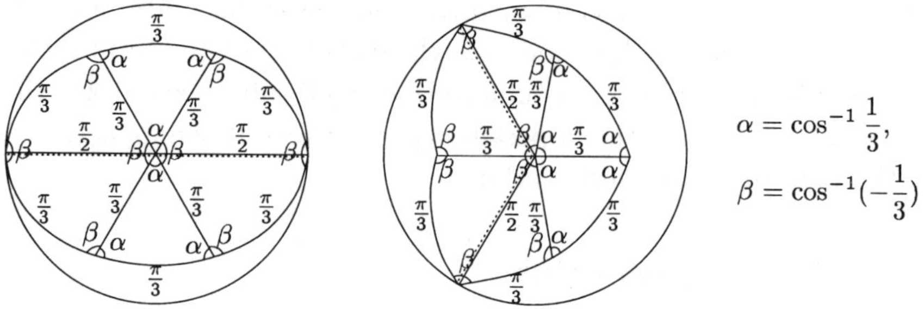

Figure 2.23

is equal to $2\cos^{-1}(-\frac{1}{3})$ *when and only when* $St(A)$ *is one of the pair as indicated in Figure* 2.23.

Proof A $6\triangle$-star has six boundary edges, $\{\widehat{B_iB_{i+1}}\}$, and six radial edges, $\{\widehat{AB_i}\}$. The fact that

$$6\cos^{-1}\frac{1}{3} = 2\pi + 1.10257\cdots \tag{15}$$

implies that at least two of its radial edges must be considerably longer than $\frac{\pi}{3}$. For otherwise, either the sum of central angles would *exceed* 2π or the convexity property will *not* hold.

Suppose that $St(A)$ contains a triangle, say $\widehat{\triangle}AB_1B_2$, with its area *less* than $\triangle_{\frac{\pi}{3}}$. Then it must be a shallow triangle with its circumradius exceeding $\frac{\pi}{3}$ and with a long radial edge. For otherwise, its central angle θ_1 must exceed the *larger angle* of the "spherical parallelogram" (cf. Eq. (70) in Sec. 2.1) with $a = \overline{AB_1}$, $b = \overline{AB_2}$ and its area equal to $2\triangle_{\frac{\pi}{3}}$, which is at least equal to $2\cos^{-1}(\frac{1}{3})$ when both a and b are equal to $\frac{\pi}{3}$. Therefore, the sum of the other five central angles of such a $6\triangle$-star must be less than $2\cos^{-1}(-\frac{1}{3})$, or even smaller when a and b are not both equal to $\frac{\pi}{3}$. However, the following lower bound estimates on central angles proves the *non-existence* of such a $6\triangle$-star.

Set the length of $\overline{AB}_j$ to be ℓ_j and the outer angle at B_j to be γ_j, $3 \leq j \leq 6$. If $\gamma_j \leq \pi$ and $\ell_j > \frac{\pi}{3}$, then one may deform the

6$\triangle$-star by pushing B_j inward *unless* $\overline{AB}_j$ is the side of a cocircular quadrilateral. On the other hand, if $\gamma_j > \pi$, then the sum of the pair of adjacent central angles, $\theta_{j-1} + \theta_j$, is *bounded below* by the angle A of Example 2.1.2 with $a = \frac{\pi}{3}$, $b = \ell_{j+1}$ and $c = \ell_{j-1}$. Therefore, the sum $\sum_{j=2}^{6} \theta_j$ is bounded below by the sum of the central angles of a cocircular pentagon and a cocircular quadrilateral (resp. its adjacent triangles), as indicated in Figure 2.24. However, it is easy to check that the above lower bound is always considerably larger than $(2\pi - \theta_1)$.

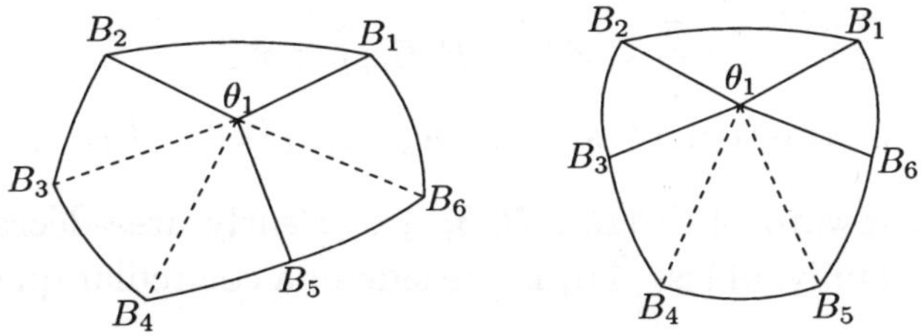

Figure 2.24

Therefore, the longest edge of such a shallow triangle must be a radial edge and the area of its adjacent triangle sharing the long radial edge must be much larger than $\triangle_{\frac{\pi}{3}}$, thus making the proof of the lower bound estimate of $2\cos^{-1}(-\frac{1}{3})$ quite obvious. Hence, it suffices to consider the other case that the area of each triangle of $St(A)$ is at least equal to $\triangle_{\frac{\pi}{3}}$. Moreover, it follows from Lemma 2.2.1 that the area of $St(A)$ is at least equal to

$$2\triangle_{\frac{\pi}{3}} + 2\square_{\frac{\pi}{3}} = 2\cos^{-1}\left(-\frac{1}{3}\right)$$
$$\left(\text{resp. } 3\triangle_{\frac{\pi}{3}} + \pentagon_{\frac{\pi}{3}} = 2\cos^{-1}\left(-\tfrac{1}{3}\right) + 0.4671\cdots\right) \tag{16}$$

if it contains at least two quadrilaterals (resp. one pentagon). Thus, it suffices to consider the remaining case that $St(A)$ contains no pentagon and at most one quadrilateral.

For the purpose of a lower bound area estimate of a certain family of star configurations, it is natural and advantageous to make use of some kinds of *area-decreasing deformations within the family* to reduce the task to that of the *non-deformable* ones in the family. The following are some simple kinds of area-decreasing deformations which provide a set of powerful tools for such a purpose.

1. First kind of belt-tightening If

$$\begin{gathered} \overset{\frown}{AB_i} > \frac{\pi}{3}, \quad \overset{\frown}{B_{i-1}B_i} + \overset{\frown}{B_iB_{i+1}} > \frac{2\pi}{3}, \\ \frac{\pi}{2} < \angle B_{i-1}B_iB_{i+1} < \pi, \end{gathered} \tag{17}$$

then the deformation of shortening $\overset{\frown}{B_{i-1}B_i} + \overset{\frown}{B_iB_{i+1}}$ simply by moving B_i inward of $\overset{\frown}{\square} AB_{i-1}B_iB_{i+1}$ is clearly area-decreasing and within the family, unless $\overline{AB}_i$ is the side of a cocircular quadrilateral.

2. Second kind of belt-tightening Suppose that both $\overset{\frown}{AB_i}$ and $\overset{\frown}{AB_{i+1}}$ are equal to $\frac{\pi}{3}$ and $\overset{\frown}{B_iB_{i+1}}$ is longer than $\frac{\pi}{3}$ but shorter than $\frac{\pi}{2}$. Then the shortening of $\overset{\frown}{B_iB_{i+1}}$ will decrease both the area of $\overset{\frown}{\triangle} AB_iB_{i+1}$ and its central angle θ_i. Moreover, the above decrement of θ_i can always be accompanied by the shearing of a suitable quadrilateral $\overset{\frown}{\square} AB_{j-1}B_jB_{j+1}$ with $\overset{\frown}{AB_j}$ considerably longer than $\frac{\pi}{3}$, so as to match the decrement of θ_i by the same amount of increment of $(\theta_{j-1} + \theta_j)$. It follows from the convexity property that the above accompanied shearing is also area-decreasing.

3. Shearing deformation along the boundary As indicated in Figure 2.25, the shearing deformation is essentially just the shearing of $\overset{\frown}{\square} B_1B_2B_3B_4$ to $\overset{\frown}{\square} B_1B_2'B_3'B_4$, namely, it fixes $\{A, B_1, B_4, B_5, B_6\}$ and preserves the length of all boundary edges. By Lemma 2.1.2, such a shearing is always *area-decreasing* if it deforms the quadrilateral further *away from the cocircular shape.* Such an area-decreasing

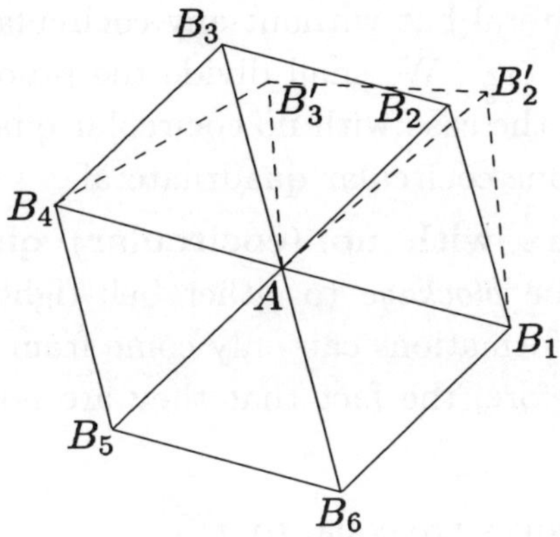

Figure 2.25

shearing is *applicable up until* it is blocked by the edge-length constraint or by the convexity property of $St(A)$.

4. Pairs of central angle compensating shearings The fourth kind of area-decreasing deformation is a coupling of a pair of shearings of two non-overlapping regions of the following two special types, so that their changes of central angles compensate each other.

(i) The shearing of $\widehat{\Box}AB_{i-1}B_iB_{i+1}$ which varies the length $\widehat{AB_i}$ between $\frac{\pi}{3}$ and the diagonal length that makes it cocircular.

(ii) The shearing of a pentagon which consists of an adjacent pair of a cocircular quadrilateral and a triangle by varying the length of their common radial edge between $\frac{\pi}{3}$ and the length that makes the pentagon itself cocircular.

To each coupled pair of such shearings, there are two directions to deform. As it turns out, there is always a direction whose total effect is monotonically area-decreasing.

With the help of the above four kinds of area-decreasing deformations, it is then quite straightforward to prove the remaining case of Lemma 2.2.2, namely, the case of 6$\triangle$-stars with at most one cocircular quadrilateral but without any cocircular pentagon or triangle smaller than $\triangle_{\frac{\pi}{3}}$. We shall divide the proof of this part into two cases, namely, the case with no cocircular quadrilateral and the case with exactly one cocircular quadrilateral.

Case 1: 6△-stars with no (cocircular) quadrilateral For such a 6△-star, the *blockage* to either belt-tightening or shearing area-decreasing deformations can only come from the $\pi/3$-bound on edge-length. Therefore, the fact that they are no longer applicable implies that

(i) all boundary edges are equal to $\pi/3$,
(ii) among each pair of *adjacent radial edges*, at least one is equal to $\pi/3$.

Therefore, such a 6△-star has at least three $\pi/3$-radial edges, while in the case of exactly three $\pi/3$-radial edges, they are separated by longer radial edges, as indicated in Figure 2.26. In fact, the two configurations of Figure 2.23 are the only such 6△-stars with at least four $\pi/3$-radial edges, both actually containing two cocircular

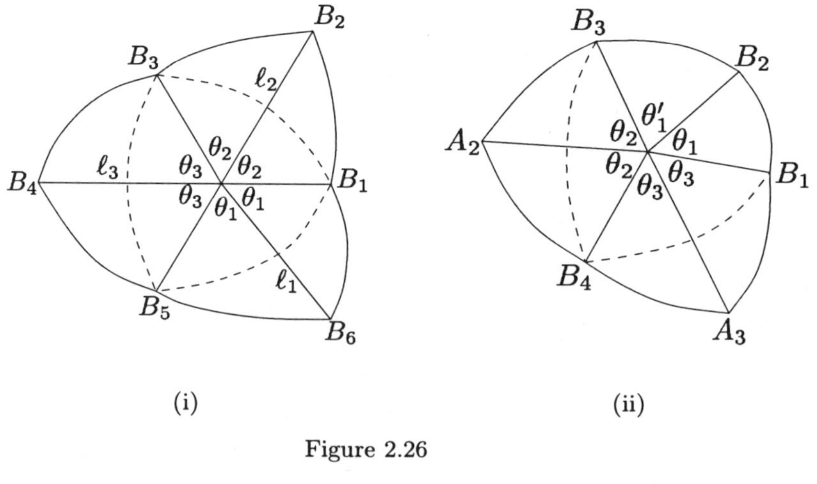

Figure 2.26

quadrilaterals. Therefore, all 6$\triangle$-stars of Case 1 always have exactly three (separated) $\pi/3$-radial edges. As indicated in Figure 2.26(i), $\widehat{AB_1} = \widehat{AB_3} = \widehat{AB_5} = \pi/3$ with their separating central angles

$$\beta < 2\theta_1, 2\theta_2, 2\theta_3 < 2\alpha \tag{18}$$

while its area is twice that of $\widehat{\triangle} B_1B_3B_5$.

By Lexell's Theorem, $\widehat{\triangle} B_1B_3B_5$ is larger than another inscribing triangle of the above $\pi/3$-circle with $\{\beta, \beta, 2\alpha\}$ as the triple of central angles, whose area is exactly equal to $\cos^{-1}(-\frac{1}{3})$. This proves that all 6$\triangle$-stars of Case 1 always have areas exceeding the bound of $2\cos^{-1}(-\frac{1}{3})$.

Case 2: 6$\triangle$-stars with exactly one (cocircular) quadrilateral Again, it suffices to consider here the ones where the belt-tightening and the shearing area-decreasing deformations are *no longer applicable.* Let $\widehat{\square} AB_1B_2B_3$ be the cocircular quadrilateral contained in such a 6$\triangle$-star. Suppose that $\widehat{AB}_1 = \widehat{AB}_3 = \pi/3$. Then the above non-applicability assumption implies that both the adjacent pairs $\{\widehat{AB}_4, \widehat{AB}_5\}$ and $\{\widehat{AB}_5, \widehat{AB}_6\}$ must contain $\pi/3$-edges. For otherwise, the shearing area-decreasing deformation would be applicable to such a 6$\triangle$-star. In fact, if the $\pi/3$-edges in the above two pairs are distinct, then the pair of triangles with only the long edge of $\{\widehat{AB}_4, \widehat{AB}_5, \widehat{AB}_6\}$ as the common edge, actually, constitute a *cocircular quadrilateral.* Hence, $\widehat{AB}_5 = \pi/3$ for those with exactly one such cocircular quadrilateral. Such 6$\triangle$-stars can, in fact, be regarded as the special case of the 6$\triangle$-ones indicated in Figure 2.26(i) with $2\theta_1 = \beta = \cos^{-1}(-\frac{1}{3})$. Therefore, it again follows from the Lexell theorem that the areas of such 6$\triangle$-stars always exceed $2\cos^{-1}(-\frac{1}{3})$.

Finally, let us study the remaining case that at least one of $\{\widehat{AB}_1, \widehat{AB}_3\}$ is longer than $\pi/3$. Set their lengths to be $2u$ and

$2t$ with $u \geq \pi/6$ and $t > \pi/6$. Then, it follows from the cocircular condition (cf. Example 2.1.2) that

$$\begin{aligned} \cos d &= \frac{\frac{1}{2}(\cos 2u + \cos 2t - 1)}{\frac{1}{2} + 2\sin u \sin t}, & d &= \widehat{B_1 B_3}, \\ \cos\varphi &= -\frac{\sin u \sin t + \cos 2u \cos 2t}{\cos u \cos t + \sin 2u \sin 2t}, & \varphi &= \angle B_1 A B_3. \end{aligned} \tag{19}$$

Moreover, it suffices to consider the ones with

$$\left(u - \frac{\pi}{6}\right) + \left(t - \frac{\pi}{6}\right) \leq 0.115. \tag{20}$$

For otherwise, it is straightforward to show that

$$\widehat{\square} AB_1B_2B_3 + \widehat{\triangle} AB_6B_1 + \widehat{\triangle} AB_3B_4 > 4\triangle_{\pi/3} + 0.52 \tag{21}$$

and hence the total area of such a $6\triangle$-star will exceed $6\triangle_{\pi/3}$ by at least 0.52, thus exceeding the lower bound of $2\cos^{-1}(-\frac{1}{3})$.

Now, let us assume that $\widehat{AB}_3 = 2t > \pi/3$ and study the shearing deformation of the adjacent pair

$$\{\widehat{\square} AB_1B_2B_3,\ \widehat{\triangle} AB_3B_4\}$$

into another pair of cocircular quadrilateral and triangle by shortening the common radial edge $\widehat{AB}_3$. Such a deformation will decrease the area but increase the central angle $(\varphi + \alpha_3)$. Therefore, such a shearing deformation should be accompanied by another shearing of quadrilaterals in the other part of the given $6\triangle$-star, such as $\widehat{\square} AB_4B_5B_6$, or $\widehat{\square} AB_5B_6B_1$, so as to *compensate* the above *increment* of central angle by exactly the same amount of *decrement* of central angle. It follows from the formulas of Example 2.1.2, that such a combined deformation is always area-decreasing. Therefore, the area of such a $6\triangle$-star is always larger than another $6\triangle$-star with a cocircular quadrilateral of $\pi/3$-side-lengths, whose area has already been proven to be at least $2\cos^{-1}(-\frac{1}{3})$. This completes the proof of Lemma 2.2.2. □

Remark Note that $(3\cos^{-1}(-\frac{1}{3}) + \cos^{-1}\frac{1}{3})$ already exceeds 2π by 0.67967. The same kind of proof will provide a lower bound estimate of the total area of a $7\triangle$-star with edge-lengths of at least $\frac{\pi}{3}$ considerably larger than

$$3\square_{\frac{\pi}{3}} + \triangle_{\frac{\pi}{3}} = 7\triangle_{\frac{\pi}{3}} + 0.770329\cdots.$$

We refer to Chap. 5 for a discussion of area estimates of $7\triangle$-star in a more general setting.

2.2.5 *Another proof of the impossibility of thirteen touching neighbors*

Suppose on the contrary that it is possible to arrange thirteen touching neighbors in a single-layer local packing. The thirteen touching points on the central sphere constitutes a set Σ of separation at least equal to $\pi/3$, whose associated spherical configuration has thirteen vertices and all edge-lengths at least equal to $\pi/3$. By Euler's formula

$$F_3 + 2F_4 = 22$$

and by adding one diagonal to each of its quadrilateral, if any, one has a triangulated spherical configuration with 22 triangles and 33 edges. Therefore, at least one of its 13 vertices has *more* than five forks and hence, the above area estimate shows that its total area is at least equal to

$$16\triangle_{\pi/3} + 2\cos^{-1}\left(-\frac{1}{3}\right) > 4\pi$$

which is, of course, a contradiction.

[illegible] The same [illegible] with [illegible] [illegible] large [illegible].

$$[illegible] = 0.97882$$

[illegible]

[illegible]

We [illegible]

Chapter 3

Circle Packings and Sphere Packings

In this chapter, we shall embark upon the journey of studying the problem of sphere packings. We shall begin with a brief section on its predecessor, the problem of circle packings, by presenting a proof which will serve as a much simplified proto-type of the proof of the major theorems on sphere packings (cf. Chaps. 1, 6 and 8).

3.1 The Problem of Circle Packings

Circle packings in $\mathbf{E}^2$ and sphere packings in $\mathbf{E}^3$ are conceptually quite similar to each other. For example, there are also three different kinds of circle packings, namely, packings with containers, finite packings *without* containers and infinite packings; the concept of *local cell* and *local density* can be defined in just the same way, while the definition of the *global density* of an *infinite circle packing* also encounters the same kind of complexity as that of infinite sphere packings discussed in Chap. 1. However, the *geometry* of circle packings is much simpler than that of sphere packings. One of the basic reasons for the simplicity of circle packings is the existence of the

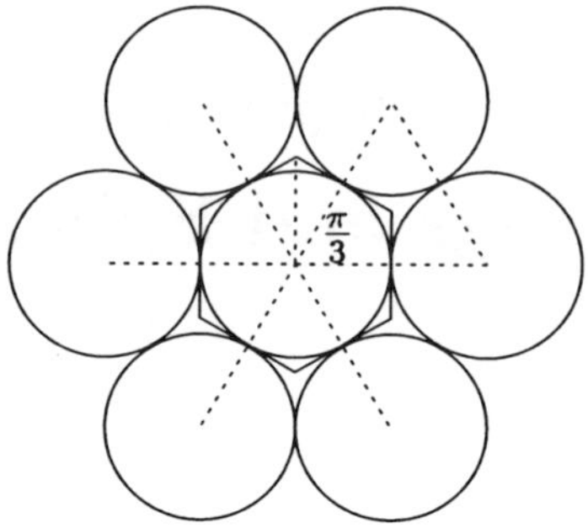

Figure 3.1

tight local packing of surrounding a circle by six touching neighbors as indicated in Figure 3.1.

The *local cell* of the above local packing is a *circumscribing regular hexagon* which is a *tiling domain* of $\mathbf{E}^2$. Therefore, just as bees stack together the hexagonal cell to build a honeycomb, one can *tile* the whole plane with identical regular hexagons and then put an inscribing circle into each of them. The circle packing so obtained, is usually referred as the *hexagonal circle packing*, $\mathcal{P}_0$, whose local cells are *uniformly* circumscribing regular hexagons. Thus, its local densities are uniformly equal to $\pi/\sqrt{12} = 0.90689968\cdots$. Moreover, the geometric tightness of the local packing as indicated in Figure 3.1 strongly suggests that it should be the *unique local packing with optimal local density*. Anyhow, the proof of the following theorem on the optimal local density of circle packing is the central problem in the study of circle packings which also enables us to settle many other issues on circle packings successfully and easily (see [Fej-1], [Thu] for its earlier proofs).

Theorem 3.1.1 *The optimal universal upper bound for the local density of circle packings in the Euclidean plane is equal to* $\pi/\sqrt{12}$, *and it can be realized as the local density when and only when the local cell is a circumscribing regular hexagon, i.e. the local packing consists of six touching neighbors.*

Remark Since the *relative density* of a finite subpacking is, by definition, a weighted average of the local densities of its individual circles, the above theorem clearly implies the following corollaries on both the relative density and the global density of circle packings.

Corollary 3.1.1 *$\pi/\sqrt{12}$ is also the optimal relative density for any finite sub-packing, $\rho(\mathcal{P}_n, \mathcal{P})$, and it is equal to $\pi/\sqrt{12}$ when and only when all its local cells are circumscribing regular hexagons.*

Corollary 3.1.2 *$\pi/\sqrt{12}$ is also the maximal value of the global density of circle packings into the Euclidean plane.*

Remark All natural or practical examples of circle packings such as bees living in a honeycomb or the cross-section of a bundle of fibre-glass optical tubes, are always packings of a finite number of circles. The infinite circle packings of the entire plane are actually the limiting geometric situation of those finite ones. Therefore, it is natural to formulate an appropriate definition of the concept of *global density* for a given finite circle packing, all by itself. We propose the following definition of a cluster of circles and the global density of such a cluster of circles.

Definition A packing of a finite number of circles is called a *cluster of circles* if any pair of them can be linked through neighboring pairs with center distances less than $2\sqrt{2}$ times the radii.

Let $\mathcal{C}$ be a given cluster of circles. Then, an extension $\mathcal{C}^*$ of $\mathcal{C}$ is called a *saturated coating* of $\mathcal{C}$ if all added circles of $\mathcal{C}^*$ are neighbors of some circles in $\mathcal{C}$ and it is impossible to add any more such neighbors to $\mathcal{C}^*$. Thus, each circle of $\mathcal{C}$ has a well-defined local cell with respect to $\mathcal{C}^*$ and hence, one can define the relative density $\rho(\mathcal{C}, \mathcal{C}^*)$ just as that of a finite sub-packing of an infinite packing.

Definition The global density of $\mathcal{C}$ is defined to be the least upper bound of $\{\rho(\mathcal{C}, \mathcal{C}^*)\}$ where $\mathcal{C}^*$ runs through all possible saturated coatings of $\mathcal{C}$.

Theorem 3.1.1′ *$\pi/\sqrt{12}$ is again the maximal possible global density of all clusters of circles, and the global density of a cluster of circles, $\mathcal{C}$, attains the above maximum of $\pi/\sqrt{12}$ when and only when $\mathcal{C}$ is a sub-cluster of circles of the hexagonal packing $\mathcal{P}_0$.*

Proof We must show that the smallest possible area for a local cell in a packing of unit circles is that of a circumscribing regular hexagon, $A_0 = 2\sqrt{3}$. The circumradius of this cell is $R = 2/\sqrt{3}$.

We will make use of an area function $A(\theta)$ defined for $0 \le \theta \le \pi$ as follows. Consider the region of the plane bounded by two radial segments at angle 2θ in the unit circle, two tangential segments to the unit circle from their end points, and in the case where $\theta > \theta_1 = \cos^{-1}(1/R)$, by an arc of the concentric circle of radius R, as indicated in Figure 3.2. Let $A(\theta)$ be the area of this region. Then we have:

The function $A(\theta)$ is convex on $0 \le \theta \le \pi$.

This follows easily from the fact that

$$A(\theta) = \begin{cases} \tan\theta\,, & \text{for } 0 \le \theta \le \theta_1\,; \\ \tan\theta_1 + R^2(\theta - \theta_1)\,, & \text{for } \theta_1 < \theta \le \dfrac{\pi}{2}\,. \end{cases} \tag{1}$$

It is sufficient to note that $A'(\theta)$ is continuous on $0 \le \theta \le \frac{\pi}{2}$, and that $A''(\theta) \ge 0$, except at $\theta = \theta_1$.

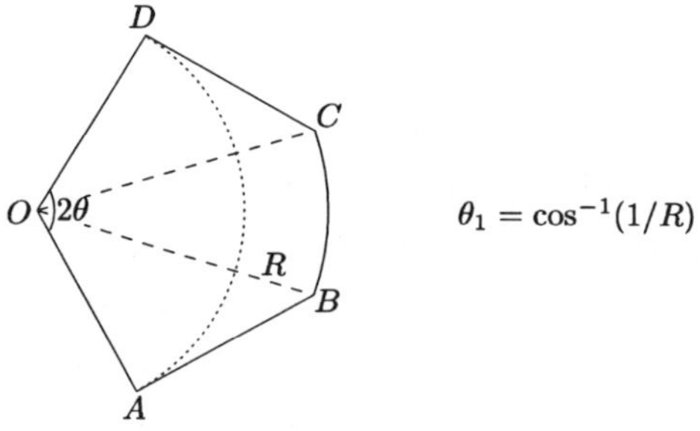

$A(\theta)$: Area$(OABCD)$.

Figure 3.2

For the purpose of this proof we define two circles in a packing to be *close neighbors* of each other if their center distance is at most $2R = 4/\sqrt{3} = 2.30940\cdots$.

Let Γ_0 be a circle in a given packing $\mathcal{P}$, and choose the center O of Γ_0 to be the origin. The angular separation θ between two close neighbors of Γ_0, where the close neighbors are tangent to each other, is a decreasing function of their center distances (as long as these are in the range of close neighbors) because if O_1 and O_2 denote the centers of the two close neighbors Γ_1 and Γ_2, then the triangle $\triangle OO_1O_2$ has angles O_1 and O_2 less than $\pi/2$ (cf. Figure 3.3).

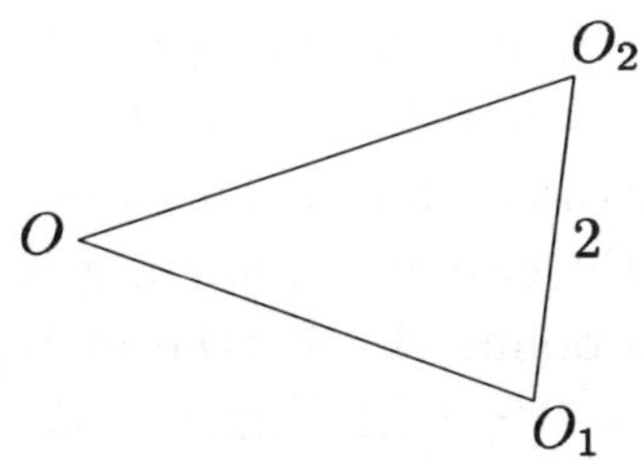

Figure 3.3

Therefore, the minimal angular separation for two close neighbors of Γ_0 occurs when they are both at the maximal distance $2R$ and touching each other. In this case the angular separations has the value

$$\hat{\theta} = 2\sin^{-1}\frac{1}{2R} = 2\sin^{-1}\frac{\sqrt{3}}{4} = 0.89566\cdots, \tag{2}$$

which is between $2\pi/7 = 0.89759\cdots$ and $2\pi/8 = 0.78539\cdots$. Hence, there is only room for at most seven close neighbors of Γ_0.

Let n be the number of close neighbors and $C = C(\Gamma_0, \mathcal{P})$ be the local cell of Γ_0 in $\mathcal{P}$. Then, we have two cases:

Case 1: $n \leq 6$. If $n = 0$, the local cell C by definition contains the circle of radius R centered at O, and therefore contains also the circumscribing regular hexagon of the unit circle. Hence, we clearly have Area $(C) \geq A_0$ in this special case.

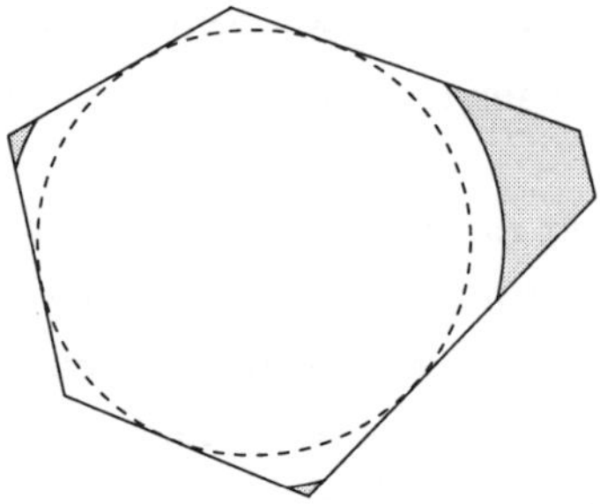

Figure 3.4

Assume therefore $1 \leq n \leq 6$. Let $\{\theta_j : 1 \leq j \leq n\}$ be the set of angles between the close neighbors of Γ_0 in cyclic order, i.e. the angular separations of the points A_j on Γ_0 defining their directions. We let $T\{A_j\}$ denote the circumscribing n-gon bounded by the tangent lines to Γ_0 at the points A_j. Furthermore, let $\bar{C}$ be this n-gon truncated at radius $R = 2/\sqrt{3}$ (cf. Figure 3.4). Then, it is clear that $\bar{C} \subseteq C$ since the non-close neighbors of Γ_0 by definition have their bisector lines with Γ_0 outside the circle of radius R centered at the origin.

By using the previous area function $A(\theta)$ and its convexity we get

$$\text{Area } (C) \geq \text{ Area } (\bar{C}) = \sum_{j=1}^{n} A\left(\frac{\theta_j}{2}\right) \geq nA\left(\frac{\pi}{n}\right) \geq 6A\left(\frac{\pi}{6}\right) = A_0 \,. \tag{3}$$

We have used the fact that $nA(\pi/n)$ is a decreasing function of n, which follows from the convexity of $A(\theta)$ since $A(0) = 0$.

Case 2: $n = 7$. From the minimum possible angular separation $\hat{\theta}$ between close neighbors of (2), it follows that a close neighbor of Γ_0 can have an angular separation from its two adjacent close neighbors of at most

$$\psi = \frac{1}{2}(2\pi - 5\hat{\theta}) = \pi - 5\sin^{-1}\frac{\sqrt{3}}{4} = 0.90243\cdots. \tag{4}$$

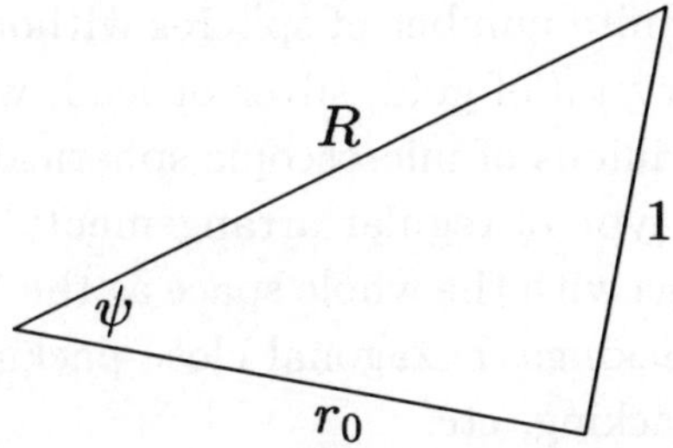

Figure 3.5

This maximum of angular space available, corresponds to a minimal possible center distance for a close neighbor of $2r_0$, where $r_0 \geq 1$ is given by the cosine relation of the triangle of Figure 3.5. Thus, we get

$$\begin{aligned} &r_0^2 - (2R\cos\psi)r_0 + R^2 - 1 = 0\,, \\ &r_0 = R\cos\psi + \sqrt{1 - R^2\sin^2\psi} = 1.13831\cdots. \end{aligned} \tag{5}$$

It follows that the local cell C contains the circle of radius r_0 centered at O, and hence

$$\text{Area}\ (C) \geq \pi r_0^2 = 4.07075\cdots > 2\sqrt{3} = A_0\,.$$

This completes the proof. □

3.2 Sphere Packings and Crystal Formations, Kepler's Conjecture and a Least Action Principle of Crystal Formation

3.2.1 *Three kinds of sphere packings and the concepts of their densities*

There are, in general, three different kinds of sphere packings, namely

(i) sphere packings with a container such as packing marbles into a jar, oranges into a box or soybeans into a silo,

(ii) packings of a finite number of spheres without a container such as a piece of crystal of gold, silver or lead, which often consists of billions of trillions of microscopic spherical atoms aggregated into a specific type of regular arrangement,

(iii) infinite packings with the whole space as the "container" such as f.c.c. lattice packings, hexagonal close packings, body centered cubic lattice packing, etc.

Let us first study the concept of density for the above three kinds of sphere packings and then proceed to discuss the problem of optimal densities for the three cases. (Some concepts and known results about these types of packings were discussed in [CFG].)

Density of sphere packings with containers The *density* of a sphere packing into a given container is defined to be the *fraction of the volume* of the container that is *occupied by the spheres.* The *least upper bound* of the densities of all possible packings of spheres of a given size into a given container is called the *optimal density* of such packings. It will, of course, depend on the *shape* of the container and the *relative sizes* of the container and the spheres. This is a rather undesirable feature of this kind of sphere packing for the mathematical study of optimal density.

Density of finite sphere packings without container First of all, unlike in the case of a finite packing with a given container, the concept of density of a finite packing *without* a container is something *yet to be defined properly*, mainly due to the lack of an obvious choice for the denominator, such as the volume of the given container in the former case. Therefore, the first thing in the study of finite packings without a container is to search for a proper definition of the (intrinsic) density. Let us first try the following which works very well in the case of finite circle packings without container.

Definition A finite sphere packing is called a *cluster of spheres* if any pair of them can be linked via neighboring pairs with center distances less than $2\sqrt{2}$ times the radii.

Definition Let $\mathcal{C}$ be a given cluster of spheres. An extension $\mathcal{C}^*$ is said to be a *single-layer coating* of $\mathcal{C}$ if all additional spheres are neighbors of some spheres in $\mathcal{C}$ and all spheres of $\mathcal{C}$ have finite volume local cells in $\mathcal{C}^*$.

Definition The *relative global density* of $\mathcal{C}$ in $\mathcal{C}^*$, denoted by $\rho(\mathcal{C}, \mathcal{C}^*)$, is defined to be the ratio between the total volume of spheres in $\mathcal{C}$ and the total volume of the local cells of spheres of $\mathcal{C}$ relative to $\mathcal{C}^*$.

Definition The *global density of the first kind* of a given cluster $\mathcal{C}$ is defined to be the least upper bound of the set of relative densities $\{\rho(\mathcal{C}, \mathcal{C}^*)\}$ with respect to all possible single-layer coatings of $\mathcal{C}$, namely

$$\rho(\mathcal{C}) = \ell.u.b.\{\rho(\mathcal{C}, \mathcal{C}^*)\}\,. \tag{1}$$

Example 3.2.1 In the very special case of a single sphere, i.e. $\mathcal{C} = \{S\}$, the above intrinsic density $\rho(\{S\})$ is exactly the *optimal local density*, while the proof of the Dodecahedron Conjecture (cf. Chap. 6) shows that

$$\rho(\{S\}) = \frac{4\pi}{3} : 20\left(1 - 2\cos\frac{2\pi}{5}\right)\tan\frac{\pi}{5} = 0.75469\cdots\,. \tag{2}$$

Example 3.2.2 Let $\mathcal{C}$ be a subcluster of spheres contained in a given hexagonal close packing, such as the packing of a piece of gold crystal. Then, there exists a single-layer coating, $\mathcal{C}^*$, of $\mathcal{C}$ which consists of all the neighbors of those spheres of $\mathcal{C}$ in the given hexagonal close packing, thus having the relative density

$$\rho(\mathcal{C}, \mathcal{C}^*) = \frac{\pi}{\sqrt{18}}\,. \tag{3}$$

Therefore, the (intrinsic) density, $\rho(\mathcal{C})$, of such a cluster is at least equal to $\frac{\pi}{\sqrt{18}}$. On the other hand, one has the following simple *upper bound estimate* of $\rho(\mathcal{C})$:

Set N to be the total number of spheres in the given subcluster $\mathcal{C}$ and N' to be the number of spheres which does *not* have twelve touching neighbors within $\mathcal{C}$. Then, it follows from the dodecahedron estimate that

$$\rho(\mathcal{C}) < \frac{N \cdot \frac{4\pi}{3}}{(N-N')4\sqrt{2}+N'V_0} = \frac{\pi}{\sqrt{18}}\left\{1-\frac{N'}{N}\left(1-\frac{V_0}{4\sqrt{2}}\right)\right\}^{-1}, \quad (4)$$

where

$$V_0 = 20\left(1-2\cos\frac{2\pi}{5}\right)\tan\frac{\pi}{5}, \quad 1-\frac{V_0}{4\sqrt{2}} = 0.0188379\cdots. \quad (4')$$

Note that N'/N becomes very, very small in the case of a piece of crystal of hexagonal close packing. Hence, its intrinsic density of the first kind is practically equal to $\pi/\sqrt{18}$.

Example 3.2.3 Let $\mathcal{C}$ be a string of N unit spheres, $\{S_i\}$, with the coordinates of their centers given by $\{(2i,0,0), 1 \le i \le N\}$. Let $\mathcal{C}^*$ be a single-layer coating of $\mathcal{C}$ which is symmetric with respect to the $\frac{2\pi}{5}$-rotation around the x-axis and consists of five common touching neighbors for each adjacent pair of spheres in $\mathcal{C}$ and additional six touching neighbors for each of the two end spheres of the string. Moreover, the arrangements of adjacent quintuples always differ by an angular-phase of $\frac{\pi}{5}$, and the angular separations between the spheres centered at $(0,0,0)$ and $(2N+2,0,0)$ and their neighboring quintuples are equal to r. Then, the volumes of the local cells $C(S_i,\mathcal{C}^*)$ can be computed as follows:

Let

$$a = \cos^{-1}\left\{\frac{3}{4}\cos\frac{2\pi}{5}+\frac{1}{4}\right\} = \cos^{-1}(0.4817627458),$$

$$b = \cos^{-1}\left\{\frac{3}{4}\cos\frac{\pi}{5}-\frac{1}{4}\right\} = \cos^{-1}(0.3567627458),$$

$$c = \cos^{-1}\left\{1 - 2\sin^2\frac{\pi}{5}\sin^2 r\right\},$$

$$d = \cos^{-1}\left\{\frac{\sqrt{3}}{2}\cos\frac{\pi}{5}\sin r - \frac{1}{2}\cos r\right\} \tag{5}$$

and $\{\sigma_i, 1 \leq i \leq 5\}$ be spherical triangles specified by their side-lengths as follows, namely

$$\begin{aligned} &\sigma_1 = \triangle\left(\frac{\pi}{3}, \frac{\pi}{3}, a\right), \quad \sigma_2 = \triangle(b, b, a), \quad \sigma_3 = \triangle(r, r, c), \\ &\sigma_4 = \triangle(d, d, c), \quad \sigma_5 = \triangle(d, d, a). \end{aligned} \tag{6}$$

Let $f(\sigma)$ be the volume formula of $T(\sigma)$ given by Lemma 4.2.1. Then,

$$\begin{aligned} &\operatorname{vol} C(S_i, \mathcal{C}^*) = 10f(\sigma_1) + 10f(\sigma_2), && 1 < i < N; \\ &\operatorname{vol} C(S_i, \mathcal{C}^*) = 5\{f(\sigma_1) + f(\sigma_3) + f(\sigma_4) + f(\sigma_5)\}, && i = 1 \text{ or } N. \end{aligned} \tag{7}$$

Straightforward computation will show that

$$10(f(\sigma_1) + f(\sigma_2)) = 5.578906172\cdots \tag{8}$$

while the volume of $C(S_1, \mathcal{C}^*)$ (resp. $C(S_N, \mathcal{C}^*)$) has a minimal value at $r = 1.13255$, namely

$$5\{f(\sigma_1) + f(\sigma_3) + f(\sigma_4) + f(\sigma_5)\} \geq 5.55802 \tag{9}$$

and the equality holds when and only when $r = 1.13255$.

Therefore, the relative density $\rho(\mathcal{C}, \mathcal{C}^*)$ of such a coating is given by

$$\begin{aligned} \rho(\mathcal{C}, \mathcal{C}^*) &= \frac{N \cdot \frac{4\pi}{3}}{(N-2)(5.578906\cdots) + 2(5.55802)} \\ &= (0.75082643) \cdot \left[1 - \frac{1}{N}(0.007487489)\right]^{-1}. \end{aligned} \tag{10}$$

3.2.1.1 *Some natural problems on the density of the first kind and a generalized Dodecahedron Conjecture*

The above definition of an intrinsic density of a given cluster of spheres is certainly rather natural. However, it is in general extremely difficult to compute. For example, the computation of such an intrinsic density of a single sphere amounts to the proof of the Dodecahedron Conjecture on the optimal local density. Therefore, the computations of the intrinsic densities of the first kind for a pair (resp. triple or quadruple) of spheres touching each other already constitute some rather challenging problems. Let us formulate here a natural problem on the densities (of the first kind) of clusters of spheres which is even more challenging.

Problem *What is the least upper bound of the densities (of the first kind) of all clusters with a given cardinality N, namely*

$$\rho_N = l.u.b.\,\{\rho(\mathcal{C});\ \#(\mathcal{C}) = N\} = \,?$$

and what are those clusters of N spheres with $\rho(\mathcal{C}) = \rho_N$?

Generalized Dodecahedron Conjecture

$$\rho_N = 0.75082643 \cdot \left[1 - \frac{1}{N}(0.007487489)\right]^{-1} \tag{11}$$

and $\rho(\mathcal{C}) = \rho_N$ when and only when $\mathcal{C}$ is the above linear string of N spheres.

Density of infinite packings (with the whole space as the container) Examples of sphere packings both in daily life and in natural occurrence are always *finite* in number even though the number of spheres can be exceedingly large such as the number of atoms in a piece of crystal. Therefore, *infinite packings* with the *whole space* as the "container" such as lattice packings and hexagonal close packings are, actually, a kind of mathematical model which should be regarded as the limiting geometric structure of the above two kinds of finite packings (i.e. with or without a container). In the very special case of lattice packings or hexagonal close packings, the local densities at all the spheres of such a packing are *uniformly the same.* Therefore, the global density of such an infinite packing is, of course, simply defined to be the uniform value of local densities. For example, the global densities of f.c.c. lattice packing (resp. all those hexagonal close packings) are equal to $\pi/\sqrt{18}$. However, in the general case of infinite packings with non-uniform local densities, the definition of their global densities, itself, becomes a highly non-trivial task which inevitably encounters several difficult issues.

Remark It is possible to have a certain sphere S_i in a certain given packing $\mathcal{P}$ with a local cell $C(S_i, \mathcal{P})$ of infinite size. However, such packings will certainly play *no* role in the study of the *optimal density* of sphere packings. Thus, we shall *always assume*, from now on, that *all local cells* of the packings that we consider, are of *finite sizes.* We shall also assume that each packing has an upper bound of the sizes of its local cells.

Let $\mathcal{P}_n = \{S_i; i \in I_n\}$ be a finite sub-packing of a given (infinite) sphere packing $\mathcal{P}$. Then, it is natural to define the relative density of $\mathcal{P}_n$ in $\mathcal{P}$ as follows.

Definition The *relative density* of $\mathcal{P}_n$ in $\mathcal{P}$, denoted by $\rho(\mathcal{P}_n, \mathcal{P})$, is defined to be

$$\begin{aligned} \rho(\mathcal{P}_n, \mathcal{P}) &= \sum_{i\in I_n} \operatorname{vol} S_i \Big/ \sum_{i\in I_n} \operatorname{vol} C(S_i, \mathcal{P}) \\ &= \sum_{i\in I_n} w_i \rho(S_i, \mathcal{P}) \Big/ \sum_{i\in I_n} w_i \,, \end{aligned} \tag{12}$$

where $w_i = \operatorname{vol} C(S_i, \mathcal{P})$.

It is natural to choose some kind of *exhaustion sequence of finite subpackings* $\{\mathcal{P}_n\}$, meaning that $\mathcal{P}_{n+1} \supseteq \mathcal{P}_n$ for all n and $\mathcal{P} = \bigcup\{\mathcal{P}_n\}$, and to define the global density of $\mathcal{P}$ as the "limit" of the relative densities $\{\rho(\mathcal{P}_n, \mathcal{P})\}$. However, such an approach will inevitably encounter the following issues which are, in general, difficult to settle properly, namely

(i) First of all, the limit of $\rho(\mathcal{P}_n, \mathcal{P})$ as $n \to \infty$ *may not* exist. Of course, one may use the *upper limit* of $\{\rho(\mathcal{P}_n, \mathcal{P})\}$ for the purpose of studying the *optimal global density* of infinite packings, thus avoiding the difficulty of possible non-existence of the limit.
(ii) However, the upper limits of the relative densities of *different exhaustion sequences* of finite subpackings *may be different.* Moreover, there is no good geometric criterion to provide a "canonical" choice of a specific family of exhaustion sequences.

So far, the commonly used definition of the global density of an infinite packing $\mathcal{P}$ makes use of the following family of exhaustion sequences which are, at least, geometrically simple and logically workable, namely

$$\mathcal{P}_n = \{S_i \in \mathcal{P}, \overline{OO_i} \leq 2nr\}\,, \tag{13}$$

where r is the common radius of the spheres of $\mathcal{P}$ and O is a chosen origin of the space. It is quite simple to show that the upper limit

$$\overline{\operatorname{Lim}_{n\to\infty}}\, \rho(\mathcal{P}_n, \mathcal{P}) \tag{14}$$

is independent of the choice of origin O for the above selection of the exhaustion sequence $\{\mathcal{P}_n\}$.

Definition The global density of $\mathcal{P}$ is defined to be the common upper limit of the relative densities of the family of exhaustion sequences of (13), namely

$$\rho(\mathcal{P}) = \overline{\operatorname{Lim}_{n\to\infty}}\, \rho(\mathcal{P}_n, \mathcal{P})\,. \tag{15}$$

Remark Due to the fact that the above definition of the global density $\rho(\mathcal{P})$ makes use of the rather specific choice of exhaustion sequences of (13), it can only be regarded as a *workable but not* *entirely satisfactory* definition of the global density of infinite packings.

Let us try another approach of selecting those exhaustion sequences of finite subpackings so that we can define the concept of global density in a canonical setting.

Definition An exhaustion sequence, $\{\mathcal{P}_n\}$, of $\mathcal{P}$ is said to be *proper* if $\mathcal{P}_{n+1}$ contains all the neighbors of those spheres of $\mathcal{P}_n$.

Definition A sphere $S_i \in \mathcal{P}_n \subset \mathcal{P}$ is called an *interior* sphere of $\mathcal{P}_n$ if all the neighbors of S_i in $\mathcal{P}$ are already contained in $\mathcal{P}_n$. Otherwise, it is called a *boundary* sphere of $\mathcal{P}_n$.

We shall denote the cardinal number of spheres (resp. interior spheres, boundary spheres) in $\mathcal{P}_n$ by $N(\mathcal{P}_n)$ (resp. $N_0(\mathcal{P}_n), N'(\mathcal{P}_n)$).

Definition An exhaustion sequence, $\{\mathcal{P}_n\}$, of $\mathcal{P}$ is said to be *admissible* if it is proper and satisfies the following condition, namely

$$\operatorname*{Lim}_{n\to\infty} \frac{N'(\mathcal{P}_n)}{N(\mathcal{P}_n)} = 0\,. \tag{16}$$

We propose the following definition of the global density of an infinite packing, namely, setting

$$\hat{\rho}(\mathcal{P}) = l.u.b.\left\{ \overline{\operatorname*{Lim}_{n\to\infty}}\, \rho(\mathcal{P}_n, \mathcal{P}_{n+1}) \right\} \tag{17}$$

where $\{\mathcal{P}_n\}$ runs through all admissible exhaustion sequences of the given infinite packing $\mathcal{P}$.

Remark Colloquially speaking, the sphere packing problem seeks to determine the *optimal density* of infinite packings with the whole space as the "container." However, one needs a precise mathematical definition of the concept of density of such an infinite packing in order to give a precise mathematical formulation of the sphere packing problem, while the definition of density of an infinite packing inevitably encounters various kinds of complexities. The above definition of $\hat{\rho}(\mathcal{P})$ represents the extremal choice which pushes the "density" to its upper limit at every stage, thus making the proof of Kepler's Conjecture the most demanding.

The following is a simple example of infinite packing which reveals some aspect of the complexities that one encounters in the definition of the density of infinite packings.

Example Let $\mathcal{P}^*$ be the h.c.p. packing of unit spheres. Choose the cartesian coordinate system of the space so that the $\{\cdots abab\cdots\}$-layer of spheres of $\mathcal{P}^*$ have their centers situated on the family of parallel planes given by $z = n \cdot \sqrt{\frac{8}{3}}$, $n \in \mathbf{Z}$. Let $\mathcal{P}$ be the subpacking of $\mathcal{P}^*$ with their centers situated on the *subfamily of planes* given by $z = n\sqrt{\frac{8}{3}}$, where n runs through all non-positive integers and all positive even integers. It is easy to see that there are three different values of the volumes of its local cells, namely

$$\operatorname{vol} C(S_i, \mathcal{P}) = \begin{cases} 4\sqrt{2} \\ 6\sqrt{2} \\ 8\sqrt{2} \end{cases} \text{if} \quad z(O_i) = \begin{cases} < 0\,, \\ = 0\,, \\ > 0\,. \end{cases} \tag{18}$$

Let $\{\mathcal{P}_n\}$ be an admissible exhaustion sequence of finite subpackings of $\mathcal{P}$. Set a_n (resp. b_n, c_n) to be the number of spheres with their centers below (resp. on, above) the (x, y)-plane. Then, the relative density $\rho(\mathcal{P}_n, \mathcal{P})$ is given by

$$\begin{aligned} \rho(\mathcal{P}_n, \mathcal{P}) &= \frac{(a_n + b_n + c_n)\frac{4\pi}{3}}{4\sqrt{2}a_n + 6\sqrt{2}b_n + 8\sqrt{2}c_n} \\ &= \left(\frac{a_n + b_n + c_n}{a_n + \frac{3}{2}b_n + 2c_n}\right)\frac{\pi}{\sqrt{18}}\,. \end{aligned} \tag{19}$$

Thus, for each given value of λ in the close interval $[\frac{1}{2}\frac{\pi}{\sqrt{18}}, \frac{\pi}{\sqrt{18}}]$, it is not difficult to construct such an admissible exhaustion sequence, $\{\mathcal{P}_n\}$, of $\mathcal{P}$ with $\operatorname{Lim}_{n\to\infty} \rho(\mathcal{P}_n, \mathcal{P}) = \lambda$.

The simple example above shows that the global density $\hat{\rho}(\mathcal{P})$ defined by (17) is, in general, far from representative of the geometric intuition of the averaged density over the whole space. Anyhow, we shall use the definition of (17) to provide a precise mathematical formulation of both the sphere packing problem and Kepler's Conjecture.

3.2.2 *The sphere packing problem and Kepler's Conjecture*

The sphere packing problem *What is the least upper bound of the global densities of all possible sphere packings in the space?*

From the above sequence of definitions, the precise mathematical definition of the "*optimal*" *density* of sphere packings with the whole space as the container is, indeed, rather involved. However, the sphere packing problem is clearly very natural and undoubtedly of fundamental importance. Of course, one might still question whether such a problem, concerning infinitely many spheres packed into the infinite expanse of the space, can have a neat answer.

Kepler's Conjecture One of the earliest publications on this problem occurred in a latin booklet of 1611 by J. Kepler, entitled "*A new year's gift — on six cornered snowflakes.*" He began with some discussions on the hexagonal structure of snowflakes and that of honeycombs and then moved on to discuss the geometry of sphere packings. For example, he examined the number of touching neighbors (for each individual sphere) and the shapes of local cells for various cases of regular packings, nowadays, called *lattice-packings.* Among them, he found that a special example of such regular packings achieved the maximal number of *twelve touching neighbors*, the one which is nowadays called *face-centered cubic lattice packing* (or simply f.c.c. packing). Then, he boldly asserted that this should be that which would achieve the "optimal density," namely, in his own words:

"The (f.c.c.) packing will be the tightest possible, so that in no other arrangements could more pellets be stuffed into the same container."

He did *not* specify the *shape* of the container, nor the *relative size* between the container and the pellets. Therefore, what he meant, was most likely the universal one of infinite size, i.e. the limiting situation with the whole space as the container. Anyhow, he *neither* gave a definitive formulation, *nor* any intuitive reasoning for the above bold assertion. He simply made such a far-reaching statement and moved on to discuss other issues. Kepler's bold assertion of 1611 amounts to the following:

Kepler's Conjecture $\pi/\sqrt{18}$ *should be the optimal density for all possible sphere packings in Euclidean* 3*-space.*

3.2.2.1 *A brief comparison between the sphere packing and the circle packing problem*

Note that the concept of global density of infinite circle packings is also defined in the same kind of complicated way. However, the problem of optimal density in the case of infinite circle packings can be settled in a rather simple manner (cf. Theorem 3.1.1). The geometric reason for the above simplicity lies in the geometric reality that the *optimal local density* of circle packing occurs when and only when the local cell is a *circumscribing regular hexagon* which is a tiling domain, thus the tightest hexagonal local packing can be globally replicated to construct the global hexagonal circle packing. Anyhow, this is the basic reason why the circle packing problem can simply be *localized* to the study of optimal local density.

The geometric reality of the sphere packing problem is much more complex than that of the circle packing problem. First of all, the local cell that realizes the *optimal local density* of sphere packing is a *circumscribing regular dodecahedron* which is *not* a tiling domain. Therefore, the optimal local density of sphere packing *cannot* be realized as the global density of infinite packings and the sphere packing problem *cannot* be settled simply by studying the geometry of single-layer local packings. Hence, one of the most important insights that one seeks in the study of the sphere packing problem, is what will be the "*essential extent of complexity*" that one has to deal with in order to solve the problem? Roughly speaking, the geometric complexity of single-layer local packings is what one has to deal with in the proof of the Dodecahedron Conjecture, while the major step in the proof of Kepler's Conjecture is the *localization scheme* which enables us to reduce its proof to the *optimal estimation* of a specific local invariant which *only depends* on the geometric complexity of *double-layer local packings*, namely, the *locally averaged density* of [Hsi-2].

3.2.3 *Sphere packings and crystal formation*

In our daily life, there are many examples of packing spheres into given containers, such as putting marbles into a jar or soybeans into a silo. However, such packings are always *with given* containers, and in the case of large number of spheres, rather random and with density much lower than $\pi/\sqrt{18} = 0.74048\cdots$.

Other than the above mentioned "man-made" sphere packings with given containers, there are many *naturally-created* sphere packings *without a container*, but with remarkable geometric regularity and often with the highest known density of $\pi/\sqrt{18}$, such as a piece of crystal of gold, silver or lead. The physical shapes of atoms can be regarded as microscopic spheres, while a small piece of a given atomic crystal often consists of billions of trillions of such microscopic spheres aggregated into a specific type of regular arrangement. Crystal structures are wonders of nature which exhibit fascinating geometric regularity and remarkable precision.

A close examination of the crystal structure of chemical elements listed in the periodic table (Table 3.1), reveals that among

Table 3.1 Crystal structure of elements.

H^1 hcp 3.75 6.12																	**He^4** hcp 3.57 5.83
Li bcc 3.491	**Be** hcp 2.27 3.59											**B** rhomb.	**C** diamond 3.567	**N** cubic 5.66 (N_2)	**O** complex (O_2)	**F**	**Ne** fcc 4.46
Na bcc 4.225	**Mg** hcp 3.21 5.21	← Crystal structure → ← a lattice parameter, in Å → ← c lattice parameter, in Å →										**Al** fcc 4.05	**Si** diamond 5.430	**P** complex	**S** complex	**Cl** complex (Cl_2)	**Ar** fcc 5.31
K bcc 5.225	**Ca** fcc 5.58	**Sc** hcp 3.31 5.27	**Ti** hcp 2.95 4.68	**V** bcc 3.03	**Cr** bcc 2.88	**Mn** cubic complex	**Fe** bcc 2.87	**Co** hcp 2.51 4.07	**Ni** fcc 3.52 6.12	**Cu** fcc 3.61	**Zn** hcp 2.66 4.95	**Ga** complex	**Ge** diamond 5.658	**As** rhomb.	**Se** hex. chains	**Br** complex (Br_2)	**Kr** fcc 5.64
Rb bcc 5.585 6.12	**Sr** fcc 6.08	**Y** hcp 3.65 5.73	**Zr** hcp 2.23 5.15	**Nb** bcc 3.30 6.12	**Mo** bcc 3.15	**Tc** hcp 2.74 4.40	**Ru** hcp 2.71 4.28	**Rh** fcc 3.80	**Pd** fcc 3.89 6.12	**Ag** fcc 4.09	**Cd** hcp 2.98 5.62	**In** tetr. 3.25 4.95	**Sn** (α) diamond 6.49	**Sb** rhomb.	**Te** hex. chains	**I** complex (I_2)	**Xe** fcc 6.13
Cs bcc 6.045	**Ba** bcc 5.02	**La** hex. 3.77 *ABAC*	**Hf** hcp 3.19 5.05	**Ta** bcc 3.30	**W** bcc 3.16	**Re** hcp 2.76 4.46	**Os** hcp 2.74 4.32	**Ir** fcc 3.84	**Pt** fcc 3.92	**Au** fcc 4.08	**Hg** rhomb.	**Tl** hcp 3.46 5.52	**Pb** fcc 4.95	**Bi** rhomb.	**Po** sc 3.34	**At** --	**Rn** --
Fr --	**Ra** --	**Ac** fcc 5.31															
			Ce fcc 5.16	**Pr** hex. 3.67 *ABAC*	**Nd** hex. 3.66	**Pm** --	**Sm** complex	**Eu** bcc 4.58	**Gd** hcp 3.63 5.78	**Tb** hcp 3.60 5.70	**Dy** hcp 3.59 5.65	**Ho** hcp 3.58 5.62	**Er** hcp 3.56 5.59	**Tm** hcp 3.54 5.56	**Yb** fcc 5.48	**Lu** hcp 3.50 5.55	
			Th fcc 5.08	**Pa** tetr. 3.92 3.24	**U** complex	**Np** complex	**Pu** complex	**Am** hex. 3.64 *ABAC*	**Cm** --	**Bk** --	**Cf** --	**Es** --	**Fm** --	**Md** --	**No** --	**Lr** --	

the eighty-one elements with stable monoatomic form, there are twenty f.c.c., twenty-four h.c.p. and four of other types of hexagonal close packings. Altogether there are forty-eight chemical elements whose crystal structures are sphere packings of the highest known density of $\pi/\sqrt{18}$. The physical intuition suggests that these forty-eight elements are the ones with rather weak inter-atomic forces, thus the *density* might play the dominant role in the formation of these crystals. As an illustration of the above physical intuition, let us examine the crystal structures of the five elements of group IV A in the periodic table, namely, Carbon (C), Silicon (Si), Germanium (Ge), Tin (Sn) and Lead (Pb). They all have four covalent bonds which are naturally arranged into four directions with uniform angular separation of $\cos^{-1}(-\frac{1}{3}) = 109.5°$. Among them Carbon has the strongest covalent bonds and the highest angular rigidity between

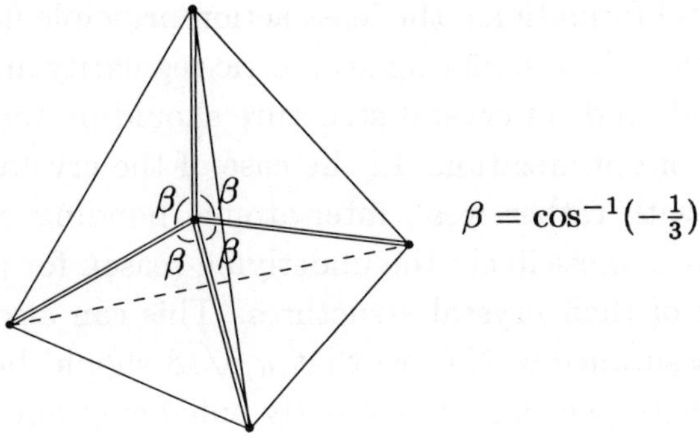

Figure 3.6

bonds. Therefore, each carbon atom in the crystal structure of diamond has only four nearest neighbors, exactly arranged in a set of directions with uniform angular separation of $\cos^{-1}(-\frac{1}{3})$, as indicated in Figure 3.6. Such crystal structure has a sphere packing density of only 0.34, which is about 46% of $\pi/\sqrt{18}$. However, diamond is renowned for having the highest stability and hardness.

The crystal structures of Silicon, Germanium and Tin are again of the same type as that of diamond. In fact, {C, Si, Ge, Sn} are the *only* four elements with the crystal structure of diamond type, while the crystal structure of lead (Pb) is the dense packing of f.c.c. type, no longer of the diamond type. Apparently, the crystal structure of diamond type is the result of the directional covalent bonding of group IV A elements. However, such covalent bonding becomes weaker and weaker as the atomic weight gets heavier and heavier. In the heaviest case of lead, the strength of covalent bonding already becomes so weak that the density emerges as the dominant factor in its crystal formation.

3.2.3.1 *The least action principle and crystal formation*

In general, the least action principle is a kind of philosophical belief that Nature always chooses the "best" among all possibilities. In the realm of crystal formations, the least action principle naturally leads to the belief that the fascinating geometric regularity and remarkable precision in all kinds of crystal structure should be the consequence of some kind of optimization. In the case of the crystal structure of those atoms with rather weak inter-atomic bonding, the optimization of density is most likely the underlying reason for the hexagonal close packing of their crystal structures. This can also be regarded as a strong assurance of Nature that $\pi/\sqrt{18}$ should be the optimal density of sphere packing, thus greatly enhancing our confidence in the validity of Kepler's Conjecture. However, in order to establish the *applicability* of the above least action principle to the crystal formation of hexagonal type, one needs to prove a much stronger result which not only proves the *optimality* of $\pi/\sqrt{18}$ for the density of finite packings, but also proves that all finite clusters of spheres with the optimal density of $\pi/\sqrt{18}$ are necessarily *hexagonal clusters.* Only when such a strong *uniqueness theorem of finite packings with optimal density* is proved, then we can say that the geometric regularity of those crystals of hexagonal close packing type is, indeed, the *consequence of the optimization of density.*

In the case of circle packings, the strong uniqueness theorem of clusters of circles with *optimal global density* (i.e. Theorem 3.1.1′) can already be deduced directly from the uniqueness aspect of the local cell with *optimal local density*, namely, it is a circumscribing regular hexagon which happens to be a *tiling domain* of the Euclidean plane. However, the *tightest local cell* in the case of sphere packing is a circumscribing *regular dodecahedron* (cf. the proof of Dodecahedron Conjecture in Chap. 6) which is *not* a tiling domain of Euclidean space. Moreover, there are various kinds of finite clusters with their global densities (of the first kind) exceeding $\pi/\sqrt{18}$, such as the string of N spheres (cf. Example 3.2.3). Therefore, the *optimization* of the *global density of the first kind cannot* be used as the least action principle for the crystal formation of those forty-eight elements of close packing type. Thus, one must search for another kind of global density that works!

3.2.4 *The least action principle of crystal formation and the localization of the proof of Kepler's Conjecture*

In the philosophical belief of the least action principle, the very existence of many crystal structures of close hexagonal packing type can be regarded as Nature's assurance of the existence of a workable global density (for finite packings without a container) while our task is just to find such a workable one! A moment of further philosophical reflection suggests that there is another message from Nature, namely, such a workable one must be *highly localizable*, meaning that it should be the *weighted average* of a certain *local invariant* which only depends on the local structure of a rather small number of layers of coating around a single sphere and moreover, the optimization of the global density so defined is always the consequence of the optimization of such a local invariant *everywhere*. Therefore, it is quite natural to begin our search of such a highly localizable, workable density in the realm of *double-layer coatings*. Anyhow, the *locally averaged density* and the *intrinsic global density of the second kind* defined in Sec. 1.4 are exactly such workable *local* (resp. *global*) *invariants* that we are searching for. The major results on sphere packings can be reduced to the *optimal estimation* of the above local (resp. global) invariants (i.e. $\frac{\pi}{\sqrt{18}}$) together with the *determination of* those *geometric structures* achieving the optimal locally averaged density (resp. optimal global density of the second kind). See Sec. 1.5 for the precise statements of the main theorems on sphere packings.

In retrospect, one of the most remarkable geometric realities on sphere packings is that both the *least action principle of crystal formation* and the *proof of Kepler's Conjecture* can be directly reduced to the *optimal estimation of a single local invariant*, namely, the *locally averaged density* of [Hsi-2] together with the determination of those *tightest double-layer local packings*. Such a clean-cut reduction by means of a single local invariant not only makes the mathematical proofs of both major results on sphere packings much simpler than otherwise, but it also reveals the basic geometric mechanism of crystal formations, namely, that *global optimization* is the *consequence* of *local optimization everywhere* while local optimization already imposes the geometric regularity that constitutes the building-block of such crystals. In fact, there is *no a priori* geometric reason for the "existence" of such a reduction of both problems to the optimal estimation of a highly localizable invariant; it is the result of Main Theorem I and it is proof that finally demonstrates the existence of such a remarkable reduction to that of the locally averaged density.

Thus, the major tasks in the study of sphere packings can be reduced to the *optimal estimates* of the *local density* (resp. the *locally averaged density*) which amounts to the optimal volume estimates of a single local cell (resp. the cluster of local cells of a central sphere and its close neighbors).

3.3 Some Basic Ideas and Crucial Understanding on which the Proofs of the Major Theorems are Based

Before we embark upon the long journey of working through the logically organized proof of the major theorems on sphere packings, let us start with a preview of some of the *basic ideas* and *crucial understanding.*

3.3.1 *Volume estimates of local cells*

3.3.1.1 *Basic geometry of local cells*

The above discussions on the central issues in the study of sphere packings, such as the least action principle of crystal formation, Kepler's Conjecture and the Dodecahedron Conjecture, clearly reveal the paramount importance of the basic *concept of local cells.* In fact, all the important issues on sphere packings are *defined in terms of the local cells*, while their proofs or solutions will rely on various kinds of *volume estimation of the local cells.* Therefore, let us begin with a brief discussion on the *geometric significance* of the concept of local cells in the study of sphere packings.

The spheres are *perfectly round* in shape, which is the *embodiment of the orthogonal symmetry* of the space, while the total expanse of the space is basically *rectilinear* in nature. The former represents the *orthogonality*, while the latter is represented by the *simply transitive* action of the *translation group.* It is intuitively clear that *roundness* and *rectilinearity cannot* fit together very well in the aspect of packing, thus there are always considerable amounts of *unfilled interstices*, no matter how you pack the round spheres into rectilinear space. The sphere packing problem seeks to determine the *minimal percentage of such interstices.* The concept of *local cells* provides a *canonical* decomposition of the whole space (in the case of infinite packings) or a portion of the space (in the case of finite packings) into *individual rectilinear pieces* (i.e. convex polyhedrons) with respect to a given packing, thus enabling us to localize the *total accounting of interstices* into that of the *unfilled portion in each local cell.* Therefore, the concept of the local cell is exactly the fundamental concept that the whole theory of sphere packings will be based upon.

The local cell of a sphere S_i in a given packing $\mathcal{P}$ is, by definition, the intersection of the S_i-sides of those half-spaces, with the *perpendicular bisectors* of the intervals, joining its center to the centers of its neighbors, as the *dividing planes.* Therefore, the local cell $C(S_i, \mathcal{P})$ and the local packing of S_i in $\mathcal{P}$ determine each other. Choose the common radii to be the unit of length, and choose the center of S_i to be the origin. Let $\{\mathbf{b}_j = \overrightarrow{OO}_j\}$ be the set of position vectors of the centers of neighbors of S_i. Then, $C(S_i, \mathcal{P})$ is given by the following set of inequalities, namely

$$C(S_i, \mathcal{P}) = \left\{ X; \overrightarrow{OX} \cdot \mathbf{b}_j \leq \frac{1}{2}|\mathbf{b}_j|^2 \right\}, \tag{1}$$

where $\mathbf{b}_j$ runs through the position vectors of the centers of the neighbors. Set A_j to be the intersection point of S_i and $\overline{OO}_j$, $\mathbf{a}_j = \overrightarrow{OA}_j$ and $|\mathbf{b}_j| = 2(1+h_j)$. Then the set of position vectors $\{\mathbf{b}_j = \overrightarrow{OO}_j\}$ can also be recorded by the set of points $\{A_j\}$ which record their directions, together with the set of numbers $\{h_j\}$ which records the center-distances between S_i and its neighbors. Thus, one has

$$C(S_i, \mathcal{P}) = \{X; \overrightarrow{OX} \cdot \overrightarrow{OA}_j \le 1 + h_j\}. \tag{2}$$

For example, if all the neighbors of S_i are touching ones, then $\{A_j\}$ is exactly the set of touching points and all $h_j = 0$, while the local cell $C(S_i, \mathcal{P})$ is exactly the circumscribing polyhedron given by

$$T(\{A_j\}) = \{X; \overrightarrow{OX} \cdot \overrightarrow{OA}_j \le 1\}. \tag{3}$$

3.3.1.2 *Basic strategy of volume estimations of local cells*

Since the local or global *densities* of sphere packings are all defined in terms of the volumes of local cells, the *volume estimation of local cells* will of course play a major role in the study of the sphere packing problem. For example, the proof of the *Dodecahedron Conjecture* amounts to achieving the *optimal lower bound estimate* of the volumes of all possible local cells, namely, the local cells of minimal volume are realized by the circumscribing regular dodecahedrons, while the central issue in the proof of Main Theorem I is a kind of *optimal, collective volume estimate of a cluster of local cells*. The task of achieving such *optimal* volume estimates is formidable and challenging, in view of the great many geometric degrees of freedom of single-layer (resp. double-layer) local packings. Thus, one needs to have a set of efficient volume estimation techniques, *specifically* developed for the local cells, in order to achieve the kind of *optimal* estimates needed for the proofs of the two major conjectures. Let us first discuss some basic strategies that lead to the development of the specific volume estimation techniques of local cells (cf. Chap. 4).

Basic strategy 1: A natural separation of the localized interstices into the rectilinear ones and the curved ones Since the volume of the unit sphere is fixed and well-known, the volume estimation of a local cell is essentially that of the interstices contained in the given local cell. However, there are in general two kinds of interstices contained in a given local cell, namely

(i) the *curved interstices* which lie between the circumscribing polyhedron $T(\{A_j\})$ and the sphere,
(ii) the *rectilinear interstices* which lie between the boundary of the given local cell and the above circumscribing polyhedron $T(\{A_j\})$. In Sec. 4.2, we shall define a canonical subdivision of the rectilinear interstices into *slabs of thickness* h_j for each *non-zero* h_j of at most 0.3, namely, the one that lies between the tangent plane of S_i at A_j and the perpendicular bisector of $\overline{OO_j}$.

The geometric intuition underlying the above strategy of separating the interstices contained in a given local cell into the rectilinear interstices and the curved interstices is the following:

The *rectilinear slabs* are interstices of a simpler kind, which can be effectively estimated by Euclidean geometry, while the curved part of interstices hugged around the sphere is intimately related to the sphere, whose effective estimation naturally relies on techniques of spherical geometry (cf. Chap. 4).

Basic strategy 2: Reduced neighbors, spherical configuration and a crucial volume formula For the sake of technical simplicity, it is advantageous to introduce the following definitions of reduced neighbors and reduced local packing.

Definition A neighbor with its center-distance at most equal to 2.6 times the radii is call a *reduced neighbor*. The subset of a given local packing $\mathcal{L}(S_0)$ consisting of S_0 and all its reduced neighbors are defined to be the *reduced packing* of $\mathcal{L}(S_0)$ and will be denoted by $\mathcal{L}'(S_0)$.

Set Σ to be the set of points on the unit sphere S_0 which records the *directions* of those *reduced neighbors*, and $T(\Sigma)$ to be the *tangent* (i.e. *circumscribing*) *polyhedron* of S_0 with Σ as the set of tangent points. Then, the part of the local cell, $C(\mathcal{L})$, *inside* of $T(\Sigma)$ is called its *core part* while the *outside part* is called its *peripheral part.* The latter consists of the rectilinear interstices of $C(\mathcal{L})$, while the former contains the curved part of the interstices as the complementary subset of S_0. For example, in the special case that all neighbors in $\mathcal{L}(S_0)$ are reduced neighbors (i.e. $\mathcal{L}'(S_0) = \mathcal{L}(S_0)$), the core part of $C(\mathcal{L})$ is exactly $T(\Sigma)$ itself. Otherwise, the core part is a subset of $T(\Sigma)$ defined by some additional inequalities corresponding to those *non-reduced* neighbors, i.e. a truncated tangent polyhedron. Anyhow, the *lower bound volume estimation of the core part* is intimately related to that of tangent polyhedra which is essentially the estimation of the isoperimetric quotient for polyhedra with a given number of faces (cf. Subsec. 2.2.1).

Geometrically speaking, the volume of $T(\Sigma)$ is, of course, an orthogonal invariant of Σ, which is a set of n unit vectors $\{\overrightarrow{OA_j}, A_j \in \Sigma\}$. Therefore, there should be a clean-cut formula expressing the volume of $T(\Sigma)$ in terms of the basic orthogonal invariants of Σ, namely, the *inner products* of the above n unit vectors. Thus, it is quite natural to begin our task of volume estimation of the core part by the search of such a volume formula of $T(\Sigma)$. First of all, the *associated spherical configuration* of Σ (cf. Sec. 2.2 for the definition of $\mathcal{S}(\Sigma)$) always provides a concise and often advantageous organization of the totality of the orthogonal invariants of Σ. In the "generic" case that all faces of $\mathcal{S}(\Sigma)$ are *triangular* and *containing their own circumcenters*, the solid angle cones corresponding to its faces subdivide $T(\Sigma)$ into pieces of *tangent subpolyhedra* as indicated in Figure 4.4 of Chap. 4, whose volumes can be neatly expressed in terms of the orthogonal invariants of individual spherical triangles, namely (cf. Lemma 4.2.1)

$$\mathrm{vol}(T(\Sigma) \cap \Gamma(\sigma)) = \frac{1}{3}\mathbf{v} \cdot \left\{ \frac{\mathbf{a} \times \mathbf{b}}{1 + \mathbf{a} \cdot \mathbf{b}} + \frac{\mathbf{b} \times \mathbf{c}}{1 + \mathbf{b} \cdot \mathbf{c}} + \frac{\mathbf{c} \times \mathbf{a}}{1 + \mathbf{c} \cdot \mathbf{a}} \right\}, \quad (4)$$

where $\Gamma(\sigma)$ is the solid angle cone spanned by a face $\sigma \in \mathcal{S}(\Sigma)$, $\{\mathbf{a}, \mathbf{b}, \mathbf{c}\}$ are the position vectors of the vertices of σ and

$$\mathbf{v} = \frac{1}{\mathbf{a} \cdot \mathbf{b} \times \mathbf{c}} \{\mathbf{a} \times \mathbf{b} + \mathbf{b} \times \mathbf{c} + \mathbf{c} \times \mathbf{a}\}. \quad (4')$$

Straightforward computation of vector algebra shows that the right-hand side of (4) is equal to

$$f(\sigma) = \frac{1}{3D} \left\{ 8 + u - \frac{4u(1 + 2u + w)}{u^2 + D^2} \right\}, \quad (5)$$

where

$$D = \det(\mathbf{a}, \mathbf{b}, \mathbf{c}) > 0,$$

$$u = 1 + \mathbf{a} \cdot \mathbf{b} + \mathbf{b} \cdot \mathbf{c} + \mathbf{c} \cdot \mathbf{a}, \quad (6)$$

$$w = (\mathbf{a} \cdot \mathbf{b})(\mathbf{b} \cdot \mathbf{c}) + (\mathbf{b} \cdot \mathbf{c})(\mathbf{c} \cdot \mathbf{a}) + (\mathbf{c} \cdot \mathbf{a})(\mathbf{a} \cdot \mathbf{b}).$$

Remarks (i) In the case that $\mathcal{L}'(S_0) = \mathcal{L}(S_0)$, the solid-angle decomposition (at the center of S_0) corresponding to $\mathcal{S}(\Sigma)$ is exactly the same as that of the Delaunay decomposition.

(ii) In the general case that some of the triangular faces of $\mathcal{S}'(\Sigma)$ (i.e. triangulation of $\mathcal{S}(\Sigma)$ via diagonal cuttings of those non-triangular faces, if any) do *not* contain their own circumcenters, the above orthogonal invariant $f(\sigma)$ of (5) is no longer equal to the volume of $T(\Sigma) \cap \Gamma(\sigma)$. However, it is a remarkable fact that the total sum is always equal to $\mathrm{vol}\, T(\Sigma)$, namely that

$$\mathrm{vol}\, T(\Sigma) = \sum_{\sigma \in \mathcal{S}'(\Sigma)} f(\sigma) \quad (7)$$

holds in general. We refer to Theorem 4.2.1 for the proof of such a remarkable formula.

3.3.1.3 *Volume estimation techniques specifically developed for local cells*

The *local cell* is a convex polyhedron canonically associated to a given (single layer) *local packing.* Thus, local cells (of sphere packings) are a rather special type of convex polyhedrons, which are intimately related to the geometry of spheres. However, due to the many degrees of freedom of local packings, there are a large number of varieties and flexibilities in the general geometric structure of local cells. Therefore, in order to achieve the kinds of *optimal estimates* needed in the proofs of both the Dodecahedron Conjecture and the Kepler Conjecture, it is necessary to develop an efficient set of volume estimation techniques, which effectively utilize the special geometric features of local cells, i.e. their intimate relations to the geometry of spheres. For example, the following Lemmas of [Hsi-2] and [Hsi-3] (cf. Chap. 4) are exactly such volume estimation techniques specifically developed for the purpose of providing clean-cut, optimal estimates of the volumes of local cells in various geometric situations.

Lemma (cf. Lemma 4.2.2) *Let* $\sigma = \overset{\frown}{\triangle} ABC$ *be a spherical triangle containing its own circumcenter,* $\{\mathbf{a}, \mathbf{b}, \mathbf{c}\}$ *and* $\triangle$ *be the position vectors of its vertices and its area, and* $\bar{\sigma}_{\triangle}$ *be the equilateral spherical triangle with* $\triangle$ *as its area. Set* $f(\sigma)$ *to be the volume function of* (5) *and* (6). *Then,*

$$f(\sigma) \geq f(\bar{\sigma}_{\triangle}) = \left(1 - 2\cos\frac{\pi + \triangle}{3}\right)\tan\frac{\pi + \triangle}{6} \tag{8}$$

and the equality holds when and only when σ *is itself equilateral.*

Lemma (cf. Lemma 4.2.3) $\varphi(\triangle) = (1 - 2\cos\frac{\pi+\triangle}{3})\tan\frac{\pi+\triangle}{6}$ *is a strictly convex function of* $\triangle$, *namely* $\varphi''(\triangle) > 0$ *for* $\triangle \in (0, 2\pi)$.

Lemma (cf. Lemma 4.3.1) *The volume of a rectilinear slab of thickness $h \leq 0.3$ is at least equal to*

$$L(h) = \frac{5}{3} \tan \frac{\pi}{5} \left(\frac{1-h}{3+h} \right) \left(\frac{1-h}{1+h} \right) \left[1 - \left(1 - \frac{h(1+h)}{1-h} \right)^3 \right] . \quad (9)$$

We refer to Chap. 4 for their proofs and further discussions, only mentioning here some direct applications of the above Lemmas.

3.3.1.4 *Some examples of volume estimation*

Example 3.3.1 Let $\mathcal{L}(S_0)$ be a single-layer local packing consisting of a central sphere S_0 and twelve touching neighbors of S_0. Set $\Sigma = \{A_j, 1 \leq j \leq 12\}$ to be the set of touching points, and $\mathcal{S}(\Sigma)$ to be the associated spherical configuration of Σ whose edge-lengths are, of course, at least equal to $\pi/3$. Moreover, the circumradii of its faces are at most equal to $\pi/4$ if and only if such a single-layer local packing is already a saturated one, meaning *no* more neighbors of the central sphere can be added to such a local packing.

Let us assume that $\mathcal{L}(S_0)$ is such a local packing. Then, the local cell $C(S_0, \mathcal{L})$ is a circumscribing dodecahedron with Σ as the set of twelve touching points. The faces in $\mathcal{S}(\Sigma)$ are either triangles or "squares" of side-lengths $\pi/3$. We shall add diagonals to those squares, if any, so as to cut them into pairs of triangles. Thus, $\mathcal{S}(\Sigma)$ consists of exactly twenty triangles, say $\{\sigma_i; 1 \leq i \leq 20\}$, each of them containing its own circumcenter, while the sum total of their areas is, of course, equal to 4π. Hence, it follows directly from Lemmas 4.2.2 and 4.2.3 that

$$\begin{aligned} \operatorname{vol} T(\Sigma) &= \sum_{i=1}^{20} f(\sigma_i) \geq \sum_{i=1}^{20} f(\bar{\sigma}_{\triangle_i}) \\ &\geq 20 f(\bar{\sigma}_{\pi/5}) = 20 \left(1 - 2\cos\frac{2\pi}{5} \right) \tan\frac{\pi}{5} . \end{aligned} \quad (10)$$

Moreover, the equality holds when and only when all triangles are regular with area $\pi/5$, namely, when $I(\Sigma)$ is a regular, spherical icosahedron.

Remark The above clean-cut estimate proves that the volume of the circumscribing regular dodecahedron is, indeed, the lower bound of the volumes of those local cells of the above special type of local packings. This, of course, only constitutes a small portion of the proof of the Dodecahedron Conjecture. However, it is an important part of the proof, because it covers a rather large vicinity of the optimal solution, i.e. the unique tightest local packing that makes the local cell have the minimal volume.

Next, let us investigate the case of more than twelve neighbors. The answer to the problem of thirteen spheres proves that a single-layer local packing $\mathcal{L}(S_0)$ with at least thirteen neighbors of S_0, must contain some non-touching neighbors of S_0. Let us first look at the following type of local packings.

Example 3.3.2 Suppose that $\mathcal{L}(S_0) = \mathcal{L}'(S_0)$ and all faces of $\mathcal{S}(\Sigma)$ are triangular and contain their own circumcenters. Then, a direct application of the above lemmas provides the following volume estimate, namely

$$\begin{aligned} \operatorname{vol} C(\mathcal{L}) &\geq \sum_{j=1}^{n} L(h_j) + \operatorname{vol} T(\Sigma) \\ &= \sum_{j=1}^{n} L(h_j) + \sum_{i=1}^{2n-4} f(\sigma_i) \\ &\geq \sum_{j=1}^{n} L(h_j) + (2n-4) f(\overline{\sigma}_{\overline{\triangle}}), \end{aligned} \tag{11}$$

where $n = \#(\Sigma)$ and $\overline{\triangle} = \frac{2\pi}{n-2}$.

Remarks (i) In the case of $n \geq 13$, the lower bound volume estimate of $T(\Sigma)$ by $(2n-4)f(\overline{\sigma}_{\overline{\triangle}})$ is no longer optimal. However, it is a simple estimate which already suffices for most cases of applications. We shall denote it simply by L_n, namely

$$L_n = (2n-4)\left(1 - 2\cos\frac{n\pi}{3n-6}\right)\tan\frac{n\pi}{6n-12}\,. \tag{12}$$

(ii) The size of the first term, i.e. $\sum_{j=1}^{n} L(h_j)$, will of course depend on the size of $\sum_{j=1}^{n} h_j$.

3.3.2 *Geometry of single-layer local packings*

Intuitively speaking, a single-layer local packing is an arrangement of a certain number of unit spheres on top of a perfectly round table of total area 4π. The geometric situation is that there is more than adequate room to arrange *twelve touching neighbors*, but far from enough room to arrange *thirteen touching neighbors.* Roughly speaking, local packings with twelve or less neighbors are always rather *untight*, while local packings with thirteen or more neighbors are always *over-crowded.* Therefore, the kind of geometric understanding of single-layer local packings that will be needed in the optimal estimates of both the local density and the locally averaged density are the quantitative measurements and estimates both on the *non-tightness* of the former and on the *over-crowdedness* of the latter.

3.3.2.1 *Quantitative refinements of the problem of thirteen spheres*

The original "problem of thirteen spheres" is a qualitative one, which asks whether it is possible to arrange thirteen touching neighbors. The negative answer of such a qualitative problem (first proved by van der Waerden and Schütte) is, just by itself, of very little use in the study of either the local density or the locally averaged density. Anyhow, what one needs is a *quantitative measurement* and an *estimate of such a quantitative measurement* of the *over-crowdedness* of local packings with thirteen or more neighbors.

For each individual neighbor S_j in a given local packing $\mathcal{L}(S_0)$, the buckling height h_j is a *quantitative measurement* of the *non-touchingness* of S_j. Therefore, the *total buckling height*, namely Σh_j, is a *quantitative measurement* of the *total amount of non-touchingness* contained in a given local packing (or reduced packing). As indicated in formula (11) of the volume estimation of Example 3.3.2, the total buckling height is closely related to the first term $\Sigma L(h_j)$ which estimates the total contribution of those rectilinear slabs contained in the local cell $C(S_0, \mathcal{L})$. Therefore, the relevant quantitative understanding of the *over-crowdedness* of local packings with thirteen or more neighbors should naturally be some kind of *lower bound estimates* of the *total buckling height* in terms of the number of neighbors or close neighbors. We refer to Chap. 5 for a systematic discussion on this topic, which will play an important role in the proofs of both the Dodecahedron Conjecture and Main Theorem I (cf. Chaps. 6 and 8).

3.3.2.2 *The non-tightness of local packings with twelve touching or almost touching neighbors*

A triple of touching neighbors clustering together in a triangular formation touching each other, represents a *tight* local formation of touching neighbors. Correspondingly, the triple of their touching points are the vertices of an equilateral triangle of side-length $\pi/3$ on the central sphere S_0 and the area of this triangle is equal to $2\tan^{-1}\frac{\sqrt{2}}{5}$. Let $\mathcal{L}(S_0)$ be a given local packing with twelve touching neighbors and Σ be the set of twelve touching points. We shall assume that the associated spherical configuration of Σ consists of twenty triangles and thirty edges, by adding diagonals to cut its quadrilateral faces, if any, into pairs of triangles. To each edge of $\mathcal{S}(\Sigma)$, say $\overset{\frown}{A_jA_{j'}}$, whose length ℓ exceeds $\pi/3$, there is a *gap* of width equal to $(4\sin\frac{\ell}{2}-2)$ between S_j and $S_{j'}$. We shall call the *difference* between the length of a given edge in $\mathcal{S}(\Sigma)$ and $\pi/3$ its *edge-excess* (resp. the difference between the area of a given triangle in $\mathcal{S}(\Sigma)$ and $2\tan^{-1}\frac{\sqrt{2}}{5}$, its *area-excess*). Notice that the *total amount of area-excess* is equal to

$$4\pi - 40\tan^{-1}\frac{\sqrt{2}}{5} = 1.54065\cdots. \tag{13}$$

Therefore, it is not difficult to use the isoperimetric inequality of spherical triangles to obtain a *lower bound estimate* of the *total amount of edge-excess*, which is closely related to the total number of gaps contained in this type of local packings. However, in the estimation of the *locally averaged density*, the *quantitative effects* of big and small gaps are quite different! Therefore, instead of working with a *total accounting* of gaps, we shall study the *distribution* of the total amount of non-tightness into the following two types of gaps, namely

(i) **gully-type**: corresponding to edges of *small* or *medium edge-excess*,
(ii) **big-hole-type**: corresponding to quadrilaterals or slightly buckled quadrilaterals, or equivalently corresponding to the *big edge-excess* of their *cutting diagonals*.

In fact, the following geometric intuition plays an important role in the overall strategy of the proof of Main Theorem I (cf. Chap. 8).

The *locally averaged density* is defined to be a specific weighted average of the local densities of the cluster of spheres consisting of a central sphere and all its close neighbors. As one shall see later, the most critical case in the optimal upper bound estimation of the locally averaged density, is the case that all spheres of $\widehat{\mathcal{L}}(S_0)$ have exactly twelve close neighbors in $\mathcal{L}^2(S_0)$. In this case, the above optimal estimate amounts to proving that the *total volume* of the thirteen local cells (i.e. that of S_0 and its twelve close neighbors) is at least equal to $52\sqrt{2}$, and moreover, the total volume is equal to the minimal value of $52\sqrt{2}$ when and only when both $\mathcal{L}(S_0)$ and $\mathcal{L}^2(S_0)$ are subclusters of a hexagonal close packing. The underlying geometric intuition of such a proof is that gaps of the gully-type always contribute a much higher proportion of volumes

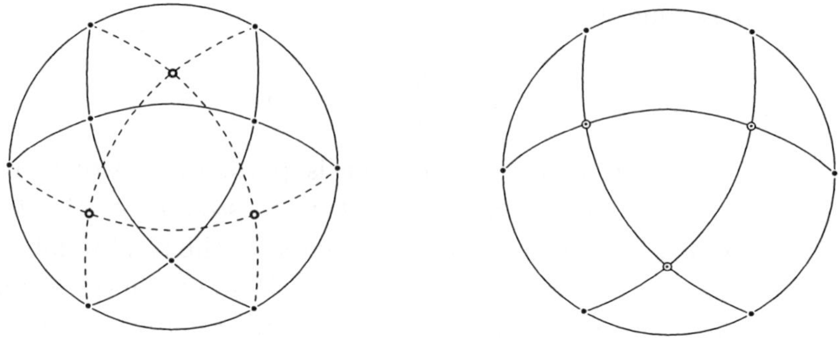

f.c.c. config. h.c.p. config.

Figure 3.7

(i.e. that of the rectilinear slabs) than that of the big-hole-type. Therefore, the total volume of the thirteen local cells will be minimal in the case that all gaps are distributed into big-holes and, if possible, without any gaps of the gully-type. Geometrically, such a local packing corresponds to a spherical configuration with maximal number of quadrilaterals and of uniform edge-lengths of $\pi/3$. The following sub-lemma shows that there are exactly two such spherical configurations.

Sublemma 3.3.1 *For spherical configurations with twelve vertices and edge-lengths of at least* $\pi/3$*, there are at most six quadrilateral faces. Moreover, there are only two such spherical configurations with six quadrilaterals, namely the f.c.c. configuration and the h.c.p. configuration, as indicated in Figure* 3.7.

Proof Let F_3 (resp. F_4) be the number of triangular (resp. quadrilateral) faces in $\mathcal{S}(\Sigma)$. As one shall see later, there cannot be any pentagonal face in such a spherical configuration. Therefore, by Euler's formula

$$F_3 + 2F_4 = 20\,. \tag{14}$$

On the other hand, it follows from Lemma 2.2.1 that the areas of triangles (resp. quadrilaterals) in such a spherical configuration are bounded below by

$$\Delta_{\pi/3} = 3\cos^{-1}\frac{1}{3} - \pi$$
$$\left(\text{resp. } \square_{\pi/3} = 4\cos^{-1}\left(-\frac{1}{3}\right) - 2\pi\right) \tag{15}$$

and their areas can be equal to the above minimal values only when they are actually equilateral triangles (resp. quadrilaterals) of side-length $\pi/3$. Note that

$$8\Delta_{\pi/3} + 6\square_{\pi/3} = 24\left[\cos^{-1}\left(\frac{1}{3}\right) + \cos^{-1}\left(-\frac{1}{3}\right)\right] - 20\pi = 4\pi\,. \tag{16}$$

Hence, $F_4 \leq 6$, and $F_4 = 6$ implies that $\mathcal{S}(\Sigma)$ consists of eight equilateral triangles and six equilateral quadrilaterals, all of side-lengths $\pi/3$. It is easy to see the two configurations indicated in Figure 3.7 are the only possible such spherical configurations with $F_4 = 6$. □

3.3.3 *Geometry of double-layer local packing*

The *invariant* that enables us to *localize* both the establishment of the least action principle and the proof of Kepler's Conjecture, namely, the *locally averaged density of* [Hsi-2] (cf. Sec. 1.4), is defined in terms of the geometry of a double-layer local packing, while their mathematical proofs can be reduced to that of the *optimal estimation* of the above local invariant. In this subsection, we shall first analyze and then outline the type of geometry on double-layer local packings that such a proof will depend upon.

1. The locally averaged density is defined solely in terms of the volumes of the local cells of the spheres of the single-layer local packing, while the spheres of the second layer *only play the role* of making the above collection of local cells *defined.* Therefore, in analyzing the actual geometric input that is needed in the optimal estimation of

the locally averaged density, it is natural to regard a double-layer local packing $\mathcal{L}^2(S_0)$ as *an extension* of the single-layer local packing $\mathcal{L}(S_0)$ it contains. Such a viewpoint focuses our attention on the geometry of $\mathcal{L}(S_0)$ and the effect of those additional spheres (i.e. the second layer ones) on the volumes of the local cells of the first layer spheres.

2. To each close neighbor of S_0, say S_j, the spheres of $\mathcal{L}(S_0)$ already provide a *partial surrounding* of S_j which is given by the local arrangement of $\mathcal{L}(S_0)$ around S_j. Thus, the restriction of the extension from $\mathcal{L}(S_0)$ to $\mathcal{L}^2(S_0)$, to the local packing of S_j amounts to a *local extension* which extends the partial surrounding of S_j in $\mathcal{L}(S_0)$ to its full-fledged local packing $\mathcal{L}(S_j, \mathcal{L}^2)$. Naturally, we shall analyze the relationship between the geometry of such a local extension and the volume of its resulting local cell, thus enabling us to pinpoint those *volume-influential geometric invariants* of such a *local extension.*

3. The relationship between the *local extensions* at individual close neighbors of S_0 and the *full extension* from $\mathcal{L}(S_0)$ to $\mathcal{L}^2(S_0)$ is such that the former are all *restrictions* of the latter, while the latter is a *compatible assemblage* of the former. Therefore, a collection of local extensions at the individual close neighbors of S_0 need to satisfy a large collection of rather intricate *compatibility conditions* in order that they can be put together without conflict of space! In general it is quite difficult to formulate precisely such compatibility conditions, or to account for their manifestations at the level of collective volume estimation for the resulting local cells. However, among the various *interlocking relationships* between the *compatible local extensions*, there is a *simple type* with rather *prominent effect* on the collective volume estimation of the local cells. Notice that two such neighboring local extensions both have to make use of some *common neighboring space*, thus forcing them to have *common spheres* in *both local extensions.*

Definition Let S' be an added sphere in a given local extension around S_j. If the center-distance between S' and *another* close neighbor S_k of S_0 is *less than* $2\sqrt{2}$, then S' is automatically also a neighbor of S_k while $\frac{1}{2}(\overline{O'O}_k - 2)$ is the *buckling height* of S' as a neighbor of S_k. The *totality* of *all such buckling height* of at most 0.3 between the *added spheres of a given local extension* around S_j and the *other close neighbors of* S_0, is defined to be the *total buckling effect* of the given local extension.

For example, certain local extensions of a given type of partial surroundings of S_j may make the resulting local cells of S_j slightly smaller than otherwise, but such a tighter local extension *needs to be accompanied by* a considerable number of total buckling height, thus *producing* an *overall increment* of the total estimates of volumes. Therefore, in the *collective estimation* of the volumes of local cells needed for the proof of Main Theorem I, it often suffices to provide some kind of lower bound volume estimates for the resulting local cells of those local extensions *which have a rather small upper bound* on their *total buckling effect* (cf. Chap. 7). This is the important geometric insight that enables us to *localize the overall collective volume estimation* for the cluster of local cells to that of *individual* local cells, *namely, to that of the local extensions* with *at most a small amount of total buckling effect*, thus making the task of proving Main Theorem I much simpler.

4. Note that the Main Theorem of [Hsi-2] not only achieves the *optimal* upper bound estimate of the locally averaged densities of all possible double-layer local packings, namely $\pi/\sqrt{18}$; but it also determines all those double-layer local packings, up to congruence and scaling, that attain the optimal locally averaged density of $\pi/\sqrt{18}$, namely, the double-layer local packings occurring in a hexagonal close packing. In fact, there are exactly the following six geometrically distinct such double-layer local packings, namely, the double layer local packing of a sphere S of the middle layer of the

following six types of close stackings of five layers of planary hexagonal packings:

$$\left.\begin{array}{ll}\text{(i)} & ababa-\text{stacking}\\ \text{(ii)} & acacb-\text{stacking}\\ \text{(iii)} & bcacb-\text{stacking}\end{array}\right\}\ \text{h.c.p. } \mathcal{L}(S)$$

$$\left.\begin{array}{ll}\text{(iv)} & abaca-\text{stacking}\\ \text{(v)} & abacb-\text{stacking}\\ \text{(vi)} & cbacb-\text{stacking}\end{array}\right\}\ \text{f.c.c. } \mathcal{L}(S)\,.$$

Moreover, it is easy to check that the number of spheres in $\mathcal{L}^2(S)$ is equal to 57 (resp. 56, 55) for the cases (i) and (iv) (resp. (ii) and (v), (iii) and (vi)).

5. One of the outstanding geometric features of the above six tightest double-layer local packings is that all their local cells contain *no rectilinear slabs*. In fact, it is not difficult to prove that they are actually the only double-layer local packings without rectilinear slabs in the local cells (defined by them). This reality heuristically suggests the following guiding philosophy which is useful in the whole proof of Main Theorem I, namely, the *collective accounting* of the *total number of the volumes of the rectilinear slabs* contained in various types of double-layer local packings, other than the above six, will always play an important role in proving that the locally averaged densities defined by such double-layer local packings, are necessarily *less than* $\pi/\sqrt{18}$! The underlying geometric intuition is that, those rectilinear slabs, if any, are always *highly volume increasing*!

Chapter 4

Geometry of Local Cells and Specific Volume Estimation Techniques for Local Cells

Local cells are the most basic type of geometric objects that are naturally associated to sphere packings, while all the other important geometric invariants of sphere packings such as the various kinds of densities are defined in terms of them. In this chapter, we shall study the geometry of single-layer local packings and their associated local cells, mainly for the purpose of developing a set of efficient volume estimation techniques, specifically for the local cells. We shall always denote the central sphere of a local packing by S_0 and denote a single-layer (resp. double-layer) local packing by $\mathcal{L}(S_0)$ (resp. $\mathcal{L}^2(S_0)$). Moreover, we shall denote the local cell of $\mathcal{L}(S_0)$ simply by $C(\mathcal{L})$ and the local cell of S_j in $\mathcal{L}^2(S_0)$ by $C(S_j, \mathcal{L}^2)$, $S_j \in \mathcal{L}(S_0)$.

4.1 Basic Geometry of Local Packings and Local Cells

It is quite easy to see that the geometric arrangement of neighbors of S_0 in a given local packing, $\mathcal{L}(S_0)$, and the geometry of its associated local cell, $C(\mathcal{L})$, determine each other completely. In this section, we shall analyze various aspects of the geometric correlations between them in order to develop a set of efficient volume estimation techniques, specifically tailored for that of the local cells.

The geometric arrangement of neighbors of S_0 in a given local packing $\mathcal{L}(S_0)$ can be concisely recorded by the set of points on S_0 (or rather, unit vectors) which records the individual directions of those neighbors, together with the set of *buckling height* which records the individual distances, namely

$$\Sigma = \{\mathbf{a}_j = \overrightarrow{OA}_j, A_j \in S_0\} \text{ and } \{h_j\}, \tag{1}$$

where the position vector of the center of the neighbor S_j is given by

$$\overrightarrow{OO}_j = 2(1 + h_j)\mathbf{a}_j . \tag{2}$$

The local cell $C(\mathcal{L})$ can then be given as

$$C(\mathcal{L}) = \{X;\ \overrightarrow{OX} \cdot \mathbf{a}_j \leq 1 + h_j, j \in I\}, \tag{3}$$

where I is the index set of neighbors in $\mathcal{L}(S_0)$.

Intuitively, it is not difficult to see that neighbors with rather large buckling height, if any, will play a rather insignificant role in the problem of lower bound volume estimation of local cells. As one shall see later, the occurrence of such further away neighbors tends to make the volume of local cell larger than otherwise, thus making the task of lower bound volume estimation easier than otherwise. Anyhow, for the sake of technical simplification, it is advantageous to introduce the concepts of *reduced neighbors* and *reduced local packing* (cf. Sec. 3.3).

Notation In this chapter, we shall always use Σ (resp. $\{h_j\}$) to denote the set of points on S_0 (resp. the set of buckling height) that records the individual directions (resp. distances) of those *reduced neighbors* of S_0, i.e. $S_j \in \mathcal{L}'(S_0)$.

The above set of geometric data, $(\Sigma, \{h_j\})$, already provides a concise bookkeeping of the arrangement of a given reduced local packing $\mathcal{L}'(S_0)$. However, in order to fully analyze the geometry of its associated local cell so as to obtain some kind of simple but already "good enough" lower bound volume estimates of its associated local cell, it is natural and useful to introduce the following system of spherical configurations and polyhedrons that can be naturally constructed from $(\Sigma, \{h_j\})$.

4.1.1 *Associated spherical configurations and polyhedrons of* $(\Sigma, \{h_j\})$

(i) $I(\Sigma)$: The *inscribing* polyhedron of S_0 with Σ as the set of vertices; it is also the *convex hull* of Σ. (We shall always assume that $I(\Sigma)$ contains O in its interior.)

(ii) $T(\Sigma)$: The *circumscribing* polyhedron of S_0 with Σ as the set of touching points. Let $\{\mathbf{a}_j\}$ be the set of unit vectors which represents the positions of points in Σ. Then $T(\Sigma)$ consists of those points whose position vectors satisfy the following set of inequalities, namely

$$\overrightarrow{OX} \cdot \mathbf{a}_j \le 1\,, \quad \mathbf{a}_j = \overrightarrow{OA_j}\,, \quad A_j \in \Sigma\,. \tag{4}$$

(iii) $\mathcal{S}(\Sigma)$: The spherical configuration associated to Σ, whose faces are, by definition, the radial projections of the faces of $I(\Sigma)$.

(iv) $\mathcal{S}^*(\Sigma)$: The dual spherical configuration of $\mathcal{S}(\Sigma)$ whose faces are the radial projections of the faces of $T(\Sigma)$, which is the dual of $I(\Sigma)$ in the sense of projective geometry. Moreover, the set of vertices of $\mathcal{S}^*(\Sigma)$ are exactly the set of circumcenters of those faces of $\mathcal{S}(\Sigma)$.

(v) $T(\Sigma, \{h_j\})$: The convex polyhedron consists of those points whose position vectors satisfy the following set of inequalities, namely

$$\overrightarrow{OX} \cdot \mathbf{a}_j \le 1 + h_j\,, \quad \mathbf{a}_j = \overrightarrow{OA_j}\,, \quad A_j \in \Sigma\,. \tag{5}$$

We shall call $T(\Sigma, \{h_j\})$ the *generalized tangent polyhedron of Σ with buckling height* $\{h_j\}$. Notice that it is equal to the local cell $C(\mathcal{L})$ if $\mathcal{L}(S_0) = \mathcal{L}'(S_0)$, i.e. if all neighbors of $\mathcal{L}$ are reduced neighbors.

Remark In the general case that $\mathcal{L}(S_0) \neq \mathcal{L}'(S_0)$, the local cell, $C(\mathcal{L})$, is a proper subset of $T(\Sigma, \{h_j\})$ obtained by the truncations of some additional inequalities, namely

$$\overrightarrow{OX} \cdot \overrightarrow{OA_\ell} \leq 1 + h_\ell \; (h_\ell > 0.3)\,, \tag{6}$$

where $\{\overrightarrow{OA_\ell}\}$ are the direction vectors of those non-reduced neighbors in $\mathcal{L}(S_0)$.

(vi) $\mathcal{S}^*(\Sigma, \{h_j\})$: The spherical configuration whose faces are, by definition, the radial projections of the faces of $T(\Sigma, \{h_j\})$.

4.1.2 *Geometric correlations between spherical configurations and associated polyhedrons*

4.1.2.1 *The correlation between $\mathcal{S}(\Sigma)$ and $T(\Sigma)$*

The projective duality between $T(\Sigma)$ and $I(\Sigma)$ naturally produces the following duality-correlations between $T(\Sigma)$ and $\mathcal{S}(\Sigma)$, which is simply the "spherical version" of $I(\Sigma)$, namely

(i) each face of $T(\Sigma)$ corresponds to a vertex of $\mathcal{S}(\Sigma)$, which is exactly the tangent point of the face,
(ii) each vertex of $T(\Sigma)$ corresponds to a face of $\mathcal{S}(\Sigma)$, $V(\sigma) \leftrightarrow \sigma$, such that $V(\sigma)$ is the point of common intersection of the tangent planes at the vertices of σ,
(iii) a pair of vertices of $T(\Sigma)$ are connected by an edge if their corresponding faces of $\mathcal{S}(\Sigma)$ are an adjacent pair, i.e. share a common edge.

The following sublemma summarizes the above duality correspondence in terms of explicit vector operations.

Sublemma 4.1.1 *Let $\sigma_1 = \widehat{\triangle}ABC$ and $\sigma_2 = \widehat{\triangle}ACD$ be a pair of adjacent faces in $\mathcal{S}(\Sigma)$ and V_1, V_2 be their corresponding vertices in $T(\Sigma)$. Let $\{\ell_i, 1 \le i \le 4\}$ be the four side-lengths of $\widehat{\square}ABCD$, in cyclic order of positive orientation. Set $d = \widehat{AC}$, $c_i = \cos \ell_i$, $x = 1 + \cos d$, $D_1 = \mathbf{a} \times \mathbf{b} \cdot \mathbf{c}$, $D_2 = \mathbf{a} \times \mathbf{c} \cdot \mathbf{d}$. Then*

$$\begin{aligned} \overrightarrow{OV_1} &= \frac{1}{D_1}\{\mathbf{a} \times \mathbf{b} + \mathbf{b} \times \mathbf{c} + \mathbf{c} \times \mathbf{a}\}, \\ \overrightarrow{OV_2} &= \frac{1}{D_2}\{\mathbf{a} \times \mathbf{c} + \mathbf{c} \times \mathbf{d} + \mathbf{d} \times \mathbf{a}\}, \\ \overrightarrow{V_1V_2} &= \left\{\frac{x - c_1 - c_2}{xD_1} + \frac{x - c_3 - c_4}{xD_2}\right\} \mathbf{a} \times \mathbf{c}. \end{aligned} \tag{7}$$

Proof (i) $\overrightarrow{OV_1}$ (resp. $\overrightarrow{OV_2}$) is characterized by the following set of equations, namely

$$\begin{gathered} \overrightarrow{OV_1} \cdot \mathbf{a} = \overrightarrow{OV_1} \cdot \mathbf{b} = \overrightarrow{OV_1} \cdot \mathbf{c} = 1 \\ (\text{resp. } \overrightarrow{OV_2} \cdot \mathbf{a} = \overrightarrow{OV_2} \cdot \mathbf{c} = \overrightarrow{OV_2} \cdot \mathbf{d} = 1). \end{gathered} \tag{8}$$

(ii) Both V_1 and V_2 are on the intersection line of T_A and T_C which is perpendicular to the plane of (OAC). Hence, $\overrightarrow{V_1V_2}$ is a multiple of $\mathbf{a} \times \mathbf{c}$. Moreover, the position vector of $T_A \cap T_C \cap (OAC)$ is given by

$$\frac{\mathbf{a} + \mathbf{c}}{1 + \mathbf{a} \cdot \mathbf{c}}. \tag{9}$$

Thus, there exist suitable constants, k_1, k_2, such that

$$\begin{aligned} \overrightarrow{OV_1} &= \frac{\mathbf{a} + \mathbf{c}}{1 + \mathbf{a} \cdot \mathbf{c}} + k_1 \mathbf{a} \times \mathbf{c}, \\ \overrightarrow{OV_2} &= \frac{\mathbf{a} + \mathbf{c}}{1 + \mathbf{a} \cdot \mathbf{c}} + k_2 \mathbf{a} \times \mathbf{c}. \end{aligned} \tag{10}$$

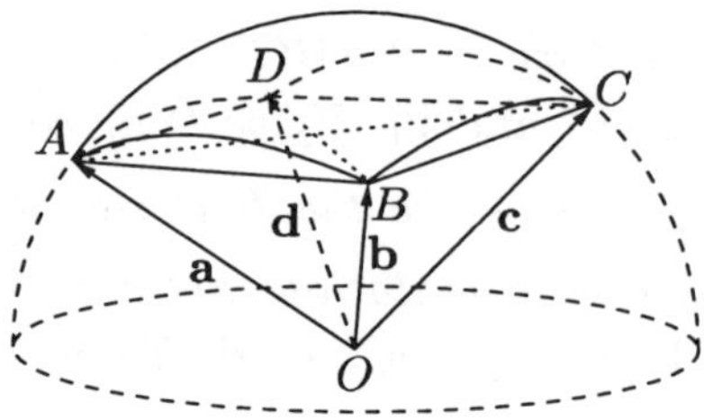

Figure 4.1

Using the conditions that $\overrightarrow{OV_1} \cdot \mathbf{b} = 1$, $\overrightarrow{OV_2} \cdot \mathbf{d} = 1$, one has

$$k_1 = \frac{-x + c_1 + c_2}{xD_1}, \quad k_2 = \frac{x - c_3 - c_4}{xD_2}. \tag{11}$$

Hence, $\overrightarrow{V_1V_2} = (k_2 - k_1)\mathbf{a} \times \mathbf{c}$ is given by (7). □

Remark Suppose that the spherical quadrilateral $\widehat{\square}ABCD$ of Figure 4.1 is decomposed into the pair of adjacent triangles $\widehat{\triangle}ABC$ and $\widehat{\triangle}ACD$ rather than the other pair $\widehat{\triangle}ABD$ and $\widehat{\triangle}BCD$. Then, it follows from the "convexity property" of $\mathcal{S}(\Sigma)$ that

$$\mathbf{a} \times \mathbf{b} \cdot \mathbf{c} + \mathbf{a} \times \mathbf{c} \cdot \mathbf{d} - \mathbf{a} \times \mathbf{b} \cdot \mathbf{d} - \mathbf{b} \times \mathbf{c} \cdot \mathbf{d} > 0. \tag{12}$$

As indicated in Figure 4.1, the above quantity is, in fact, equal to six times the volume of the tetrahedron with $\{A, B, C, D\}$ as its four vertices.

4.1.2.2 *The basic geometry of $T(\Sigma, \{h_j\})$*

Conceptually, the only difference between $T(\Sigma)$ and $T(\Sigma, \{h_j\})$ is that the defining inequalities of the latter are modified to be $\overrightarrow{OX} \cdot \overrightarrow{OA_j} \leq 1 + h_j$, $A_j \in \Sigma$, instead of uniformly equal to 1 for the right-hand side in the case of $T(\Sigma)$. For example, in the special case that all $\{h_j\}$ are uniformly the same, $T(\Sigma, \{h\})$ is simply the $(1 + h)$-magnification of $T(\Sigma)$ and hence, the radial projections of the faces of $T(\Sigma, \{h\})$ and that of $T(\Sigma)$ are an identical set of spherical polygons, namely, $\mathcal{S}^*(\Sigma, \{h\}) = \mathcal{S}^*(\Sigma)$ in the special case of $h_j = h$ for all j. However, in the non-uniform case, $\mathcal{S}^*(\Sigma, \{h_j\})$ and $\mathcal{S}^*(\Sigma)$ are, in general, different. Moreover, in a certain geometric situation, even the combinatorial structures of $\mathcal{S}^*(\Sigma, \{h_j\})$ and $\mathcal{S}^*(\Sigma)$ can be different. Anyhow, let us analyze the local geometric situation of $T(\Sigma, \{h_j\})$ associated to the above spherical quadrilateral $\widehat{\square}ABCD$ as follows.

Set $\{h_j, 1 \leq j \leq 4\}$ to be the respective buckling height of $\{S_A, S_B, S_C, S_D\}$. Then, the four corresponding planes are given by $\overrightarrow{OX} \cdot \overrightarrow{OA} = 1 + h_1$, $\overrightarrow{OX} \cdot \overrightarrow{OB} = 1 + h_2$, etc. and they determine four points of triple-intersections whose position vectors can be explicitly given as follows, namely

$$\overrightarrow{O\bar{V}_1} = \frac{1}{\mathbf{a} \times \mathbf{b} \cdot \mathbf{c}} \{(1 + h_3)\mathbf{a} \times \mathbf{b} + (1 + h_1)\mathbf{b} \times \mathbf{c} + (1 + h_2)\mathbf{c} \times \mathbf{a}\},$$

$$\overrightarrow{O\bar{V}_2} = \frac{1}{\mathbf{a} \times \mathbf{c} \cdot \mathbf{d}} \{(1 + h_4)\mathbf{a} \times \mathbf{c} + (1 + h_1)\mathbf{c} \times \mathbf{d} + (1 + h_3)\mathbf{d} \times \mathbf{a}\},$$

$$\overrightarrow{O\bar{V}_1'} = \frac{1}{\mathbf{a} \times \mathbf{b} \cdot \mathbf{d}} \{(1 + h_4)\mathbf{a} \times \mathbf{b} + (1 + h_1)\mathbf{b} \times \mathbf{d} + (1 + h_2)\mathbf{d} \times \mathbf{a}\},$$

$$\overrightarrow{O\bar{V}_2'} = \frac{1}{\mathbf{b} \times \mathbf{c} \cdot \mathbf{d}} \{(1 + h_4)\mathbf{b} \times \mathbf{c} + (1 + h_2)\mathbf{c} \times \mathbf{d} + (1 + h_3)\mathbf{d} \times \mathbf{b}\}. \tag{13}$$

In the special case that $h_1 = h_2 = h_3 = h_4 = 0$, the above four points $\{\bar{V}_1, \bar{V}_2; \bar{V}_1', \bar{V}_2'\}$ become $\{V_1, V_2; V_1', V_2'\}$, while the inequality (12) determines that $\{V_1, V_2\}$ should be the pair that occur as vertices of $T(\Sigma)$.

Sublemma 4.1.2 $\{\bar{V}_1, \bar{V}_2\}$ *(resp.* $\{\bar{V}_1', \bar{V}_2'\}$*) are the pair of points that belong to the set of vertices of* $T(\Sigma, \{h_j\})$ *if and only if*

$$(1 + h_4)\mathbf{a} \times \mathbf{b} \cdot \mathbf{c} + (1 + h_2)\mathbf{a} \times \mathbf{c} \cdot \mathbf{d} - (1 + h_3)\mathbf{a} \times \mathbf{b} \cdot \mathbf{d} - (1 + h_1)\mathbf{b} \times \mathbf{c} \cdot \mathbf{d} > 0 \ (\textit{resp.} \ < 0). \tag{14}$$

Proof Using (13), one has

$$\overrightarrow{O\bar{V}_1} \cdot \mathbf{d} < 1 + h_4 \quad \text{and} \quad \overrightarrow{O\bar{V}_2} \cdot \mathbf{b} < 1 + h_2 \tag{15}$$

if and only if (14) is positive, while

$$\overrightarrow{O\bar{V}_1'} \cdot \mathbf{c} < 1 + h_3 \quad \text{and} \quad \overrightarrow{O\bar{V}_2'} \cdot \mathbf{a} < 1 + h_1 \tag{15$'$}$$

if and only if (14) is negative. □

Remarks (i) $\{\bar{V}_1, \bar{V}_2, \bar{V}_1', \bar{V}_2'\}$ become cocircular if (14) happens to be equal to zero.

(ii) The second case of (14) (i.e. the negative case) happens when

$$h_3\ \mathbf{a} \times \mathbf{b} \cdot \mathbf{d} + h_1\ \mathbf{b} \times \mathbf{c} \cdot \mathbf{d} - h_4\ \mathbf{a} \times \mathbf{b} \cdot \mathbf{c} - h_2\ \mathbf{a} \times \mathbf{c} \cdot \mathbf{d}$$
$$> \mathbf{a} \times \mathbf{b} \cdot \mathbf{c} + \mathbf{a} \times \mathbf{c} \cdot \mathbf{d} - \mathbf{a} \times \mathbf{b} \cdot \mathbf{d} - \mathbf{b} \times \mathbf{c} \cdot \mathbf{d}. \tag{16}$$

Such a geometric situation can occur only when the volume of the tetrahedron spanned by $\{A, B, C, D\}$ is quite small (meaning that $\{A, B, C, D\}$ are almost co-planar) and $h_1 + h_3$ is considerably larger than $h_2 + h_4$, so as to make the left-hand side of (16) exceed the right-hand side of (16).

(iii) The local combinatorial structures of $\mathcal{S}^*(\Sigma)$ and $\mathcal{S}^*(\Sigma, \{h_h\})$ are the same in the first case, while in the second case, there occurs a "switch" of local combinatorial structures as indicated in Figure 4.2.

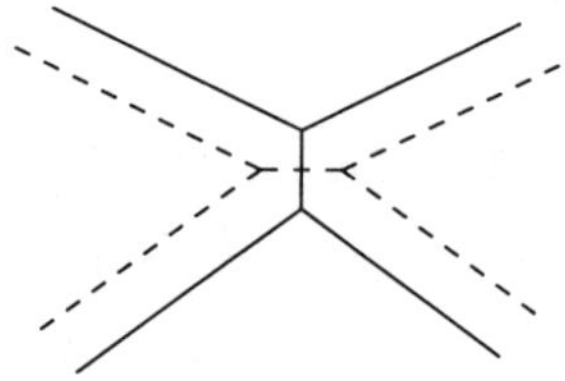

Figure 4.2

Sublemma 4.1.3

$$\begin{aligned}
\overrightarrow{O\bar{V}_1} &= \alpha\mathbf{a} + \gamma\mathbf{c} + \bar{k}_1\mathbf{a}\times\mathbf{c}\,,\\
\overrightarrow{O\bar{V}_2} &= \alpha\mathbf{a} + \gamma\mathbf{c} + \bar{k}_2\mathbf{a}\times\mathbf{c}\,,\\
\overrightarrow{O\bar{V}_1'} &= \beta\mathbf{b} + \delta\mathbf{d} + \bar{m}_1\mathbf{b}\times\mathbf{d}\,,\\
\overrightarrow{O\bar{V}_2'} &= \beta\mathbf{b} + \delta\mathbf{d} + \bar{m}_2\mathbf{b}\times\mathbf{d}\,.
\end{aligned} \tag{17}$$

Then,

$$\begin{aligned}
\alpha &= \frac{1}{1+\mathbf{a}\cdot\mathbf{c}} + \frac{1}{2}\left\{\frac{h_1+h_3}{1+\mathbf{a}\cdot\mathbf{c}} + \frac{h_1-h_3}{1-\mathbf{a}\cdot\mathbf{c}}\right\},\\
\beta &= \frac{1}{1+\mathbf{b}\cdot\mathbf{d}} + \frac{1}{2}\left\{\frac{h_2+h_4}{1+\mathbf{b}\cdot\mathbf{d}} + \frac{h_2-h_4}{1-\mathbf{b}\cdot\mathbf{d}}\right\},\\
\gamma &= \frac{1}{1+\mathbf{a}\cdot\mathbf{c}} + \frac{1}{2}\left\{\frac{h_1+h_3}{1+\mathbf{a}\cdot\mathbf{c}} + \frac{h_3-h_1}{1-\mathbf{a}\cdot\mathbf{c}}\right\},\\
\delta &= \frac{1}{1+\mathbf{b}\cdot\mathbf{d}} + \frac{1}{2}\left\{\frac{h_2+h_4}{1+\mathbf{b}\cdot\mathbf{d}} + \frac{h_4-h_2}{1-\mathbf{b}\cdot\mathbf{d}}\right\},\\
\bar{k}_1 &= \frac{-1}{\mathbf{a}\times\mathbf{b}\cdot\mathbf{c}}\{1+h_2-\alpha\mathbf{a}\cdot\mathbf{b}-\gamma\mathbf{c}\cdot\mathbf{b}\},\\
\bar{k}_2 &= \frac{1}{\mathbf{a}\times\mathbf{c}\cdot\mathbf{d}}\{1+h_4-\alpha\mathbf{a}\cdot\mathbf{d}-\gamma\mathbf{c}\cdot\mathbf{d}\},\\
\bar{m}_1 &= \frac{1}{\mathbf{a}\times\mathbf{b}\cdot\mathbf{d}}\{1+h_1-\beta\mathbf{b}\cdot\mathbf{a}-\delta\mathbf{d}\cdot\mathbf{a}\},\\
\bar{m}_2 &= \frac{-1}{\mathbf{b}\times\mathbf{c}\cdot\mathbf{d}}\{1+h_3-\beta\mathbf{b}\cdot\mathbf{c}-\delta\mathbf{d}\cdot\mathbf{c}\}.
\end{aligned} \tag{17'}$$

Proof The pair of points $\{\bar{V}_1, \bar{V}_2\}$ (resp. $\{\bar{V}_1', \bar{V}_2'\}$) lie on the intersection line of the pair of planes given by

$$\begin{aligned} &\overrightarrow{OX}\cdot\mathbf{a}=1+h_1,\ \overrightarrow{OX}\cdot\mathbf{c}=1+h_3 \\ &(\text{resp. } \overrightarrow{OX}\cdot\mathbf{b}=1+h_2,\ \overrightarrow{OX}\cdot\mathbf{d}=1+h_4)\,. \end{aligned} \tag{18}$$

Therefore, $\{\alpha,\gamma\}$ (resp. $\{\beta,\delta\}$) of (17) are determined by the following system of linear equations:

$$\begin{cases} \alpha+\gamma\mathbf{a}\cdot\mathbf{c}=1+h_1 \\ \alpha\mathbf{a}\cdot\mathbf{c}+\gamma=1+h_3 \end{cases} \left(\text{resp. } \begin{cases} \beta+\delta\mathbf{b}\cdot\mathbf{d}=1+h_2 \\ \beta\mathbf{b}\cdot\mathbf{d}+\delta=1+h_4 \end{cases}\right). \tag{19}$$

Hence, simple computation will show that $\{\alpha,\beta,\gamma,\delta\}$ and $\{\bar{k}_1,\bar{k}_2,\bar{m}_1,\bar{m}_2\}$ of (17) are given by (17′). □

4.1.2.3 *The relationships between $T(\Sigma)$ and $T(\Sigma,\{h_j\})$ (resp. $\mathcal{S}^*(\Sigma)$ and $\mathcal{S}^*(\Sigma,\{h_j\})$)*

Recall that $T(\Sigma)$ and $T(\Sigma,\{h_j\})$ are respectively defined by similar systems of inequalities, namely

$$\overrightarrow{OX}\cdot\overrightarrow{OA_j}\le 1 \ (\text{resp. } 1+h_j)\,, \quad A_j\in\Sigma\,.$$

It is clear that $T(\Sigma)$ and $T(\Sigma,\{h_j\})$ have pairwise parallel faces with a distance of h_j between the jth pair. The radial projections of the faces of $T(\Sigma)$ (resp. $T(\Sigma,\{h_j\})$) constitutes the spherical configuration (i.e. a decomposition of S_0 into spherical polygons) $\mathcal{S}^*(\Sigma)$ (resp. $\mathcal{S}^*(\Sigma,\{h_j\})$). Notice that $I(\Sigma)$ and $T(\Sigma)$ are projective dual to each other. Thus, their radial projections, i.e. $\mathcal{S}(\Sigma)$ and $\mathcal{S}^*(\Sigma)$, are dual to each other in the following spherical geometry setting, namely

(i) to each face σ of $\mathcal{S}(\Sigma)$, the corresponding vertex of $\mathcal{S}^*(\Sigma)$ is exactly the circumcenter of σ,

(ii) to each edge of $\mathcal{S}(\Sigma)$, its corresponding edge of $\mathcal{S}^*(\Sigma)$ is exactly the spherical arc joining the circumcenters of the pair of faces of $\mathcal{S}(\Sigma)$ sharing the given edge, thus it is situated on the perpendicular bisector of the given edge,

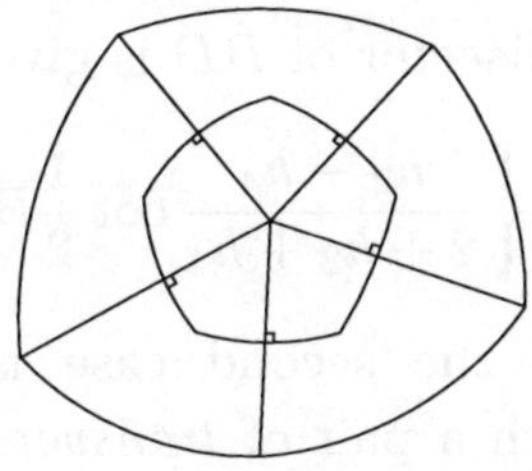

Figure 4.3

(iii) to a given vertex A_j of $\mathcal{S}(\Sigma)$, its corresponding face of $\mathcal{S}^*(\Sigma)$ is the spherical polygon enclosed by the perpendicular bisectors of the radial edges of $St(A)$ in $\mathcal{S}(\Sigma)$, as indicated in Figure 4.3:

Now, let us discuss the geometric relationship between $\mathcal{S}^*(\Sigma)$ and $\mathcal{S}^*(\Sigma, \{h_j\})$. First of all, corresponding to each $A_j \in \Sigma$, both $\mathcal{S}^*(\Sigma)$ and $\mathcal{S}^*(\Sigma, \{h_j\})$ have exactly one face containing A_j in its interior. However, in the case of edges, there are two kinds of geometric situations according to whether

$$\{(1+h_4)\mathbf{a}\times\mathbf{b}\cdot\mathbf{c}+(1+h_2)\mathbf{a}\times\mathbf{c}\cdot\mathbf{d}-(1+h_3)\mathbf{a}\times\mathbf{b}\cdot\mathbf{d}-(1+h_1)\mathbf{b}\times\mathbf{c}\cdot\mathbf{d}\}$$

and

$$\{\mathbf{a}\times\mathbf{b}\cdot\mathbf{c}+\mathbf{a}\times\mathbf{c}\cdot\mathbf{d}-\mathbf{a}\times\mathbf{b}\cdot\mathbf{d}-\mathbf{b}\times\mathbf{c}\cdot\mathbf{d}\}$$

are of the same sign or of the opposite signs. If they are of the same sign, then the pair of corresponding edges of $\mathcal{S}^*(\Sigma)$ and $\mathcal{S}^*(\Sigma, \{h_j\})$ are both perpendicular to an edge of $\mathcal{S}(\Sigma)$, such as $\widehat{AC}$ in Sublemma 4.1.2, and their longitudinal difference is given by

$$\tan^{-1}\left\{\frac{h_1 - h_3}{2 + h_1 + h_3}\cot\frac{1}{2}\widehat{AC}\right\}. \tag{20}$$

On the other hand, if they are of the opposite signs, then the pair of corresponding edges are respectively perpendicular to the two diagonals of $\widehat{\square ABCD}$, i.e. one perpendicular to $\widehat{AC}$ and the other perpendicular to $\widehat{BD}$; the longitudinal difference between the latter and the perpendicular bisector of $\widehat{BD}$ is given by

$$\tan^{-1}\left\{\frac{h_2 - h_4}{2 + h_2 + h_4}\cot\frac{1}{2}\widehat{BD}\right\}. \qquad (20')$$

The local geometry of the second case is as indicated in Figure 4.2. Notice that such a pair of *transversal* corresponding edges of $\mathcal{S}^*(\Sigma)$ and $\mathcal{S}^*(\Sigma, \{h_j\})$ can occur only when both are rather short, (cf. Remarks following Sublemma 4.1.2).

4.1.2.4 *The peripheral part and its rectilinear slabs*

Definition The complementary set of $T(\Sigma)$ in $T(\Sigma, \{h_j\})$ is defined to be the *peripheral part* of $C(\mathcal{L}')$. The *peripheral part* of $C(\mathcal{L})$ is defined to be the intersection of $C(\mathcal{L})$ and the peripheral part of $C(\mathcal{L}')$, i.e. its restriction to $C(\mathcal{L})$.

Definition Corresponding to each neighbor $S_j \in \mathcal{L}'(S_0)$ with $h_j > 0$, both $T(\Sigma)$ and $T(\Sigma, \{h_j\})$ have a face perpendicular to $\overrightarrow{OA_j}$ and with a distance of h_j between them. The *convex hull* of such a pair of parallel faces is defined to be the *rectilinear slab in* $C(\mathcal{L}')$ *associated to* S_j. Its restriction to $C(\mathcal{L})$ is called the rectilinear slab in $C(\mathcal{L})$ associated to S_j.

Let σ_j^* and $\bar{\sigma}_j^*$ be the pair of corresponding faces of $\mathcal{S}^*(\Sigma)$ and $\mathcal{S}^*(\Sigma, \{h_j\})$ which contain A_j in their interior and F_j, $\bar{F}_j$ be the pair of corresponding faces of $T(\Sigma)$ and $T(\Sigma, \{h_j\})$ which are both perpendicular to $\overrightarrow{OA_j}$. In case no switching (cf. Sublemma 4.1.2) occurs at those boundary edges of both σ_j^* and $\bar{\sigma}_j^*$, F_j and $\bar{F}_j$ have *matching-pairs of parallel edges.* On the other hand, if some switchings occur, such as the one described in Sublemma 4.1.2 with negative sign for (14), then there are edges of either F_j or $\bar{F}_j$ without their matching parallel ones. In such a case, it is natural to regard their matching parallel ones as being degenerate to vertices. In the former case, all the *side-faces of the rectilinear slab* with F_j and $\bar{F}_j$ as its bottom and top faces are trapezoids, while in the latter case some of the side-faces become triangles.

From the above description of the geometric structures of rectilinear slabs of $C(\mathcal{L}')$, it is not difficult to see that they constitute a collection of *non-overlapping* pieces of the *peripheral part* of $C(\mathcal{L}')$. However, the union of all the rectilinear slabs, in general, may not be equal to the totality of the peripheral part. In fact, corresponding to each occurrence of a switch such as the one described in Sublemma 4.1.2, the tetrahedron spanned by $\{V_1, V_2, \bar{V}_1', \bar{V}_2'\}$ will be such a missing piece and the *complementary set of* the above union of rectilinear slabs in the peripheral part is exactly the union of those small tetrahedrons corresponding to the occurrence of switches.

4.1.3 *Basic strategies of volume estimation of local cells*

In concluding this section, we shall discuss the basic strategies that lead to the development of an effective system of volume estimation techniques, specifically tailored for that of the local cells. As one shall see later, such a system of volume estimation techniques constitutes a powerful set of geometric lemmas in the study of various kinds of densities of sphere packings. However, the following basic strategies are the decisive starting points that also provide the proper setting for those volume estimation lemmas.

4.1.3.1 *The separation of a local cell into its core part and its peripheral part*

Definition The part of a given local cell, $C(\mathcal{L})$, inside of the tangent polyhedron $T(\Sigma)$ is defined to be its *core part*, while the outside part is defined to be the *peripheral part.*

The underlying geometric intuition of the above strategy of separating a local cell into its core part and peripheral part is that the latter consists of the *rectilinear portion* of interstices which can be estimated by techniques of Euclidean geometry, while the former contains the *curved* portion of interstices and hugs around the sphere that naturally relies on techniques of spherical geometry to provide an efficient estimate.

Remark Note that the above tangent polyhedron $T(\Sigma)$ is, by definition, the one associated to the *reduced local packing* $\mathcal{L}'(S_0)$ rather than the local packing $\mathcal{L}(S_0)$ itself; the set of points Σ only records the directions of those neighbors of at most 0.3 buckling height. Such a choice of using the reduced local packing $\mathcal{L}'(S_0)$ rather than the fully-fledged one in the above strategic separation of $C(\mathcal{L})$ into the core part and the peripheral part is actually based upon the following geometric insight and technical consideration, namely, the effect of having some neighbors with buckling height exceeding 0.3 amounts to some rather shallow truncations of some tips of $C(\mathcal{L}')$ peaking beyond 0.3 "above sea level." The remaining portion of such a peak always has a considerable amount of volume, thus making the task of lower bound volume estimation of such a local cell (resp. core part) easier than otherwise. Anyhow, it is technically more advantageous to treat the effect of having a *non-reduced* neighbor as a truncation rather than producing a thick rectilinear slab with a quite small top face.

4.1.3.2 *The subdivision of the peripheral part into rectilinear slabs*

Geometrically, the peripheral part of a given local cell $C(\mathcal{L})$ (or $C(\mathcal{L}')$) can be decomposed into *non-overlapping* pieces of rectilinear slabs plus some (small) pieces of tetrahedrons if some switches of local combinatorial structures between $\mathcal{S}^*(\Sigma)$ and $\mathcal{S}^*(\Sigma, \{h_j\})$ occur. Based upon the above decomposition of the peripheral part, it is quite natural to try the following simple approach for obtaining a kind of simple, but already good enough, lower bound volume estimate of the peripheral part:

(i) Because those small pieces of tetrahedrons may not occur at all, and their total contribution of volume is very small anyway, it is natural to discount such an amount, if any.

(ii) By studying the *localized geometry* of the surroundings of a pair of spheres $\{S, S'\}$ with $2(1+h)$ center–distance, one first works out the localized task of lower bound volume estimate of an individual rectilinear slab solely in terms of its thickness h (cf. Lemma 4.3.1).

(iii) The total sum of the above kind of lower bound volume estimates for individual rectilinear slabs will, of course, be a lower bound volume estimate of the peripheral part. However, the real power of such an estimate will depend on the strength of the estimates on the *total buckling height* that we shall discuss in Chap. 5.

4.1.3.3 *Spherical configuration and a basic volume formula of $T(\Sigma)$, the basic strategy for the estimation of the core part*

The *core part* of a given local cell $C(\mathcal{L})$ is, by definition, the part inside of the tangent polyhedron $T(\Sigma)$ associated to the *reduced* local packing $\mathcal{L}'(S_0)$. Thus, it is equal to $T(\Sigma)$ if $\mathcal{L}(S_0) = \mathcal{L}'(S_0)$. Otherwise, it is equal to the subset of $T(\Sigma)$ with some tips peaking beyond "0.3 above sea level" being truncated by the perpendicular bisectors corresponding to those *non-reduced* neighbors. Anyhow, the core part of a local cell is either a tangent polyhedron or a tangent polyhedron with some *rather shallow truncations.* Intuitively, one expects to make extensive use of the close connection between such a polyhedron and the unit sphere S_0 in order to provide some efficient lower bound estimates of its volume. Therefore, it is natural that the spherical geometry of its "*generating point set*" Σ will play an important role in developing the volume estimation techniques for that of the core part. In the critical "generic" case that all the faces of $\mathcal{S}(\Sigma)$ are triangular and contain their own circumcenters, the decisive first step is, again, the choice of an advantageous subdivision of the given tangent polyhedron $T(\Sigma)$ into individual pieces, which enables us to reduce the task of lower bound volume estimation of the core part to that of individual pieces, thus greatly simplifying the technical complexity from the *global geometry of* $T(\Sigma)$ to the *localized geometry of individual pieces.* Notice that only the most suitable subdivision will enable us to have the *advantages in both ways*, namely, on the one hand, we can localize the estimation task to that of individual pieces; and on the other hand, the *sum total* of such simplified individual estimates will still produce the *optimal lower bound estimates for those crucial cases.* As it turns out, the subdivision of $T(\Sigma)$ provided by the subdivision of the *total solid angle at the origin* given by $\mathcal{S}(\Sigma)$ is exactly such an advantageous one! This leads to the discovery of an advantageous volume formula of $T(\Sigma)$ that we shall prove in Subsec. 4.2.2.

In retrospect, the strategic separation of a local cell into its *peripheral part* and its *core part* and the above crucial choice of subdivision of the core part into subpolyhedrons using the solid angle decomposition of $\mathcal{S}(\Sigma)$ are probably some of the decisive steps that radically contribute to the solving of the sphere packing problem [Hsi-2]. Anyhow, the above crucial subdivision naturally leads to the development of the volume estimation lemmas of [Hsi-2] that we are going to discuss in the next section.

4.2 Technique of Volume Estimation of the Core Part

In this section, we shall develop a system of volume estimation techniques specifically tailored for that of the core part [Hsi-2]. As it has already been pointed out in Sec. 4.1, the crucial starting point is the canonical subdivision of $T(\Sigma)$ according to the subdivision of the total solid angle at the center of S_0 given by $\mathcal{S}(\Sigma)$. Set n to be the number of vertices of $\mathcal{S}(\Sigma)$, i.e. the cardinal number of Σ (resp. reduced neighbors). In the generic case, all the faces of $\mathcal{S}(\Sigma)$ are triangular. Thus, the number of edges is equal to $\frac{3}{2}$-times the number of faces, the reason being that each triangle has three edges while each edge is shared by two triangles. Therefore, Euler's formula becomes

$$F - E + V = F - \frac{3}{2}F + n = 2 \;\Rightarrow\; F = 2n - 4\,. \tag{1}$$

Thus, a triangulated spherical configuration with n vertices always consists of $2n - 4$ triangular faces.

Let us first consider the case that all faces of $\mathcal{S}(\Sigma)$ are *triangular* and each triangular face *contains its own circumcenter*. Let $\sigma = \widehat{\triangle ABC}$ be such a face of $\mathcal{S}(\Sigma)$ and $\Gamma(\sigma)$ be the solid angle cone spanned by σ. Then the subpolyhedron $T(\Sigma) \cap \Gamma(\sigma)$ is given by

$$T(\sigma) = \{X \in \Gamma(\sigma) \text{ and } \overrightarrow{OX} \cdot \overrightarrow{OA},\ \overrightarrow{OX} \cdot \overrightarrow{OB},\ \overrightarrow{OX} \cdot \overrightarrow{OC} \le 1\}\,. \tag{2}$$

We shall call it the *tangent subpolyhedron of* σ.

4.2.1 *The volume function of tangent subpolyhedron*

Let $\sigma = \widehat{\triangle ABC}$ be a spherical triangle containing its own circumcenter. Set T_A, T_B, T_C to be the tangent planes of S_0 at A, B and C respectively and V, $\hat{A}$, $\hat{B}$ and $\hat{C}$ to be the intersection points of the following triples of planes, namely

$$\begin{aligned} &V = T_A \cap T_B \cap T_C\,, \quad \hat{A} = T_B \cap T_C \cap (OBC)\,, \\ &\hat{B} = T_A \cap T_C \cap (OAC)\,, \quad \hat{C} = T_A \cap T_B \cap (OAB)\,. \end{aligned} \tag{3}$$

Then, the above tangent subpolyhedron of σ, $T(\sigma)$, is exactly the *convex hull* of $\{O, A, B, C, V, \hat{A}, \hat{B}, \hat{C}\}$, while the condition on circumcenter means that V is inside of the solid angle cone $\Gamma(\sigma)$ (cf. Figure 4.4).

In the special case that $\sigma = \widehat{\triangle ABC}$ has three right-angles, the above $T(\sigma)$ is exactly a unit cube. In general, it is a "twisted" cube as indicated in Figure 4.4. Naturally, the first step to exploring the true advantage of the above subdivision is to derive a volume formula of the above specific type of polyhedrons.

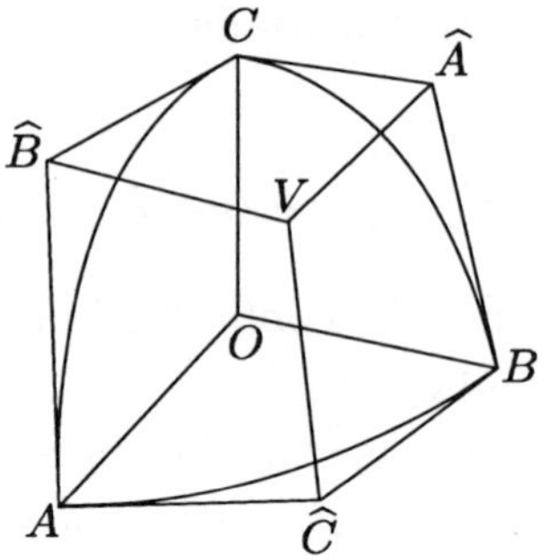

Figure 4.4

4.2.1.1 *A basic volume formula*

Lemma 4.2.1 *Let $\sigma = \widehat{\triangle} ABC$ be a spherical triangle containing its own circumcenter. Then,*

$$\operatorname{vol} T(\sigma) = \frac{1}{3D}\left\{8 + u - \frac{4u(1+2u+w)}{u^2+D^2}\right\}, \tag{4}$$

where $D = \det(\mathbf{a}, \mathbf{b}, \mathbf{c}) > 0$,

$$\begin{aligned} u &= 1 + \mathbf{a}\cdot\mathbf{b} + \mathbf{b}\cdot\mathbf{c} + \mathbf{c}\cdot\mathbf{a}\,, \\ w &= (\mathbf{a}\cdot\mathbf{b})(\mathbf{b}\cdot\mathbf{c}) + (\mathbf{b}\cdot\mathbf{c})(\mathbf{c}\cdot\mathbf{a}) + (\mathbf{c}\cdot\mathbf{a})(\mathbf{a}\cdot\mathbf{b})\,. \end{aligned} \tag{5}$$

Proof We choose the orientation of $\widehat{\triangle} ABC$ so that $\det(\mathbf{a}, \mathbf{b}, \mathbf{c}) = \mathbf{a}\cdot(\mathbf{b}\times\mathbf{c}) > 0$. It is straightforward to check that

$$\begin{aligned} &\overrightarrow{O\hat{A}} = \frac{\mathbf{b}+\mathbf{c}}{1+\mathbf{b}\cdot\mathbf{c}}, \quad \overrightarrow{O\hat{B}} = \frac{\mathbf{c}+\mathbf{a}}{1+\mathbf{c}\cdot\mathbf{a}}, \quad \overrightarrow{O\hat{C}} = \frac{\mathbf{a}+\mathbf{b}}{1+\mathbf{a}\cdot\mathbf{b}}, \\ &\mathbf{v} = \overrightarrow{OV} = \frac{1}{D}(\mathbf{a}\times\mathbf{b} + \mathbf{b}\times\mathbf{c} + \mathbf{c}\times\mathbf{a}) \end{aligned} \tag{6}$$

namely, the above vectors satisfy their respective vector equations of (3). On the other hand, the polyhedron $T(\sigma)$ can be decomposed into three cones with V as their common vertex and the quadrilaterals $\square OA\hat{C}B$, $\square OB\hat{A}C$ and $\square OC\hat{B}A$ as their bases, respectively.

Note that the oriented normal vectors of the above quadrilaterals with lengths equal to their areas are respectively given as

$$\frac{\mathbf{a}\times\mathbf{b}}{1+\mathbf{a}\cdot\mathbf{b}},\ \frac{\mathbf{b}\times\mathbf{c}}{1+\mathbf{b}\cdot\mathbf{c}},\ \frac{\mathbf{c}\times\mathbf{a}}{1+\mathbf{c}\cdot\mathbf{a}}. \tag{7}$$

Hence, the volume of $T(\sigma)$, which is the sum of the volumes of the above three cones, can be computed by vector algebra as

$$\begin{aligned}\operatorname{vol} T(\sigma) &= \frac{1}{3}\mathbf{v}\cdot\left\{\frac{\mathbf{a}\times\mathbf{b}}{1+\mathbf{a}\cdot\mathbf{b}}+\frac{\mathbf{b}\times\mathbf{c}}{1+\mathbf{b}\cdot\mathbf{c}}+\frac{\mathbf{c}\times\mathbf{a}}{1+\mathbf{c}\cdot\mathbf{a}}\right\}\\ &= \frac{1}{3D}\left\{8+u-\frac{4u(1+2u+w)}{u^2+D^2}\right\}. \end{aligned}\tag{8}$$

□

Corollary 4.2.1 *Let* $\triangle$ *(resp.* R*) be the area (resp. circumradius) of* $\sigma = \widehat{\triangle ABC}$*. Then,*

$$\operatorname{vol} T(\sigma) = \frac{1}{6}\sin\triangle\left\{1+\frac{8}{u}-\sec^2 R\right\} = \frac{1}{6}\sin\triangle\left\{\frac{8}{u}-\tan^2 R\right\}. \tag{9}$$

Proof Recall that (cf. Sec. 2.2)

$$\sin\triangle = \frac{2uD}{u^2+D^2},\quad \sec^2 R = |\mathbf{v}|^2 = \frac{1}{D^2}\{4(1+w)-u^2\}. \tag{10}$$

Hence, straightforward algebraic computation will show that

$$\frac{1}{6}\sin\triangle\left\{1+\frac{8}{u}-\sec^2 R\right\} = \frac{1}{3D}\left\{8+u-\frac{4u(1+2u+w)}{u^2+D^2}\right\}. \tag{11}$$

□

Corollary 4.2.2 *Let* $\bar{\sigma}_\triangle$ *be the equilateral spherical triangle with* $\triangle$ *as its area. Then,*

$$\operatorname{vol} T(\bar{\sigma}_\triangle) = \left(1-2\cos\frac{\pi+\triangle}{3}\right)\tan\frac{\pi+\triangle}{6}. \tag{12}$$

Proof The three angles of such an equilateral spherical triangle are, of course, all equal to $\frac{1}{3}(\pi+\triangle)$. Therefore, by the cosine law, the cosines of its three sides are all given by

$$\cos a = \cos b = \cos c = \frac{\cos\frac{1}{3}(\pi+\triangle)}{1-\cos\frac{1}{3}(\pi+\triangle)}. \tag{13}$$

Set it to be t. Then, $u = 1+3t$, $w = 3t^2$,

$$D^2 = 1 + 2t^3 - 3t^2 = (1-t)^2(1+2t) \tag{14}$$

and hence,

$$\operatorname{vol} T(\bar{\sigma}_\triangle) = \frac{1}{\sqrt{1+2t}}\frac{1-t}{1+t} = \left(1 - 2\cos\frac{\pi+\triangle}{3}\right)\tan\frac{\pi+\triangle}{6}. \tag{15}$$

□

Remark If the circumcenter of $\overset{\frown}{\triangle}ABC$ happens to be outside of $\overset{\frown}{\triangle}ABC$, then the intersection point of the three tangent planes, V, is situated outside of the solid angle cone spanned by $\overset{\frown}{\triangle}ABC$. As indicated in Figure 4.5, V is located at the other side of (OBC). Then,

$$\frac{1}{3}\mathbf{v}\cdot\frac{\mathbf{b}\times\mathbf{c}}{1+\mathbf{b}\cdot\mathbf{c}} \tag{16}$$

becomes *negative* and the right-hand side of (8) is equal to the volume of the convex hull of $\{O, A, B, C, D, E, \hat{B}, \hat{C}\}$ subtracted by that of the tetrahedron of $\{\hat{A}, D, E, V\}$, where D and E are the cross-intersection of $\{\overline{V\hat{B}}, \overline{\hat{A}C}\}$ and $\{\overline{V\hat{C}}, \overline{\hat{A}B}\}$, respectively.

4.2.2 *A volume formula of $T(\Sigma)$*

In the "generic" case that all the faces of $\mathcal{S}(\Sigma)$ are triangular and each of them contains its own circumcenter, the portion of $T(\Sigma)$ inside of the solid angle cone, $\Gamma(\sigma_i)$, spanned by $\sigma_i \in \mathcal{S}(\Sigma)$ is exactly $T(\sigma_i)$ of Lemma 4.2.1. Therefore, one has the following formula for

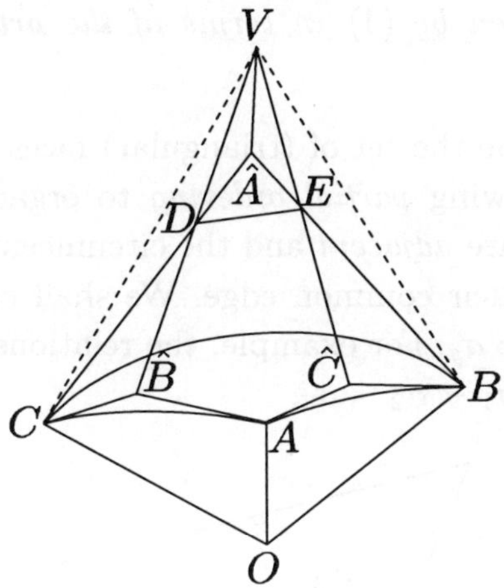

Figure 4.5

the volume of such a circumscribing polyhedron $T(\Sigma)$:

$$\operatorname{vol} T(\Sigma) = \sum \{f(\sigma_i), \sigma_i \in \mathcal{S}(\Sigma)\}, \tag{17}$$

where $f(\sigma_i)$ is given by (4) in terms of the orthogonal invariants of σ_i.

At first glance, the above remark seems to indicate that there may have been some difficulty in using the geometric organization of $\mathcal{S}(\Sigma)$ and the function $f(\sigma)$ of (4) to produce a clean-cut formula of the volume of $T(\Sigma)$ *in general*. However, a careful analysis of the geometry of both $T(\Sigma)$ and $\mathcal{S}(\Sigma)$ shows that the formula of (17) actually hold *in general*, namely, *without* the above condition that all the faces of $\mathcal{S}(\Sigma)$ are triangular and each contains its own circumcenter.

Theorem 4.2.1 *Let $\mathcal{S}(\Sigma)$ be the associated spherical configuration of Σ and $\mathcal{S}'(\Sigma) = \{\sigma_i\}$ be a triangulation of $\mathcal{S}(\Sigma)$ which subdivides each non-triangular faces of $\mathcal{S}(\Sigma)$, if any, into triangles by adding diagonals. Then,*

$$\operatorname{vol} T(\Sigma) = \sum \{f(\sigma_i), \sigma_i \in \mathcal{S}'(\Sigma)\}, \tag{17'}$$

where $f(\sigma_i)$ is given by (4) in terms of the orthogonal invariants of σ_i.

Proof Let $\{\sigma_i\}$ be the set of (triangular) faces of $\mathcal{S}'(\Sigma)$. We shall introduce the following *partial ordering* to organize them, namely, $\sigma_i \prec \sigma_j$ if $\{\sigma_i, \sigma_j\}$ are *adjacent* and the circumcenter of σ_i is situated at the σ_j side of their common edge. We shall call such a relation: σ_i is subordinate to σ_j. For example, the relationship as indicated in Figure 4.6 is that $\sigma_1 \prec \sigma_2 \prec \sigma_3$.

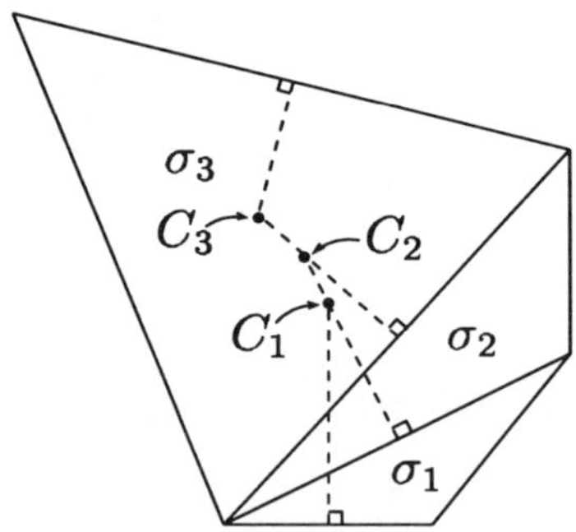

Figure 4.6

Let $\{\sigma_{i_1}, \sigma_{i_2}, \ldots, \sigma_{i_k}\}$ be the set of maximal elements with respect to the above partial ordering, and $\{F_1, F_2, \ldots, F_k\}$ be the decomposition of $\mathcal{S}'(\Sigma)$ into sub-collections of faces, each of them consisting a single maximal element and all its subordinates, if any. Set

$$\Omega_j = \bigcup\{\sigma_i; \sigma_i \in F_j\}, \quad 1 \leq j \leq k \tag{18}$$

and $\Gamma(\Omega_j)$ to be the solid angle cone spanned by Ω_j. Of course, F_j may consist of a single element σ_{i_j} and $\Omega_j = \sigma_{i_j}$ in such a case.

Let F be a sub-collection of faces of $\mathcal{S}'(\Sigma)$ satisfying the condition that it also includes all the *superiors* of $\sigma_i \in F$. Set $\Omega = \bigcup\{\sigma_i \in F\}$, $\Gamma(\Omega)$ to be the solid angle cone of Ω and

$$T(F) = \{X \in \Gamma(\Omega); \overrightarrow{OX} \cdot \overrightarrow{OA}_j \leq 1\}, \tag{19}$$

where A_j runs through all the vertices of $\sigma_i \in F$. Then, it is easy to see that

$$T(\Sigma) \cap \Gamma(\Omega) \subseteq T(F)$$

and moreover

$$T(\Sigma) \cap \Gamma(\Omega) = T(F)$$

if and only if F also satisfies another condition that it includes all the *subordinates* of $\sigma_i \in F$. In particular, the above sub-collections $\{F_j, 1 \leq j \leq k\}$ all satisfy both conditions and hence,

$$T(\Sigma) \cap \Gamma(\Omega_j) = T(F_j)\,, \quad \operatorname{vol} T(\Sigma) = \sum_{j=1}^{k} \operatorname{vol} T(F_j)\,. \tag{20}$$

Therefore, for the proof of Theorem 4.2.1, it suffices to prove that

$$\operatorname{vol} T(F) = \sum \{f(\sigma_i), \sigma_i \in F\} \tag{21}$$

for sub-collections F satisfying the first condition by induction on the number of faces in F.

If $F = \{\sigma\}$ consists of a single triangle, then σ has *no* superior (i.e. σ contains its own circumcenter) and (21) is exactly Lemma 4.2.1.

For the proof of the induction step, set $F = \{\sigma_i, 1 \leq i \leq \ell\} = F' \cup \{\sigma_\ell\}$ such that σ_ℓ is a minimal element of F. Then F' also satisfies the first condition and hence,

$$\operatorname{vol} T(F') = \sum_{i=1}^{\ell-1} f(\sigma_i) \tag{21'}$$

by the induction assumption. If σ_ℓ contains its own circumcenter, then it again follows from Lemma 4.2.1 and (21′) that

$$\operatorname{vol} T(F) = \operatorname{vol} T(F') + \operatorname{vol} T(\sigma_\ell) = \sum_{i=1}^{\ell} f(\sigma_i)\,. \tag{21''}$$

Otherwise, the immediate superior of σ_ℓ belongs to F', say $\sigma_{\ell-1}$. It follows from the *convexity* of $I(\Sigma)$ that the circumradii $r(\sigma_{\ell-1}) \geq$

$r(\sigma_\ell)$ and hence, the difference between $T(F')$ and $T(F) \cap \Gamma(\Omega')$, $\Omega' = \bigcup\{\sigma_i, 1 \le i \le \ell - 1\}$, is that a tetrahedron such as $T(\widehat{A}DEV)$ of Figure 4.5 has been cut away from the former, i.e. by the only additional tangent plane. Therefore,

$$\begin{aligned}
\operatorname{vol} T(F) &= \operatorname{vol} T(F) \cap \Gamma(\Omega') + \operatorname{vol} T(F) \cap \Gamma(\sigma_\ell) \\
&= \operatorname{vol} T(F') - \operatorname{vol} T(\widehat{A}DEV) + \operatorname{vol} T(F) \cap \Gamma(\sigma_\ell) \\
&= \sum_{i=1}^{\ell-1} f(\sigma_i) + f(\sigma_\ell) = \sum_{i=1}^{\ell} f(\sigma_i) \,. \qquad (22)
\end{aligned}$$

This completes the inductive proof of (21) and hence also the proof of Theorem 4.2.1. □

Remarks (i) The volume of $T(\Sigma)$ is, of course, a function of the n position vectors of the set of n touching points, i.e. $\{\mathbf{a}_j = \overrightarrow{OA}_j, A_j \in \Sigma\}$, which should be expressible in terms of their basic orthogonal invariants such as $\{\mathbf{a}_i \cdot \mathbf{a}_j, 1 \le i, j \le n\}$. However, the formula (17′) is much simpler than one might have originally expected, thus providing a powerful tool in the study of the volumes of circumscribing polyhedra.

(ii) In the above discussion, the canonical decomposition of the total solid angle actually *only* provides a natural geometric way of discovery of the formula (17′) in the "generic" case that all faces of $\mathcal{S}(\Sigma)$ are triangular and each of them contains its own circumcenter. In essence, it is the canonical, concise organization of the orthogonal invariants of Σ (or rather $\{\mathbf{a}_j\}$) provided by $\mathcal{S}(\Sigma)$ that plays the crucial role in proving the volume formula (17′). Moreover, the "convexity" property of $\mathcal{S}(\Sigma)$ plays an important role for the validity of the volume formula (17′).

(iii) In the application of the above volume formula of $T(\Sigma)$ to the study of the isoperimetric problem of polyhedra, it is useful to combine Lemma 4.2.1 and Theorem 4.2.1 to derive a formula for its *partial gradient*, i.e. with respect to a single vector variable.

4.2.2.1 *A gradient formula of* $\operatorname{vol} T(\Sigma)$

Let $T(\Sigma)$ be a given circumscribing polyhedron of the unit sphere with Σ as its tangent points, and $T(\Sigma_t)$ be a given deformation of $T(\Sigma)$ within the realm of circumscribing polyhedra. Then, the initial velocity vectors of the deformation $\Sigma \to \Sigma_t$ consists of a collection of its individual velocity vectors which constitutes a *vector field defined on* Σ. We shall denote it simply by $\mathcal{V}(\Sigma)$. Geometrically, the *gradient* of the volume function of Theorem 4.2.1, i.e. (17′), at Σ is another vector field defined on Σ, say denoted by $\nabla \operatorname{vol}(\Sigma)$, such that

$$\frac{d}{dt}\operatorname{vol} T(\Sigma_t)\bigg|_{t=0} = \langle \nabla \operatorname{vol}(\Sigma), \mathcal{V}(\Sigma)\rangle \stackrel{(\text{def})}{=} \sum \{\nabla \operatorname{vol}_A \cdot \mathcal{V}_A ; A \in \Sigma\}\,. \tag{23}$$

As a direct application of the volume formula (17′), let us proceed to compute the formula of $\nabla \operatorname{vol}(\Sigma)$ (namely, the tangent vector $\nabla \operatorname{vol}_A$ at each $A \in \Sigma$) by straightforward differentiation. Let $St(C)$ be the star configuration of $\mathcal{S}'(\Sigma)$ at a given vertex C and $\{\sigma_1, \ldots, \sigma_\ell\}$ be the set of triangular faces of $St(C)$. Then, it follows from (17′) that

$$\nabla \operatorname{vol}_C = \sum_{i=1}^{\ell} \nabla f(\sigma_i)\bigg|_C\,. \tag{24}$$

Therefore, it suffices to compute the gradient of $f(\mathbf{a}, \mathbf{b}, \mathbf{c})$ with respect to $\mathbf{c}$ as follows. Set

$$d\mathbf{x} = d\xi(\mathbf{a} - (\mathbf{a}\cdot\mathbf{c})\mathbf{c}) + d\eta(\mathbf{b} - (\mathbf{b}\cdot\mathbf{c})\mathbf{c})\,,$$

$$\begin{aligned}\triangle f &= f(\mathbf{a}, \mathbf{b}, \mathbf{c} + d\mathbf{x}) - f(\mathbf{a}, \mathbf{b}, \mathbf{c})\\ &\equiv \Phi d\xi + \Psi d\eta \quad (\text{modulo higher order terms of } \{d\xi, d\eta\}).\end{aligned} \tag{25}$$

We shall proceed to compute the explicit formulas of the above pair of functions $\{\Phi, \Psi\}$ in terms of the basic orthogonal invariants of $\{\mathbf{a}, \mathbf{b}, \mathbf{c}\}$, namely, $\{\mathbf{a}\cdot\mathbf{b}, \mathbf{b}\cdot\mathbf{c}, \mathbf{c}\cdot\mathbf{a}\}$. In the following computations,

$\equiv$ always means *equality modulo higher order terms in* $\{d\xi, d\eta\}$.

$$\begin{aligned}
d\mathbf{v} &\equiv \frac{1}{\mathbf{a}\times\mathbf{b}\cdot(\mathbf{c}+d\mathbf{x})}(\mathbf{a}\times\mathbf{b}+(\mathbf{b}-\mathbf{a})\times(\mathbf{c}+d\mathbf{x}))-\mathbf{v}\\
&\equiv D^{-1}\left(1-\frac{\mathbf{a}\times\mathbf{b}\cdot d\mathbf{x}}{D}\right)(D\mathbf{v}+(\mathbf{b}-\mathbf{a})\times d\mathbf{x})-\mathbf{v}\\
&\equiv -\frac{\mathbf{a}\times\mathbf{b}\cdot d\mathbf{x}}{D}\mathbf{v}+\frac{1}{D}(\mathbf{b}-\mathbf{a})\times d\mathbf{x}\\
&\equiv (\mathbf{a}\cdot\mathbf{c}\,d\xi+\mathbf{b}\cdot\mathbf{c}\,d\eta)\mathbf{v}-\frac{\mathbf{a}\times\mathbf{b}}{D}(d\xi+d\eta)\\
&\quad -\frac{1}{D}(\mathbf{a}\cdot\mathbf{c}\,d\xi+\mathbf{b}\cdot\mathbf{c}\,d\eta)(D\mathbf{v}-\mathbf{a}\times\mathbf{b})\\
&\equiv -\{(1-\mathbf{a}\cdot\mathbf{c})d\xi+(1-\mathbf{b}\cdot\mathbf{c})d\eta\}\frac{\mathbf{a}\times\mathbf{b}}{D}\,. \qquad (26)
\end{aligned}$$

Set $x=\mathbf{b}\cdot\mathbf{c}$, $y=\mathbf{a}\cdot\mathbf{c}$, $z=\mathbf{a}\cdot\mathbf{b}$ and

$$\mathbf{w}=\frac{\mathbf{a}\times\mathbf{b}}{1+\mathbf{a}\cdot\mathbf{b}}+\frac{\mathbf{b}\times\mathbf{c}}{1+\mathbf{b}\cdot\mathbf{c}}+\frac{\mathbf{c}\times\mathbf{a}}{1+\mathbf{c}\cdot\mathbf{a}}\,. \qquad (27)$$

Then,

$$\begin{aligned}
d\mathbf{w} &= \frac{\mathbf{a}\times\mathbf{b}}{1+z}+\frac{\mathbf{b}\times(\mathbf{c}+d\mathbf{x})}{1+x+\mathbf{b}\cdot d\mathbf{x}}+\frac{(\mathbf{c}+d\mathbf{x})\times\mathbf{a}}{1+y+\mathbf{a}\cdot d\mathbf{x}}-\mathbf{w}\\
&\equiv \frac{1}{1+x}\left(1-\frac{\mathbf{b}\cdot d\mathbf{x}}{1+x}\right)(\mathbf{b}\times\mathbf{c}+\mathbf{b}\times d\mathbf{x})-\frac{\mathbf{b}\times\mathbf{c}}{1+x}\\
&\quad +\frac{1}{1+y}\left(1-\frac{\mathbf{a}\cdot d\mathbf{x}}{1+y}\right)(\mathbf{c}\times\mathbf{a}+d\mathbf{x}\times\mathbf{a})-\frac{\mathbf{c}\times\mathbf{a}}{1+y}\\
&\equiv -\frac{\mathbf{b}\cdot d\mathbf{x}}{(1+x)^2}\mathbf{b}\times\mathbf{c}+\frac{1}{1+x}\mathbf{b}\times d\mathbf{x}\\
&\quad -\frac{\mathbf{a}\cdot d\mathbf{x}}{(1+y)^2}\mathbf{c}\times\mathbf{a}+\frac{1}{1+y}d\mathbf{x}\times\mathbf{a}\\
&\equiv -\frac{\mathbf{a}\times\mathbf{b}}{1+x}d\xi-\frac{\mathbf{b}\times\mathbf{c}}{(1+x)^2}\\
&\quad \times\{(z-xy)d\xi+(1-x^2)d\eta+(1+x)(yd\xi+xd\eta)\}
\end{aligned}$$

$$
\begin{aligned}
&-\frac{\mathbf{a}\times\mathbf{b}}{1+y}d\eta-\frac{\mathbf{c}\times\mathbf{a}}{(1+y)^2}\\
&\times\{(1-y^2)d\xi+(z-xy)d\eta+(1+y)(yd\xi+xd\eta)\}\\
&\equiv-\left(\frac{d\xi}{1+x}+\frac{d\eta}{1+y}\right)(\mathbf{a}\times\mathbf{b})-\frac{(y+z)d\xi+(1+x)d\eta}{(1+x)^2}(\mathbf{b}\times\mathbf{c})\\
&-\frac{(1+y)d\xi+(x+z)d\eta}{(1+y)^2}\mathbf{c}\times\mathbf{a}\,. \qquad (28)
\end{aligned}
$$

Therefore,

$$
\begin{aligned}
df&=\Phi d\xi+\Psi d\eta=\frac{1}{3}(\mathbf{v}\cdot d\mathbf{w}+\mathbf{w}\cdot d\mathbf{v})=-\frac{1}{3}\left\{\left(\frac{d\xi}{1+x}+\frac{d\eta}{1+y}\right)\mathbf{a}\times\mathbf{b}\cdot\mathbf{v}\right.\\
&+\frac{(y+z)d\xi+(1+x)d\eta}{(1+x)^2}\mathbf{b}\times\mathbf{c}\cdot\mathbf{v}+\frac{(1+y)d\xi+(x+z)d\eta}{(1+y)^2}\mathbf{c}\times\mathbf{a}\cdot\mathbf{v}\\
&\left.+\frac{(1-y)d\xi+(1-x)d\eta}{D}\mathbf{a}\times\mathbf{b}\cdot\mathbf{w}\right\}. \qquad (29)
\end{aligned}
$$

Moreover, straightforward computations of vector algebra show that

$$
\begin{aligned}
(\mathbf{a}\times\mathbf{b})\cdot\mathbf{v}&=\frac{1}{D}(1-z)(1+z-x-y)\,,\\
(\mathbf{b}\times\mathbf{c})\cdot\mathbf{v}&=\frac{1}{D}(1-x)(1+x-y-z)\,,\\
(\mathbf{c}\times\mathbf{a})\cdot\mathbf{v}&=\frac{1}{D}(1-y)(1+y-x-z)\,,\\
(\mathbf{a}\times\mathbf{b})\cdot\mathbf{w}&=1-z+\frac{zx-y}{1+x}+\frac{zy-x}{1+y}\,.
\end{aligned}
\qquad (30)
$$

Hence, direct substitutions and algebraic simplifications will show that

$$
\begin{aligned}
\Phi=-\frac{1}{3D}&\left\{\frac{1}{1+x}[(1+z)(1-xy)-(1-z)(x+y)+y^2-z^2]\right.\\
&\left.+\frac{1-y}{1+y}(yz-2x-z)+(1-y)(2-z)-\frac{(y+z)^2(1-x)}{(1+x)^2}\right\},
\end{aligned}
$$

$$\Psi = -\frac{1}{3D}\left\{\frac{1}{1+y}[(1+z)(1-xy)-(1-z)(x+y)+x^2-z^2]\right.$$
$$\left.+\frac{1-x}{1+x}(xz-2y-z)+(1-x)(2-z)-\frac{(x+z)^2(1-y)}{(1+y)^2}\right\}. \tag{31}$$

Let $\nabla_{\mathbf{c}} f$ be the *partial gradient* of $f(\mathbf{a}, \mathbf{b}, \mathbf{c})$ with respect to $\mathbf{c}$, namely

$$\nabla_{\mathbf{c}} f \cdot d\mathbf{x} = \Phi d\xi + \Psi d\eta . \tag{32}$$

Set $\mathbf{a}' = \mathbf{a} - (\mathbf{a}\cdot\mathbf{c})\mathbf{c}$, $\mathbf{b}' = \mathbf{b} - (\mathbf{b}\cdot\mathbf{c})\mathbf{c}$ and $\nabla_{\mathbf{c}} f = k_1\mathbf{a}' + \hat{k}_2\mathbf{b}'$. Then, one has

$$\begin{cases} k_1\mathbf{a}'\cdot\mathbf{a}' + \hat{k}_2\mathbf{b}'\cdot\mathbf{a}' = \Phi\,, \\ k_1\mathbf{a}'\cdot\mathbf{b}' + \hat{k}_2\mathbf{b}'\cdot\mathbf{b}' = \Psi\,, \end{cases} \tag{33}$$

where

$$\mathbf{a}'\cdot\mathbf{a}' = 1-y^2\,, \quad \mathbf{b}'\cdot\mathbf{b}' = 1-x^2\,,$$
$$\mathbf{a}'\cdot\mathbf{b}' = z-xy\,, \quad \begin{vmatrix} \mathbf{a}'\cdot\mathbf{a}' & \mathbf{b}'\cdot\mathbf{a}' \\ \mathbf{a}'\cdot\mathbf{b}' & \mathbf{b}'\cdot\mathbf{b}' \end{vmatrix} = D^2\,. \tag{33$'$}$$

Hence,

$$\begin{cases} k_1 = \dfrac{1}{D^2}\{(1-x^2)\Phi + (xy-z)\Psi\}\,, \\ \hat{k}_2 = \dfrac{1}{D^2}\{(1-y^2)\Psi + (xy-z)\Phi\}\,. \end{cases} \tag{34}$$

Set $\{\mathbf{a}_i, 1 \le i \le \ell\}$ to be the position vectors of the other vertices of $St(C)$ in cyclic order, $x_i = \mathbf{c}\cdot\mathbf{a}_i$, $z_i = \mathbf{a}_i\cdot\mathbf{a}_{i+1}$ and $\mathbf{a}'_i = \mathbf{a}_i - (\mathbf{a}_i\cdot\mathbf{c})\mathbf{c}$. Then,

$$\nabla_{\mathbf{c}} f(\sigma_i) = k_i\mathbf{a}'_i + \hat{k}_{i+1}\mathbf{a}'_{i+1}\,, \tag{35}$$

where

$$k_i = \frac{1}{D_i^2}\{(1-x_{i+1}^2)\Phi_i + (x_ix_{i+1}-z_i)\Psi_i\}$$

$$\hat{k}_{i+1} = \frac{1}{D_i^2}\{(1-x_i^2)\Psi_i + (x_i x_{i+1} - z_i)\Phi_i\}$$

$$\begin{aligned}\Phi_i = \frac{-1}{3D_i}\Bigg\{ &\frac{1}{1+x_{i+1}}[(1+z_i)(1-x_i x_{i+1}) \\ &- (1-z_i)(x_i + x_{i+1}) + x_i^2 - z_i^2] \\ &+ \frac{1-x_i}{1+x_i}(x_i z_i - 2x_{i+1} - z_i) \\ &+ (1-x_i)(2-z_i) - \frac{(x_i+z_i)^2(1-x_{i+1})}{(1+x_{i+1})^2}\Bigg\}\end{aligned}$$

$$\begin{aligned}\Psi_i = \frac{-1}{3D_i}\Bigg\{ &\frac{1}{1+x_i}[(1+z_i)(1-x_i x_{i+1}) \\ &- (1-z_i)(x_i + x_{i+1}) + x_{i+1}^2 - z_i^2] \\ &+ \frac{1-x_{i+1}}{1+x_{i+1}}(x_{i+1} z_i - 2x_i - z_i) \\ &+ (1-x_{i+1})(2-z_i) - \frac{(x_{i+1}+z_i)^2(1-x_i)}{(1+x_i)^2}\Bigg\}. \qquad (35')\end{aligned}$$

We state the final result of the above computation as the following corollary of Theorem 4.2.1.

Corollary 4.2.3 *The gradient vector field of the volume function of* (17′) *is given by*

$$\nabla \operatorname{vol}_C = \sum_{i=1}^{\ell}(k_i + \hat{k}_i)\mathbf{a}_i', \qquad (36)$$

where k_i, $\hat{k}_i$, $1 \le i \le \ell$, *are given by* (35) *and* (35′).

4.2.3 *Two basic lemmas of volume estimation*

Lemma 4.2.2 *Let* $\sigma = \widehat{\triangle ABC}$ *be a (spherical) triangle with area* $\triangle$, *containing its own circumcenter, and* $\bar{\sigma}_\triangle$ *be the equilateral triangle* *with the same area* $\triangle$*. Then,*

$$\operatorname{vol} T(\sigma) \geq \operatorname{vol} T(\bar{\sigma}_\triangle) \tag{37}$$

and the above equality holds if and only if $\sigma = \bar{\sigma}_\triangle$*.*

Proof (i) Set $x = 1 + \cos a$, $y = 1 + \cos b$ and $z = 1 + \cos c$. Then,

$$\begin{cases} x + y + z = u + 2 \\ xy + yz + zx = 1 + 2u + w \quad (0 \leq x, y, z \leq 2) \\ xyz = \dfrac{1}{2}(u^2 + D^2) \end{cases} \tag{38}$$

constitutes a set of *basic invariants* for the *moduli space* $\mathcal{M}$ of congruence classes of spherical triangles. Recall that $\tan \frac{\triangle}{2} = \frac{D}{u}$. Therefore, it is easy to show that, under the constraint of $u = \text{constant} > 0$ (resp. < 0), $\triangle$ is maximal (resp. minimal) when and only when the triangle is regular. Hence, the image of $\mathcal{M}$ in the $(u, \triangle)$-space is given as follows (cf. Figure 4.7):

$$\mathcal{R} = \mathcal{R}_+ \cup \{(0, \pi)\} \cup \mathcal{R}_- \,, \tag{39}$$

where

$$\mathcal{R}_+ = \left\{ (u, \triangle) | 0 < u \leq 4, 0 \leq \frac{\triangle}{2} < \tan^{-1} \frac{(4-u)\sqrt{1+2u}}{3\sqrt{3}u} \right\},$$

$$\mathcal{R}_- = \left\{ (u, \triangle) | -1/2 \leq u < 0, \pi + \tan^{-1} \frac{(4-u)\sqrt{1+2u}}{3\sqrt{3}u} \leq \frac{\triangle}{2} \leq \pi \right\}. \tag{39'}$$

The boundary points on the graph of

$$\tan \frac{\triangle}{2} = \frac{(4-u)\sqrt{1+2u}}{3\sqrt{3}u}, \quad u \in \left[-\frac{1}{2}, 0 \right) \cup (0, 4] \tag{40}$$

are exactly the image points of the regular triangles.

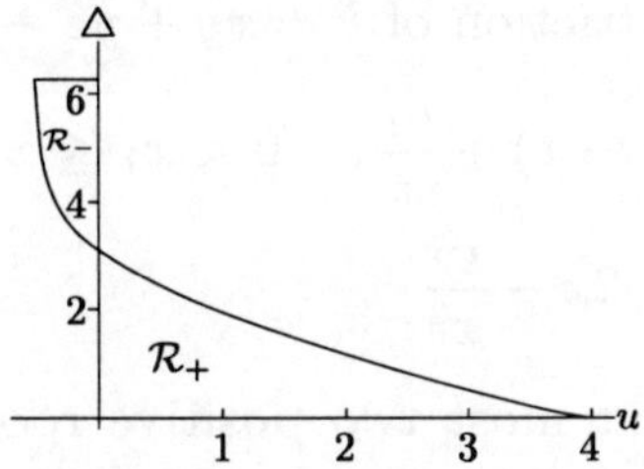

Figure 4.7

(ii) Let $p : \mathcal{M} \to \mathcal{R}$ be the above projection, $(u, \triangle)$ be a given interior point of $\mathcal{R}$ (i.e. $\tan\frac{\triangle}{2} \neq \frac{(4-u)\sqrt{1+2u}}{3\sqrt{3u}}$) and $p^{-1}(u, \triangle)$ be its *inverse image*. We claim that there are exactly two isosceles triangles, say $\sigma_1(u, \triangle)$ and $\sigma_2(u, \triangle)$, in $p^{-1}(u, \triangle)$ such that the evaluation map of $\tilde{v} = xy + yz + zx$ is a *bijection* between $p^{-1}(u, \triangle)$ and the closed interval $[\tilde{v}(\sigma_1), \tilde{v}(\sigma_2)]$, namely

$$\tilde{v}|_{p^{-1}(u,\triangle)} : p^{-1}(u, \triangle) \to [\tilde{v}(\sigma_1), \tilde{v}(\sigma_2)]\,. \tag{41}$$

If one lifts back to the (x, y, z)-space, then the preimage of $p^{-1}(u, \triangle)$ is the algebraic curve defined by

$$\Gamma : \begin{cases} x + y + z = u + 2 = c_1\,, \\ xyz = \dfrac{1}{2}(u^2 + D^2) = \dfrac{1}{2}u^2\left(1 + \tan^2\dfrac{\triangle}{2}\right) = c_2\,, \end{cases} \tag{42}$$

which is a curve in the plane of $x+y+z = c_1$, and it is *reflectionally symmetric* with respect to the three lines of $x = y$, $y = z$ and $z = x$. For example, the line of $y = z$ cuts the above curve Γ into two symmetric branches $\Gamma_\pm$, where

$$\Gamma_\pm : \begin{cases} y = \dfrac{1}{2}\left\{(c_1 - x) \pm \sqrt{(c_1 - x)^2 - \dfrac{4c_2}{x}}\right\}, \\ z = \dfrac{1}{2}\left\{(c_1 - x) \mp \sqrt{(c_1 - x)^2 - \dfrac{4c_2}{x}}\right\}. \end{cases} \tag{42'}$$

Set $g(x)$ to be the restriction of $\tilde{v} = xy + yz + zx$ on $\Gamma_\pm$. Then,

$$\begin{aligned} g(x) &= x(c_1 - x) + \frac{c_2}{x}, \quad 0 < x_1 \le x \le x_2 < 2, \\ g'(x) &= c_1 - 2x - \frac{c_2}{x^2}. \end{aligned} \tag{43}$$

Therefore, $g'(x)$ has at most two positive roots, and hence by the symmetric property of Γ it has *exactly* two positive roots, thus proving the monotonicity of the value of $\tilde{v}$ on a *fundamental arc*. This proves the assertion that the following evaluation map is bijective:

$$\tilde{v}|_{p^{-1}(u,\triangle)} : p^{-1}(u, \triangle) \to [\tilde{v}(\sigma_1), \tilde{v}(\sigma_2)]. \tag{41}$$

(iii) Let $\mathcal{M}'_k$ be the subset of $\mathcal{M}$ defined by the constraints:

$$\begin{gathered} \frac{D}{u} = \tan\frac{\triangle}{2}(= k), \\ \mathrm{Min}\{2 + x - y - z, 2 + y - z - x, 2 + z - x - y\} \ge 0. \end{gathered} \tag{44}$$

Set $F(u, \tilde{v})$ to be the restriction of the formula (4) to the above subset $\mathcal{M}'_k$, namely

$$\begin{aligned} F(u, \tilde{v}) &= \frac{1}{3D}\left\{8 + u - \frac{4u\tilde{v}}{u^2 + D^2}\right\} \\ &= \frac{1}{3k}\left\{\frac{8}{u} + 1 - \frac{4\tilde{v}}{(1 + k^2)u^2}\right\}, \quad \left(k = \frac{D}{u}\right). \end{aligned} \tag{45}$$

Set λ_0 (resp. λ_1) to be the cosine of the side-length of $\bar{\sigma}_\triangle$ (resp. the square with $2\triangle$ as its area). Then,

$$\lambda_0 = \frac{\cos\frac{1}{3}(\pi + \triangle)}{1 - \cos\frac{1}{3}(\pi + \triangle)}. \tag{46}$$

The cosines of the diagonals of the above square are equal to $2\lambda_1 - 1$ and hence, by applying cosine law,

$$\lambda_1 = \frac{1 + \cos\frac{1}{2}(\pi + \triangle)}{1 - \cos\frac{1}{2}(\pi + \triangle)}. \tag{47}$$

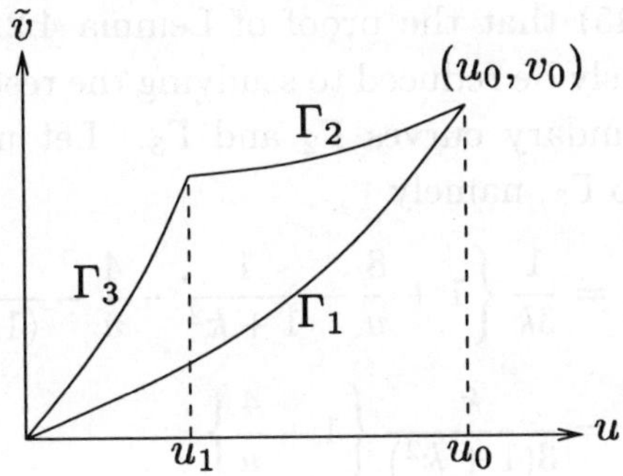

Figure 4.8

Set

$$u_0 = 1 + 3\lambda_0 \quad \text{and} \quad u_1 = 4\lambda_1 \,. \tag{48}$$

Let us first consider the case that $\triangle < \pi$, i.e. both u and k are positive. It follows from (iii) that $\mathcal{M}'_k$ can be bijectively represented by its image set in the $(u, \tilde{v})$-space, which is the domain Ω indicated in Figure 4.8:

$$\begin{cases} \Gamma_1 = \{(u, \tilde{v}(\sigma_1(u, \triangle))); 0 < u \le u_0\} \,, \\ \Gamma_2 = \{(u, \tilde{v}(\sigma_2(u, \triangle))); u_1 \le u \le u_0\} \,, \\ \Gamma_3 = \left\{ \left(u, \dfrac{1}{4}u^2 + (2 + k^2)u \right); 0 < u \le u_1 \right\} . \end{cases} \tag{49}$$

Geometrically, the above boundary curves Γ_1, Γ_2 and Γ_3 are respectively the image sets of $\{\sigma_1(u, \triangle); 0 < u \le u_0\}$, $\{\sigma_2(u, \triangle); u_1 \le u \le u_0\}$ and those triangles with their circumcenters located on their long sides. Notice that the algebraic condition for the latter is that

$$\text{Min}\{2 + x - y - z, 2 + y - z - x, 2 + z - x - y\} = 0 \,, \tag{50}$$

which implies that

$$\tilde{v} = xy + yz + zx = \frac{1}{4}u^2 + (2 + k^2)u \,. \tag{51}$$

If follows from (45) that the proof of Lemma 4.2.2 for the case of $\Delta < \pi$ can obviously be reduced to studying the restriction of $F(u, \tilde{v})$ to the upper boundary curves Γ_2 and Γ_3. Let us first verify the restriction of F to Γ_3, namely

$$\begin{aligned} F(u,\tilde{v})|_{\Gamma_3} &= \frac{1}{3k}\left\{1+\frac{8}{u}-\frac{1}{1+k^2}-\frac{4}{u}-\frac{4}{(1+k^2)u}\right\} \\ &= \frac{k}{3(1+k^2)}\left\{1+\frac{4}{u}\right\} \\ &\geq \frac{k}{3(1+k^2)}\left\{1+\frac{4}{u_1}\right\} \\ &> F(u_0,\tilde{v}_0)\,. \end{aligned} \tag{52}$$

Next let us verify the restriction of F to Γ_2 as follows. In the range $u < u_0$, the following cubic equation, namely

$$y^2(u+2-2y) - \frac{1}{2}(1+k^2)u^2 = 0 \tag{53}$$

has exactly two positive solutions $0 < \eta_1(u) < \eta_2(u)$. Set

$$\psi(u) = \frac{2}{u} - \frac{\tilde{v}(u)}{(1+k^2)u^2}\,, \tag{54}$$

where

$$\begin{aligned} \tilde{v}(u) &= 2\eta_2(u)[u+2-2\eta_2(u)] + \eta_2(u)^2 \\ &= 2(u+2)\eta_2(u) - 3\eta_2(u)^2\,. \end{aligned} \tag{54$'$}$$

Straightforward differentiation gives

$$\begin{aligned} \frac{d\psi}{du} &= -\frac{2}{u^2} - \frac{1}{(1+k^2)u^2}\frac{d\tilde{v}}{du} + \frac{2\tilde{v}}{(1+k^2)u^3}\,, \\ \frac{d\tilde{v}}{du} &= 2\eta_2 + 2(u+2-3\eta_2)\frac{d\eta_2}{du}\,, \\ \eta_2^2 &+ 2\eta_2(u+2-3\eta_2)\frac{d\eta_2}{du} = (1+k^2)u\,. \end{aligned} \tag{55}$$

Therefore,

$$\frac{d\tilde{v}}{du} = \eta_2 + (1+k^2)\frac{u}{\eta_2} \tag{56}$$

and it follows from (56) and the defining equation of η_2 that

$$(1+k^2)u^3\frac{d\psi}{du} = -2(1+k^2)u + \eta_2 u + 4\eta_2 - 2\eta_2^2 \,. \tag{57}$$

It is easy to check that

$$\begin{aligned} (1+k^2)u^3\frac{d\psi}{du}\bigg|_{u_0} &= -\frac{1+\lambda_0}{1+3\lambda_0}(1-\lambda_0)^2 < 0\,, \\ (1+k^2)u^3\frac{d\psi}{du}\bigg|_{u_1} &= 0\,. \end{aligned} \tag{58}$$

The following computation shows that $\frac{d\psi}{du} < 0$ for all $u_1 < u \le u_0$. Solving u in terms of η_2 from the following equation

$$-2(1+k^2)u + \eta_2 u + 4\eta_2 - 2\eta_2^2 = 0 \tag{59}$$

and substituting it into the defining equation of η_2, namely

$$2\eta_2^2(u + 2 - 2\eta_2) - (1+k^2)u^2 = 0\,, \tag{60}$$

one obtains

$$\begin{gathered} k^2\eta_2^2 - 4k^2(1+k^2)\eta_2 + 4k^2(1+k^2) = 0 \\ (\text{or } \eta_2^2 - 4(1+k^2)\eta_2 + 4(1+k^2) = 0)\,. \end{gathered} \tag{61}$$

Therefore, $\frac{d\psi}{du}$ has no zero in $(u_1, u_0]$ and hence $\psi(u)$ is monotonically decreasing in $(u_1, u_0]$, thus verifying that

$$F(u,\tilde{v})|_{\Gamma_2} \ge F(u_0, \tilde{v}_0)\,. \tag{62}$$

(iv) Finally, let us consider the proof of the remaining cases of $\triangle = \pi$ and $\triangle > \pi$. If $\triangle = \pi$, then $u \equiv 0$ and D is maximal when and only when the triangle σ is the regular one. Therefore, $\operatorname{vol} T(\sigma) = \frac{8}{3D}$

attains its unique minimal value of $2\sqrt{3}$ when σ is the regular triangle of side-length $\cos^{-1}(-\frac{1}{3})$.

If $\triangle > \pi$, then such a triangle automatically contains its own circumcenter because the area of its circumscribing circle is at most 2π. Hence, by Corollary 4.2.1 and $\triangle > \pi$

$$\operatorname{vol} T(\sigma) = \frac{1}{6}|\sin\triangle|\left\{\frac{8}{|u|} + |\mathbf{v}|^2 - 1\right\}, \tag{63}$$

which provides an easy proof of Lemma 4.2.2 for the case of $\triangle > \pi$. □

Lemma 4.2.3 *The volume of $T(\bar{\sigma}_\triangle)$, namely*

$$\varphi(\triangle) = \operatorname{vol} T(\bar{\sigma}_\triangle) = \left(1 - 2\cos\frac{\pi+\triangle}{3}\right)\tan\frac{\pi+\triangle}{6} \tag{64}$$

is a convex function of $\triangle$, i.e. $\varphi''(\triangle) > 0$, $0 < \triangle < 2\pi$.

Proof

$$\left(1 - 2\cos\frac{\pi+\triangle}{3}\right)\tan\frac{\pi+\triangle}{6} = 3\tan\frac{\pi+\triangle}{6} - 2\sin\frac{\pi+\triangle}{3}, \tag{65}$$

which is a sum of two strictly convex functions, and hence strictly convex on $0 < \triangle < 2\pi$. □

Remark The circumcenter of $\widehat{\triangle ABC}$ is situated on the edge $\widehat{BC}$ if and only if $2+x-y-z=0$ (or $1+\cos a = \cos b + \cos c$). It is easy to check that $f(\sigma)$ still exceeds $f(\bar{\sigma}_\triangle)$ by a considerable amount in such a "boundary case" for the estimate of Lemma 4.2.2. Therefore, the conclusion of Lemma 4.2.2, namely, $f(\sigma) \geq f(\bar{\sigma}_\triangle)$ will still hold until the circumcenter of σ is further away from its boundary by a considerable amount (cf. Sublemma 4.4.2).

4.2.4 *Some direct applications*

In this subsection, we shall discuss some direct applications of the above basic lemmas of volume estimation.

Example 4.2.1 Let $\mathcal{L}(S_0)$ be a local packing consisting of twelve touching neighbors with Σ as the set of touching points. If the circumradii of all faces of $\mathcal{S}(\Sigma)$ are at most equal to $\pi/4$, then $\mathcal{L}(S_0)$ is already a saturated local packing and the volume of its local cell is at least equal to that of the regular circumscribing dodecahedron.

Proof Notice that $\pi/4$ is exactly the circumradius of a "square" (i.e. equilateral cocircular quadrilateral) of side-length $\pi/3$. Thus, the faces of $\mathcal{S}(\Sigma)$ are either triangles or $\frac{\pi}{3}$–squares and hence, all of them contain their own circumcenters. Let V_j (resp. $r(\sigma_j)$) be the corresponding vertex (resp. circumradius) of a face σ_j of $\mathcal{S}(\Sigma)$. Then,

$$|\overrightarrow{OV_j}| = \sec r(\sigma_j) \leq \sec \frac{\pi}{4} = \sqrt{2} \tag{66}$$

while the center-distance between S_0 and any additional sphere in the direction of $\Gamma(\sigma_j)$ is at least equal to $4\cos r(\sigma_j)(\geq 2\sqrt{2})$. Hence, it *cannot* be an additional neighbor. Thus, $\mathcal{L}(S_0)$ is already a saturated local packing. We shall cut the squares in $\mathcal{S}(\Sigma)$, if any, into pairs of triangles by adding diagonals to them so that $\mathcal{S}(\Sigma)$ consists of exactly twenty triangular faces $\{\sigma_j, 1 \leq j \leq 20\}$, all of them containing their own circumcenters. Therefore, Lemmas 4.2.2 and 4.2.3 are directly applicable. Set $\triangle_j$ to be the area of σ_j, one has

$$\begin{aligned}\operatorname{vol} T(\Sigma) &= \sum_{j=1}^{20} \operatorname{vol} T(\sigma_j) = \sum_{j=1}^{20} f(\sigma_j) \geq \sum_{j=1}^{20} f(\bar{\sigma}_{\triangle_j}) \\ &\geq 20 f(\bar{\sigma}_{\pi/5}) = 20\left(1 - 2\cos\frac{2\pi}{5}\right)\tan\frac{\pi}{5} = 5.550291\cdots.\end{aligned} \tag{67}$$

Moreover, $\operatorname{vol} T(\Sigma)$ is equal to $20(1 - 2\cos\frac{2\pi}{5})\tan\frac{\pi}{5}$ when and only when all σ_j are themselves regular spherical triangles of area $\pi/5$, namely, $\mathcal{L}(S_0)$ is the local packing with *symmetrically arranged twelve touching neighbors.* □

Remark Although the above volume estimate only verifies the dodecahedron conjecture for a rather specific subfamily of local packings, it already covers a *vicinity of quite comfortable size* of the unique tightest local packing (i.e. the one with the smallest local cell) in the moduli space of all possible single-layer local packings. Intuitively speaking, local packings which are small deformations of the tightest one are exactly the ones whose local cells have their volumes close to that of the regular circumscribing dodecahedron. Therefore, as the local packings get closer and closer to the tightest one, the volume estimates must be more and more accurate in order to prove the *optimal* estimate, i.e. the Dodecahedron Conjecture. Hence, it is exactly in the vicinity of the tightest one that the accuracy of volume estimates becomes most demanding. Thus, the volume estimate of (67) for such a rather large neighborhood of the tightest one (in the critical case of twelve touching neighbors), not only accomplishes the above task in a clean-cut way, but it also demonstrates the fit of the volume estimation techniques developed in [Hsi-2] for that of the local cells of sphere packing.

Example 4.2.2 Let $T(\Sigma)$ be the tangent polyhedron with Σ as the set of touching points and $\mathcal{S}(\Sigma)$ be its associated spherical configuration. If all the faces of $\mathcal{S}(\Sigma)$ are triangles containing their own circumcenters, then one has the following lower bound estimate of its volume in terms of its number of faces, namely

$$\operatorname{vol} T(\Sigma) \geq (2n-4)\left(1 - 2\cos\frac{n\pi}{3n-6}\right)\tan\frac{n\pi}{6n-12}, \tag{68}$$

where n is the number of faces of $T(\Sigma)$.

Proof Note that n is also the number of points in Σ. Thus, $\mathcal{S}(\Sigma)$ consists of $(2n-4)$ triangular faces $\{\sigma_j\}$. Set $\triangle_j$ to be the area of σ_j and $\bar{\triangle} = \frac{4\pi}{2n-4}$, one has by Lemmas 4.2.1, 4.2.2 and 4.2.3 and the above assumption:

$$\begin{aligned}\operatorname{vol} T(\Sigma) &= \sum_{j=1}^{2n-4} \operatorname{vol} T(\sigma_j) = \sum_{j=1}^{2n-4} f(\sigma_j) \geq \sum_{j=1}^{2n-4} f(\bar{\sigma}_{\triangle_j}) \\ &\geq (2n-4) f(\bar{\sigma}_{\bar{\triangle}}) \\ &= (2n-4)\left(1 - 2\cos\frac{n\pi}{3n-6}\right)\tan\frac{n\pi}{6n-12}\,. \qquad (69)\end{aligned}$$

□

Remark The above lower bounds are optimal *only when* there exist spherical configurations with $(2n-4)$ equilateral triangles of area $\bar{\triangle}$. Thus, the above lower bounds, L_n, are optimal only for the three cases of $n = 4$, 6 and 12. For all the other cases, the estimates, L_n, are just some useful lower bounds which still have room for further improvement. We list some of their numerical values as follows:

n	12	13	14	15	16
L_n	5.550291	5.418351	5.309697	5.218668	5.141300

Lemma 4.2.2 provides a useful lower bound volume estimate for the tangent subpolyhedrons of those triangular faces of $\mathcal{S}(\Sigma)$ *containing their own circumcenters.* Suppose that σ is a *non-triangular* cocircular polygon containing its own circumcenter and $T(\sigma)$ is defined to be the portion of the solid angle cone $\Gamma(\sigma)$ bounded by the tangent planes at the vertices of σ. For the purpose of obtaining the same lower bound volume estimate for $T(\Sigma)$ whose associated spherical configuration happens to contain such non-triangular faces, it suffices to generalize Lemma 4.2.2 to the following estimate, namely

$$\operatorname{vol} T(\sigma) \geq (k-2) f(\bar{\sigma}_{\bar{\triangle}})\,,$$

where $(k-2)\bar{\triangle}$ is equal to the area of the given k-gon σ.

Sublemma 4.2.1 *Let σ be a cocircular k-gon, $k \geq 4$, containing its own circumcenter C. Set R (resp. $A(\sigma)$ and B_i, $1 \leq i \leq k$) to be its circumradius (resp. area and vertices in cyclic order). Assume that all the circumcenters of $\{\widehat{\triangle} CB_iB_{i+1}\}$ are also within σ. Set $A(k, R)$ to be the area of the cocircular equilateral k-gon with circumradius R and*

$$\begin{aligned}E(k,R) = &\frac{k}{3}\tan\frac{\pi}{k}(\sec R - 1)\tan^2\frac{R}{2}\\ &+ k\left(1 - 2\cos\frac{1}{3}\left(\pi + \frac{A(k,R)}{k}\right)\right)\tan\frac{1}{6}\left(\pi + \frac{A(k,R)}{k}\right)\\ &- (k-2)\left(1 - 2\cos\frac{1}{3}\left(\pi + \frac{A(k,R)}{k-2}\right)\right)\\ &\times \tan\frac{1}{6}\left(\pi + \frac{A(k,R)}{k-2}\right). \end{aligned} \tag{70}$$

Then, one has the following estimate, namely

$$\operatorname{vol} T(\sigma) - (k-2)f(\bar{\sigma}_{\bar{\triangle}}) \geq E(k,R) > 0, \ \bar{\triangle} = \frac{A(\sigma)}{k-2}. \tag{71}$$

Proof Set $\sigma_i = \widehat{\triangle} CB_iB_{i+1}$, $1 \leq i \leq k$. From the assumption that all their circumcenters are inside of σ, it is easy to see that each σ_j contains its own circumcenter. The tangent plane T_C cuts $T(\sigma)$ into two pieces, one of them is a cone of height $\sec R - 1$, while the other can be decomposed into the union of $\{T(\sigma_i), 1 \leq i \leq k\}$. The total volume of the latter can be estimated by a direct application of Lemmas 4.2.2 and 4.2.3, namely, for $\triangle_i = A(\sigma_i)$,

$$\begin{aligned}\sum_{i=1}^{k} \operatorname{vol} T(\sigma_i) &= \sum_{i=1}^{k} f(\sigma_i) \geq \sum_{i=1}^{k} f(\bar{\sigma}_{\triangle_i})\\ &\geq k\left(1 - 2\cos\frac{1}{3}\left(\pi + \frac{A(\sigma)}{k}\right)\right)\tan\frac{1}{6}\left(\pi + \frac{A(\sigma)}{k}\right). \end{aligned} \tag{72}$$

On the other hand, the former is a cone whose base is a circumscribing k-gon to a circle of radius $\tan\frac{R}{2}$. Thus, its volume is bounded below by that of such a cone with the regular circumscribing k-gon as its base, namely

$$\text{vol(Cone part of } T(\sigma)) \geq \frac{k}{3}\tan\frac{\pi}{k}(\sec R - 1)\tan^2\frac{R}{2}\,. \tag{73}$$

Hence,

$$\operatorname{vol} T(\sigma) - (k-2)f(\bar{\sigma}_{\bar{\Delta}}) \geq \frac{k}{3}\tan\frac{\pi}{k}(\sec R - 1)\tan^2\frac{R}{2} + \psi_k(A(\sigma))\,, \tag{74}$$

where

$$\begin{aligned}\psi_k(A) = {} & k\left(1 - 2\cos\frac{1}{3}\left(\pi + \frac{A}{k}\right)\right)\tan\frac{1}{6}\left(\pi + \frac{A}{k}\right) \\ & - (k-2)\left(1 - 2\cos\frac{1}{3}\left(\pi + \frac{A}{k-2}\right)\right)\tan\frac{1}{6}\left(\pi + \frac{A}{k-2}\right).\end{aligned} \tag{75}$$

It is straightforward to check that the above $\psi_k(A)$ is a (negative value) decreasing function of A. Therefore, it follows from the fact $A(\sigma) \leq A(k, R)$ that

$$\operatorname{vol} T(\sigma) - (k-2)f(\bar{\sigma}_{\bar{\Delta}}) > E(k, R), \ \bar{\Delta} = \frac{A(\sigma)}{k-2}\,. \tag{76}$$

Finally, the positivity of $E(k, R)$ follows from straightforward numerical computation using the following explicit formula of $A(k, R)$, namely

$$A(k, R) = 2k\alpha - (k-2)\pi\,, \quad \sin\alpha = \frac{\sin a}{\sin R}, \quad \tan a = \cos\frac{\pi}{k}\tan R\,. \tag{77}$$

Notice that an equilateral cocircular spherical k-gon with circumradius R can be subdivided into $2k$ spherical triangles as indicated in Figure 4.9. □

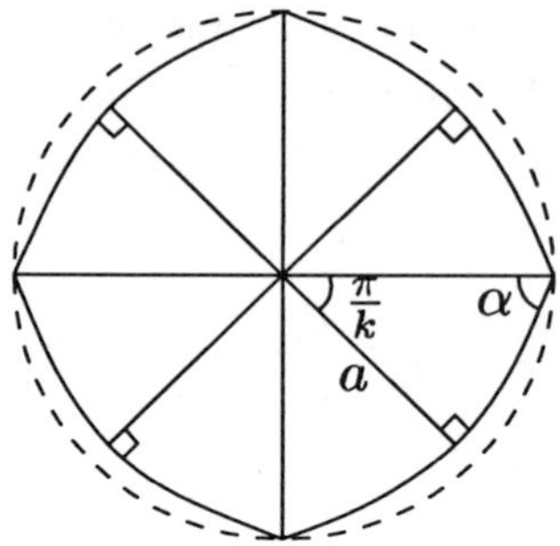

Figure 4.9

4.3 Volume Estimation of a Rectilinear Slab

Let h be the buckling height of a reduced neighbor, S', of S. Then the local cells $C(\mathcal{L}')$ (resp. $C(\mathcal{L})$) both contain a rectilinear slab of thickness h in their peripheral parts, while the latter is simply the restriction of the former to $C(\mathcal{L})$. In this section, we shall prove a *lower bound volume estimate* of such a rectilinear slab solely in terms of its thickness h [Hsi-2] [Hsi-3]. Geometrically, the above h-slab of $C(\mathcal{L}')$ is defined to be the convex hull of the pair of faces of $T(\Sigma, \{h_j\})$ and $T(\Sigma)$ *perpendicular to* $\overline{OO'}$ (cf. Sec. 4.1), which are determined by the geometry of the local surrounding of the given pair $\{S, S'\}$ in $\mathcal{L}'(S)$. Therefore, it is natural to start with a subsection on the localized geometry surrounding such a pair of spheres $\{S, S'\}$ with center–distance $2(1+h)$.

4.3.1 *The local geometry surrounding a given neighbor of $\mathcal{L}'(S_0)$*

The lower bound volume estimation of a single rectilinear slab, all by itself, naturally leads to the study of the local geometry surrounding a given neighbor of $\mathcal{L}'(S_0)$. In this subsection, we shall denote the central sphere simply by S and a given neighbor of $\mathcal{L}'(S)$ by S'. The local surrounding of such a pair of spheres $\{S, S'\}$ in a given

$\mathcal{L}'(S)$ consists of those common neighbors of both S and S', say $\{S_j, 1 \le j \le k\}$. Set the intersection of $\overline{OO'}$ with S (resp. S') to be A (resp. A') and that of $\overline{OO_j}$ (resp. $\overline{O'O_j}$) with S (resp. S') to be A_j (resp. A'_j). Then, the pair of star configurations, $St(A)$ on S and $St(A')$ on S', constitutes a natural set of geometric data that determines the geometric structure of such a local surrounding. Let $\{\ell_j\}$ (resp. $\{\ell'_j\}$) be the lengths of the radial edges of $St(A)$ (resp. $St(A')$) and $\{\theta_j\}$ be the dihedral angles between $\{\triangle OO'O_j, 1 \le j \le k\}$ which are also the central angles of both $St(A)$ and $St(A')$. Then, the geometry of a such a local surrounding of $\{S, S'\}$ is completely determined by the above three sets of angular invariants, namely, $\{\ell_j\}$, $\{\ell'_j\}$ and $\{\theta_j\}$, $1 \le j \le k$. For example, the relative geometry of the triple $\{S, S', S_j\}$ is determined by the pair $\{\ell_j, \ell'_j\}$.

As indicated in Figure 4.10, set C_j (resp. R_j, $\triangle_j$) to be the circumcenter (resp. circumradius, area) of $\triangle OO'O_j$, M and M_j to be the middle point of $\overline{OO'}$ and $\overline{OO_j}$, and D_j to be the point of intersection of the tangent lines at A and A_j.

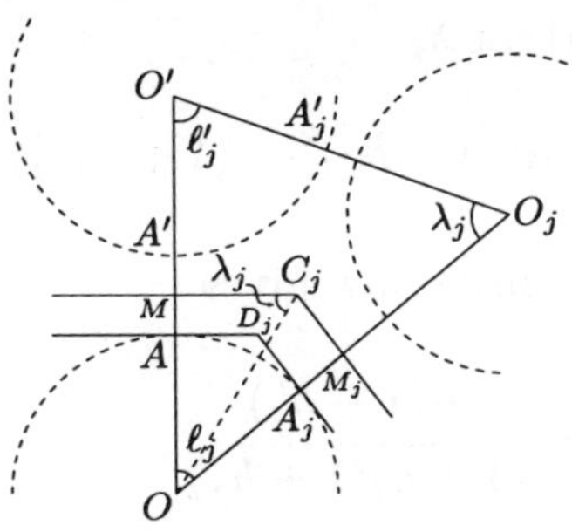

Figure 4.10

Sublemma 4.3.1 *Set* $\overline{AD}_j = f_1(h, h_j, h'_j)$, $\overline{MC}_j = f_2(h, h_j, h'_j)$ *and* $\triangle_j$ *to be the area of* $\triangle OO'O_j$. *Then,*

$$\begin{aligned} f_1(h, h_j, h'_j) &= \triangle_j^{-1}((1 + h'_j)^2 - (h - h_j)^2)\,, \\ f_2(h, h_j, h'_j) &= \triangle_j^{-1}(1 + h)((1 + h'_j)^2 + (h_j - h)(2 + h_j + h))\,. \end{aligned} \tag{1}$$

$$\begin{aligned}\frac{\partial f_1}{\partial h'_j} &= \frac{1}{(1+\cos\ell_j)(1+h)\sin\ell'_j}\,,\\ \frac{\partial f_2}{\partial h'_j} &= \frac{\cos\ell'_j}{\sin\ell_j\sin\ell'_j\sin\lambda_j}\,.\end{aligned}\tag{2}$$

$$\begin{aligned}\frac{\partial f_1}{\partial h_j} &= \frac{-\cot\lambda_j}{(1+\cos\ell_j)(1+h_j)}\,,\\ \frac{\partial f_2}{\partial h_j} &= \frac{\cos\ell_j}{\sin\ell_j\sin\ell'_j\sin\lambda_j}\,.\end{aligned}\tag{3}$$

Proof Note that $\angle AOD_j = \frac{\ell_j}{2}$ and $\angle MC_jO = \angle O'O_jO = \lambda_j$. Therefore, it follows from the cosine law and the area formula of $\triangle OO'O_j$ that

$$\begin{aligned}f_1 &= \overline{AD}_j = \tan\frac{\ell_j}{2} = \frac{1-\cos\ell_j}{\sin\ell_j} = \triangle_j^{-1}\{(1+h'_j)^2-(h-h_j)^2\}\,,\\ f_2 &= \overline{MC}_j = (1+h)\cot\lambda_j \\ &= \triangle_j^{-1}\{(1+h'_j)^2+(1+h_j)^2-(1+h)^2\}(1+h)\,.\end{aligned}\tag{4}$$

The differentiation of cosine law gives

$$\begin{aligned}-\sin\ell_j\frac{\partial\ell_j}{\partial h'_j} &= \frac{-(1+h'_j)}{(1+h)(1+h_j)}\,,\\ -\sin\ell_j\frac{\partial\ell_j}{\partial h_j} &= \frac{-(1+h)^2+(1+h_j)^2+(1+h'_j)^2}{2(1+h)(1+h_j)^2}\,,\\ -\sin\lambda_j\frac{\partial\lambda_j}{\partial h_j} &= \frac{(1+h)^2+(1+h_j)^2-(1+h'_j)^2}{2(1+h_j)^2(1+h'_j)}\,,\\ -\sin\lambda_j\frac{\partial\lambda_j}{\partial h'_j} &= \frac{(1+h)^2+(1+h'_j)^2-(1+h_j)^2}{2(1+h_j)(1+h'_j)^2}\,.\end{aligned}\tag{5}$$

Hence,

$$\frac{\partial \ell_j}{\partial h'_j} = \frac{(1+h'_j)}{(1+h)(1+h_j)\sin\ell_j} = \frac{1}{(1+h)\sin\ell'_j},$$

$$\begin{aligned}\frac{\partial \ell_j}{\partial h_j} &= \frac{-((1+h_j)^2+(1+h'_j)^2-(1+h)^2)}{2(1+h)(1+h_j)^2\sin\ell_j}\\ &= \frac{-\cos\lambda_j}{(1+h)\sin\ell'_j} = \frac{-\cot\lambda_j}{1+h_j},\end{aligned} \tag{6}$$

$$\frac{\partial \lambda_j}{\partial h_j} = \frac{-((1+h)^2+(1+h_j)^2-(1+h'_j)^2)}{2(1+h_j)^2(1+h'_j)\sin\lambda_j} = \frac{-\cot\ell_j}{1+h_j},$$

$$\frac{\partial \lambda_j}{\partial h'_j} = \frac{-((1+h)^2+(1+h'_j)^2-(1+h_j)^2)}{2(1+h_j)(1+h'_j)^2\sin\lambda_j} = \frac{-\cot\ell'_j}{1+h'_j}.$$

Therefore,

$$\begin{aligned}\frac{\partial f_1}{\partial h'_j} &= \frac{1}{2}\sec^2\frac{\ell_j}{2}\frac{\partial \ell_j}{\partial h'_j} = \frac{1}{(1+\cos\ell_j)(1+h)\sin\ell'_j},\\ \frac{\partial f_2}{\partial h'_j} &= (1+h)(-\csc^2\lambda_j)\frac{\partial\lambda_j}{\partial h'_j}\\ &= \frac{(1+h)\cot\ell'_j}{(1+h'_j)\sin^2\lambda_j} = \frac{\cos\ell'_j}{\sin\ell_j\sin\ell'_j\sin\lambda_j},\\ \frac{\partial f_1}{\partial h_j} &= \frac{1}{2}\sec^2\frac{\ell_j}{2}\frac{\partial \ell_j}{\partial h_j} = \frac{-\cot\lambda_j}{(1+\cos\ell_j)(1+h_j)},\\ \frac{\partial f_2}{\partial h_j} &= (1+h)(-\csc^2\lambda_j)\frac{\partial\lambda_j}{\partial h_j}\\ &= \frac{(1+h)\cot\ell_j}{(1+h_j)\sin^2\lambda_j} = \frac{\cos\ell_j}{\sin\ell_j\sin\ell'_j\sin\lambda_j}.\end{aligned} \tag{7}$$

□

Set

$$\begin{aligned} \delta_1(h, h_j, h'_j) &= f_1(h, 0, h'_j) - f_1(h, h_j, h'_j)\,, \\ \delta_2(h, h_j, h'_j) &= f_2(h, h_j, h'_j) - f_2(h, 0, h'_j)\,. \end{aligned} \tag{8}$$

Then, by the mean-value theorem, there exists $0 < \xi < h_j$ such that

$$\frac{\delta_2(h, h_j, h'_j)}{\delta_1(h, h_j, h'_j)} = \left(\frac{\partial f_2}{\partial h_j} \Big/ - \frac{\partial f_1}{\partial h_j} \right)_{h_j = \xi}$$

$$= \frac{(1 + \cos \ell_j(h, \xi, h'_j)) \cos \ell_j(h, \xi, h'_j)(1 + \xi)}{\sin \ell_j(h, \xi, h'_j) \sin \ell'_j(h, \xi, h'_j) \cos \lambda_j(h, \xi, h'_j)}\,. \tag{9}$$

In case that $h'_j \leq h$, one has

$$\begin{aligned} \lambda_j(h, \xi, h'_j) &\geq \ell_j(h, \xi, h'_j)\,, \\ \cos \lambda_j(h, \xi, h_j) &\leq \cos \ell_j(h, \xi, h'_j) \end{aligned} \tag{10}$$

and hence, the right-hand side of (9) is bounded below by

$$\frac{(1 + \cos \ell_j(h, \xi, h'_j))(1 + \xi)}{\sin \ell_j(h, \xi, h'_j) \sin \ell'_j(h, \xi, h'_j)}$$

$$= \cot \frac{1}{2} \ell_j(h, \xi, h'_j) \csc \ell'_j(h, \xi, h'_j)(1 + \xi)\,. \tag{11}$$

As one shall see later, the above analysis on δ_1 and δ_2 provides the needed quantitative analysis on the geometry of the given rectilinear slab further *restricted* to the *normal section of* $\triangle OO'O_j$.

Next let us analyze the central angles $\{\theta_j\}$ which are exactly the dihedral angles between the family of triangles $\{\triangle OO'O_j,\ 1 \leq j \leq k\}$. Notice that θ_j is clearly bounded below by the one when $\overline{O_j O_{j+1}}$ reaches its lower bound of 2, namely, $\theta_j \geq \hat{\theta}_j$ while $\hat{\theta}_j$ is given by

$$\sin \ell'_j \sin \ell'_{j+1} \cos \hat{\theta}_j = \cos \hat{a}_j - \cos \ell'_j \cos \ell'_{j+1} ,$$
$$\cos \hat{a}_j = \frac{(1+h'_j)^2 + (1+h'_{j+1})^2 - 1}{2(1+h'_j)(1+h'_{j+1})} . \tag{12}$$

Notice that the set of central angles of $St(A)$ and $St(A')$ are always the same. Thus, it suffices to use the geometry of just one of the above two star configurations to analyze the central angles and there should be no essential difference between them. However, it is advantageous to use $St(A')$ on S' rather than $St(A)$ on S, because it suffices to consider the case that $\{h'_j\}$ are *all quite small* for the purpose of proving Lemma 4.3.1 (cf. Subsec. 4.3.2).

For the sake of notational simplicity, set $j = 1$ and denote $\hat{\theta}_1$ by $\hat{\theta}$ in the following discussion. Again, let us first analyze the special but also the most crucial case that both h'_1 and h'_2 are equal to zero, namely, the case that

$$\sin \ell'_1 \sin \ell'_2 \cos \hat{\theta} = \frac{1}{2} - \cos \ell'_1 \cos \ell'_2 ,$$
$$\cos \ell'_1 = \frac{(1+h)^2 - h_1(2+h_1)}{2(1+h)} , \tag{13}$$
$$\cos \ell'_2 = \frac{(1+h)^2 - h_2(2+h_2)}{2(1+h)} ,$$

which defines $\hat{\theta}$ as a function of $\{h, h_1, h_2\}$, symmetric in h_1 and h_2. Thus, it suffices to consider the region $0 < h \leq 0.3$ and $0 \leq h_1 \leq h_2 \leq 0.3$. Geometrically, it amounts to analyzing the angle $\hat{\theta}$ of the spherical triangle $\widehat{\triangle A' A'_1 A'_2}$ as indicated in Figure 4.11.

The angle α_2 is always less than $\pi/2$ and hence $\hat{\theta}$ is a decreasing function in h_2 while $\hat{\theta}$ is a decreasing (resp. increasing) function in h_1 if $\alpha_1 < \pi/2$ (resp. $\alpha_1 > \pi/2$), or equivalently, if

$$\cos \ell'_2 - \frac{1}{2} \cos \ell'_1 > 0 \quad (\text{resp. } < 0) . \tag{14}$$

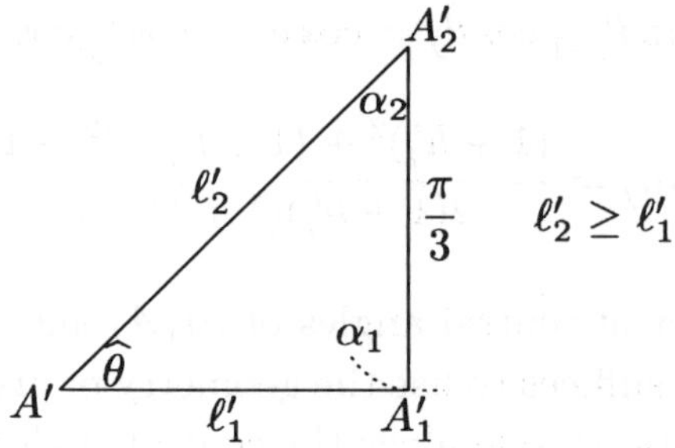

Figure 4.11

For a given h, it is quite easy to see that the minimum of the above $\hat{\theta}$ occurs at $h_1 = 0$ and $h_2 = 0.3$, namely

$$\cos \ell_1' = \frac{1+h}{2}, \quad \sin \ell_1' = \frac{1}{2}\sqrt{4-(1+h)^2},$$

$$\cos \ell_2' = \frac{(1+h)^2 - 0.69}{2(1+h)}, \quad \sin \ell_2' = \frac{1}{2}\sqrt{5.38-(1+h)^2 - \frac{0.4761}{(1+h)^2}}. \tag{15}$$

Set $u = (1+h)^2$ and $\mu(h) = \hat{\theta}(h, 0, 0.3)$. Then,

$$\mu(h) = \cos^{-1} \frac{2.69-u}{\sqrt{4-u}\sqrt{5.38-u-0.4761/u}}, \quad u = (1+h)^2, \tag{16}$$

which is a monotonically increasing function of h with $\mu(0) = \frac{\pi}{3} + 0.007111$ and $\mu(0.3) = \frac{\pi}{3} + 0.159197$. The above simple analysis on the central angles of the local surrounding of a given pair $\{S, S'\}$ in $\mathcal{L}'(S)$ shows that such a surrounding with six common neighbors must be accompanied by a certain amount of $\{h_j', 1 \leq j \leq 6\}$. Moreover, the ones with smaller total amount, i.e. $\sum_{j=1}^{6} h_j'$, can only be achieved by alternating arrangements of $\{h_j, 1 \leq j \leq 6\}$ with three of them close to zero while the other three are close to 0.3. As we shall see later, the occurrence of even a small amount of $\sum h_j'$ already makes the proof of the volume estimate of h-slab simpler than otherwise, while an amount of at least 0.12 will make such a proof almost automatic (cf. the proof of Lemma 4.3.1). Therefore, it suffices to analyze $\hat{\theta}(h, 0, 0.3, h'_1, h'_2)$ for the case that $\{h, h'_1, h'_2\}$ are all quite small, as follows.

The initial value, $\hat{\theta}(0, 0, 0.3, 0, 0)$, is equal to $\frac{\pi}{3} + 0.007111$. Therefore, it suffices to compute $\frac{\partial\hat{\theta}}{\partial h}, \frac{\partial\hat{\theta}}{\partial h'_1}, \frac{\partial\hat{\theta}}{\partial h'_2}$, at the initial point $p_0 = (0, 0, 0.3, 0, 0)$ in order to control the value of $\hat{\theta}$ in its small vicinity. Straightforward differentiations using (12) and (16) will show that

$$\left.\frac{\partial\hat{\theta}}{\partial h}\right|_{p_0} = 0.4065648\,,$$

$$\left.\frac{\partial\hat{\theta}}{\partial h'_1}\right|_{p_0} = -0.7572102\,, \tag{17}$$

$$\left.\frac{\partial\hat{\theta}}{\partial h'_2}\right|_{p_0} = -0.1803851\,.$$

Therefore, up to the first order of $\{h, h'_1, h'_2\}$, one has

$$\begin{aligned}\hat{\theta}(h, 0, 0.3, h'_1, h'_2) \equiv{}& \frac{\pi}{3} + 0.007111 + 0.4065648h \\ & - 0.7572102h'_1 - 0.1803851h'_2\end{aligned} \tag{18}$$

which can be regarded as a lower bound estimate of θ_1 up to the first order of $\{h, h'_1, h'_2\}$. By adding such lower bound estimates of $\{\theta_i, 1 \le i \le 6\}$, one has the following collective estimate of $\{h'_i, 1 \le i \le 6\}$, namely

$$\begin{aligned}&0.7572102(h'_1 + h'_3 + h'_5) + 0.1803851(h'_2 + h'_4 + h'_6) \\ &\quad \ge 3(0.007111 + 0.4065648h)\end{aligned} \tag{19}$$

up to the first order of $\{h, h'_i, 1 \le i \le 6\}$.

The above analysis on the central angles of a local surrounding of $\{S, S'\}$ with six common neighbors shows that the total amount, $\sum_{i=1}^{6} h'_i$, will be minimal in the geometric arrangement with

$$h_1 = h_3 = h_5 = 0\,, \quad h_2 = h_4 = h_6 = 0.3\,, \quad h'_1 = h'_3 = h'_5 = 0 \quad (20)$$

while $(h'_2 + h'_4 + h'_6)$ is approximately bounded below by 16.63 $\times\,(0.007111 + 0.4065648h)$.

4.3.2 *Lower bound volume estimates of rectilinear slabs*

In this subsection, we shall prove the following lower bound volume estimate of an h-slab [Hsi-2] [Hsi-3].

Lemma 4.3.1 *The volume of a rectilinear slab of thickness $h \le 0.3$ is at least equal to*

$$L(h) = \frac{5}{3}\tan\frac{\pi}{5}\left(\frac{1-h}{3+h}\right)\left(\frac{1-h}{1+h}\right)\left\{1-\left[1-\frac{h(1+h)}{1-h}^{3}\right]\right\}. \quad (21)$$

Remarks (i) Let us first explain the geometric meaning of the above lower bound, $L(h)$, before we proceed to give its proof. Suppose that the given pair of spheres $\{S, S'\}$ with center–distance $2(1+h)$ can, in fact, be surrounded by five common neighbors *touching both.* Then the rectilinear slab is simply a frustum associated to a cone whose base is a *circumscribing pentagon* to a circle of radius equal to

$$\tan\frac{R}{2} = \sqrt{\frac{1-h}{3+h}}\,, \quad \left(R = \cos^{-1}\frac{1+h}{2}\right) \quad (22)$$

while its height is equal to

$$\sec R - 1 = \frac{1-h}{1+h}\,. \quad (23)$$

Therefore, its base area is at least equal to that of the regular circumscribing pentagon, namely

$$A \geq 5 \tan \frac{\pi}{5} \tan^2 \frac{R}{2} = 5 \tan \frac{\pi}{5} \left(\frac{1-h}{3+h} \right) \tag{24}$$

and hence the volume of such a frustum is at least equal to (21), where $[1 - \frac{h(1+h)}{1-h}]$ is exactly the ratio between the height of the cone and that of the truncated cone.

(ii) In fact, it will be quite clear from the proof that the lower bound estimate of (21) is optimal when and only when it is possible to arrange five common neighbors touching both S and S'. Geometrically, it corresponds to the possibility of arranging five points on the spherical circle of radius $R = \cos^{-1} \frac{1+h}{2}$ with separation of $\pi/3$, namely

$$R \geq \sin^{-1} \frac{1}{2 \sin \frac{\pi}{5}} = 1.017222\,, \quad h \leq 0.05146\,. \tag{25}$$

Thus, the proof of Lemma 4.3.1 amounts to show that the resulting rectilinear slab of any other local surroundings of such a pair $\{S, S'\}$ always have larger volume than the above frustum.

Proof of Lemma 4.3.1 Set $R = \cos^{-1} \frac{1+h}{2}$. Note that the bottom face of the rectilinear slabs described in Remark (i) is a circumscribing pentagon of a circle of radius $\tan \frac{R}{2}$ while the *regular* one is the one with *minimal area.* Let us begin with the following simple fact of plane geometry. As indicated by the dotted one in Figure 4.12, the *regular circumscribing* pentagon of the unit circle is also the inscribing pentagon of the concentric circle of radius $\sec \frac{\pi}{5} = \sqrt{5}-1$, while all the other circumscribing pentagons always have some vertices inside and some vertices outside of the circle of radius $(\sqrt{5}-1) = 1.236068$. By the convexity of $y = \tan x$, $0 < x < \pi/2$, it is easy to show that the regular one has the minimal area among all such circumscribing pentagons. However, an even stronger statement is still true, namely, it is also the one with *minimal area* even among those circumscribing

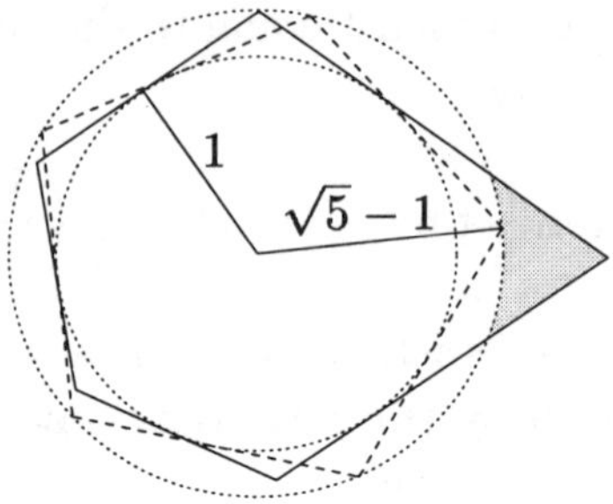

Figure 4.12

pentagons with tips truncated by the circle of radius $\sqrt{5}-1$, as indicated in Figure 4.12.

The proof of the above fact is the same as that of the convexity of $A(\theta)$ used in the proof of Theorem 3.1.1′.

In view of the above fact and Remark (i), it is natural to introduce the following upper bound, namely, set

$$\ell(h)=\cos^{-1}\left\{\frac{-3+2\sqrt{5}+(7-2\sqrt{5})h}{9-2\sqrt{5}+(2\sqrt{5}-5)h}\right\} \tag{26}$$

or equivalently

$$\tan\frac{1}{2}\ell(h)=(\sqrt{5}-1)\tan\frac{R}{2}=(\sqrt{5}-1)\sqrt{\frac{1-h}{3+h}}\,. \tag{27}$$

In case S_j is a common neighbor with $\ell_j \geq \ell(h)$, then the distance between A and the line of $T_A \cap T_{A_j}$ is at least equal to $(\sqrt{5}-1)\tan\frac{R}{2}$. Thus, the effect of such a common neighbor can simply be treated as a truncation which does *not* affect the lower bound estimation of the area of the bottom face. In particular, the cuttings of those planes that produce the "*switchings*," if any, will definitely have no essential role in the proof of this lemma, except making it much simpler than otherwise. Similarly, the occurrence of a non-reduced neighbor (of S) adjacent to S' will only produce a cutting of the slab considerably further away than that of a reduced neighbor with

$\ell_j \approx \ell(h)$. Therefore, it suffices to focus our attention on those S_j with $\ell_j < \ell(h)$. We shall call them the *crucial common neighbors* of $\{S, S'\}$. We shall label them by $\{S_j, 1 \leq j \leq k\}$ in cyclic order and divide the proof of this lemma into two cases according to $k \leq 5$ and $k \geq 6$.

Case 1: $k \leq 5$. Let us first look at the subcase that all ℓ_j are at least equal to $\cos^{-1} \frac{1+h}{2}$. In such a subcase, the rectilinear slab contains a frustum with a *truncated* circumscribing k-gon, $k \leq 5$, of the circle of radius $\tan \frac{R}{2} = \sqrt{\frac{1-h}{3+h}}$ as its base, while the truncation lines, if any, are always outside of the circle of radius $(\sqrt{5}-1)\sqrt{\frac{1-h}{3+h}}$. Hence, the volume of the rectilinear slab is at least equal to the volume of such a frustum which is at least equal to $L(h)$. Thus, it suffices to consider the remaining subcase that some of ℓ_j are less than $\cos^{-1} \frac{1+h}{2}$. Let $\ell_2 = \overset{\frown}{AA_2}$ be such a one that is less than $\cos^{-1} \frac{1+h}{2}$. Set A_2^* to be the point on the great circle of AA_2 such that $\overset{\frown}{AA_2^*} = \cos^{-1} \frac{1+h}{2}$ and $T_{A_2^*}$ to be the tangent plane of S at A_2^*. The intersection of $T_{A_2^*}$ with T_A (resp. the plane of $\overrightarrow{OX} \cdot \mathbf{a} = 1 + h$) are both parallel lines of the sides of the bottom face and the top face corresponding to A_2, namely, they all are perpendicular to $\triangle OO'O_2$. As indicated in Figure 4.13, this pair of

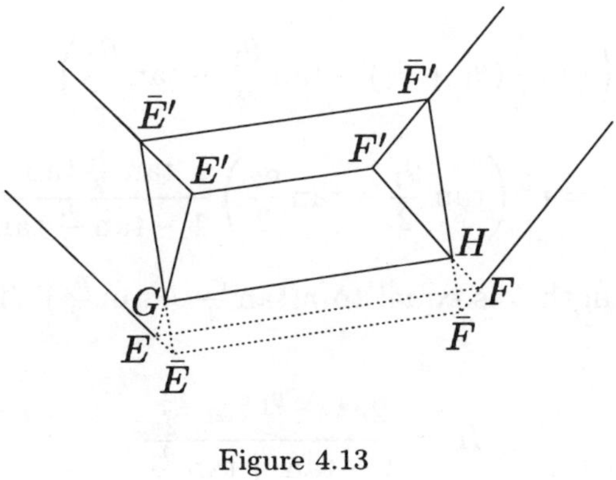

Figure 4.13

intersection lines define a new pair of parallel faces and the *convex hull* of such a new pair of parallel faces provides a *new rectilinear slab*. The only difference between the old one and the new one is that the latter adds a triangular prism outlined by the dotted-line edges while it subtracts the triangular prism outlined by the solid-line edges.

The following Sublemma computes the difference between the volumes of such a pair of triangular prisms (with a tilted cut at their ends) for the most crucial case.

Sublemma 4.3.2 *In the case of $h_1 = h_3 = h_1' = h_3' = 0$ the difference between the volume of the subtracted prism and that of the added one (cf. Figure 4.13) is given by*

$$\frac{h}{6}\left(\tan\frac{\theta_1}{2}+\tan\frac{\theta_2}{2}\right)\left\{\sqrt{\frac{1-h}{3+h}}\left[3(\delta_2-\delta_1)-\frac{h(1+h)}{1-h}(2\delta_2-\delta_1)\right]\right.$$

$$\left.-\frac{1-\tan\frac{\theta_1}{2}\tan\frac{\theta_2}{2}}{2\tan\frac{\theta_1}{2}\tan\frac{\theta_2}{2}}(\delta_2^2-\delta_2\delta_1+\delta_1^2)\right\}, \tag{28}$$

where δ_2, δ_1 are the pair of functions discussed in the previous subsection (cf. (8)).

Proof of Sublemma 4.3.2 As indicated in Figure 4.14, the area of the triangle bounded by three tangent lines of a circle of radius r is given by

$$r^2\left(\tan\frac{1}{2}(\theta_1+\theta_2)-\tan\frac{\theta_1}{2}-\tan\frac{\theta_2}{2}\right)$$

$$=r^2\left(\tan\frac{\theta_1}{2}+\tan\frac{\theta_2}{2}\right)\frac{\tan\frac{\theta_1}{2}\tan\frac{\theta_2}{2}}{1-\tan\frac{\theta_1}{2}\tan\frac{\theta_2}{2}}. \tag{29}$$

The base length b is equal to $r(\tan\frac{\theta_1}{2}+\tan\frac{\theta_2}{2})$. Thus, its height H is given by

$$H=\frac{2r\tan\frac{\theta_1}{2}\tan\frac{\theta_2}{2}}{1-\tan\frac{\theta_1}{2}\tan\frac{\theta_2}{2}}. \tag{30}$$

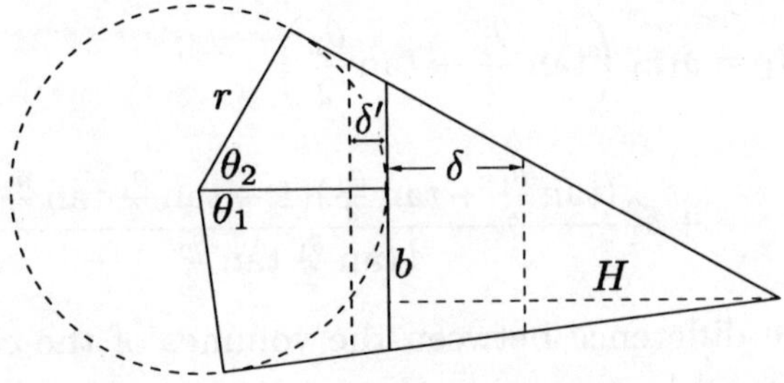

Figure 4.14

Therefore, the area of the trapezoid with width δ is equal to

$$\left\{1 - \left(\frac{H-\delta}{H}\right)^2\right\}\frac{1}{2}bH = \delta b - \frac{\delta^2 b}{2H}, \tag{31}$$

while in the case of "$-\delta'$", one has

$$\left\{\left(\frac{H+\delta'}{H}\right)^2 - 1\right\}\frac{1}{2}bH = \delta' b + \frac{\delta'^2 b}{2H}. \tag{32}$$

Set

$$\begin{aligned} r_1 &= \tan\frac{R}{2} = \sqrt{\frac{1-h}{3+h}}, \\ r_2 &= r_1 \cdot \left(1 - \frac{h(1+h)}{1-h}\right), \end{aligned} \tag{33}$$

and δ_1, δ_2 to be the pair of functions of $\{h, h_2, h_2'\}$ of (8). Then, the area of the top face (resp. bottom face) of the subtracted (resp. added) prism is given by

$$B_2 = \delta_2 r_2 \left(\tan\frac{\theta_1}{2} + \tan\frac{\theta_2}{2}\right)$$
$$- \delta_2^2 \frac{(\tan\frac{\theta_1}{2} + \tan\frac{\theta_2}{2})(1 - \tan\frac{\theta_1}{2}\tan\frac{\theta_2}{2})}{4\tan\frac{\theta_1}{2}\tan\frac{\theta_2}{2}}$$

$$\left(\text{resp. } B_1 = \delta_1 r_1 \left(\tan\frac{\theta_1}{2} + \tan\frac{\theta_2}{2}\right)\right.$$

$$\left. + \delta_1^2 \frac{(\tan\frac{\theta_1}{2} + \tan\frac{\theta_2}{2})(1 - \tan\frac{\theta_1}{2}\tan\frac{\theta_2}{2})}{4\tan\frac{\theta_1}{2}\tan\frac{\theta_2}{2}}\right). \tag{34}$$

Therefore, the difference between the volumes of the corresponding cones of trapezoid bases is given by

$$\frac{1}{3}B_2 \cdot \frac{\delta_2 h}{\delta_2 + \delta_1} - \frac{1}{3}B_1 \frac{\delta_1 h}{\delta_2 + \delta_1}$$

$$= \frac{h}{3}\left(\tan\frac{\theta_1}{2} + \tan\frac{\theta_2}{2}\right)\left\{r_1\left[(\delta_2 - \delta_1) - \frac{\delta_2^2}{\delta_2 + \delta_1}\frac{h(1+h)}{1-h}\right]\right.$$

$$\left. - (\delta_2^2 - \delta_2\delta_1 + \delta_1^2)\left(\frac{1 - \tan\frac{\theta_1}{2}\tan\frac{\theta_2}{2}}{4\tan\frac{\theta_1}{2}\tan\frac{\theta_2}{2}}\right)\right\}. \tag{35}$$

Next let us compute the difference between the volumes of the two triangular cones. The length of $\overline{GH}$ is given by

$$\ell = r_1\left(\tan\frac{\theta_1}{2} + \tan\frac{\theta_2}{2}\right)\left[1 - \frac{\delta_1}{\delta_2 + \delta_1}\frac{h(1+h)}{1-h}\right] \tag{36}$$

while the difference in volume is equal to

$$\frac{1}{6}\ell \cdot \left[\frac{\delta_2^2 h}{\delta_2 + \delta_1} - \frac{\delta_1^2 h}{\delta_2 + \delta_1}\right] = \frac{1}{6}h\ell(\delta_2 - \delta_1) = \frac{h}{6}r_1\left(\tan\frac{\theta_1}{2} + \tan\frac{\theta_2}{2}\right)$$

$$\times \left[1 - \frac{\delta_1}{\delta_2 + \delta_1}\frac{h(1+h)}{1-h}\right](\delta_2 - \delta_1). \tag{37}$$

Hence, the difference in the volumes of the two prisms is equal to the sum of (35) and (37), which is equal to the formula of (28). □

Remarks (i) By the estimate of (11), $\delta_2 : \delta_1$ is always larger than 2. Thus, the volume of the new rectilinear slab is always smaller than the original one. Therefore, the volume of the given one is always

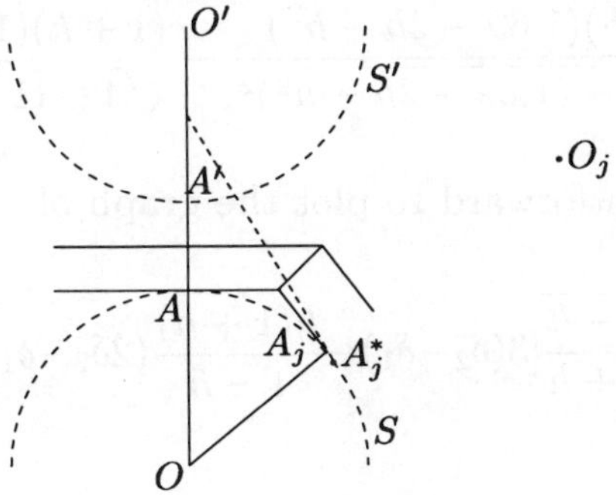

Figure 4.15

bounded below by that of the frustum associated to $\{T_{A_j^*}, 1 \le j \le k \le 5\}$, which is at most equal to $L(h)$. This proves the lemma for the case of at most five crucial common neighbors of $\{S, S'\}$. The cross–sectional geometry of $(OO'O_j)$ with $\ell_j = \widehat{AA_j}$ less than $\widehat{AA_j^*} = \cos^{-1}\frac{1+h}{2}$ is as indicated in Figure 4.15.

(ii) In the general case of non-zero $\{h_1, h_3, h_1', h_3'\}$, the tilted cutting planes at both ends of the two triangular prisms are, in general, no longer the tangent planes $T_{A_1^*}$ and $T_{A_3^*}$. Instead, they are respectively replaced by the trapzoidal side-faces of the given rectilinear slab, such as $\square EFF'E'$ in Figure 4.13, but for $j = 1$ and 3 instead of $j = 2$. Anyhow, it again follows from the estimate of (11) on $\delta_2 : \delta_1$ applying to $j = 1$ and 3 instead of 2 that such a changing of the tilted cutting planes at both ends will only make the difference between the volumes of the two prisms larger than the previous case. Thus, the one given in (28) is actually a *lower bound* for the general case.

Example 4.3.1 Set $h \le 0.1$, $h_2 = 0.3$, $h_2' = 0$ and $\theta_1 = \theta_2 = \frac{\pi}{3}$. Then,

$$\delta_1 = \sqrt{\frac{1-h}{3+h}} - \sqrt{\frac{(0.91) + (0.6)h - h^2}{(4.29) + (4.6)h + h^2}}$$

$$\delta_2 = \frac{(1+h)(1.69-2h-h^2)}{\sqrt{6.76-(1.69-2h-h^2)^2}} - \frac{(1+h)(1-2h-h^2)}{\sqrt{4-(1-2h-h^2)^2}}$$

Hence, it is straightforward to plot the graph of

$$F(h) = \frac{1}{3\sqrt{3}}\left\{\sqrt{\frac{1-h}{3+h}}[3(\delta_2-\delta_1)-\frac{h(1+h)}{1-h}(2\delta_2-\delta_1)]-(\delta_2^2-\delta_2\delta_1+\delta_1^2)\right\}$$

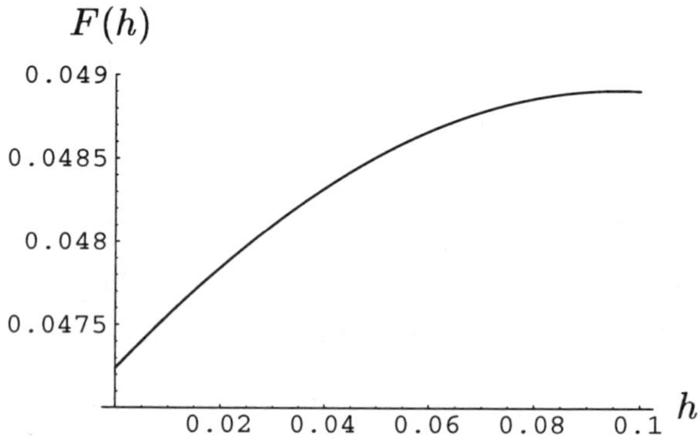

Case 2: $k \geq 6$. Let us first consider the case of $k = 6$. Of course one may also carry out the same kind of construction as before. Therefore, the volume of the given rectilinear slab is bounded below by the sum of those lower bounds of Sublemma 4.3.2 for $1 \leq j \leq 6$ and that of the frustum associated to $\{T_{A_j^*}, 1 \leq j \leq 6\}$. However, the base area of such a frustum is, in general, bounded below by $(2\sqrt{3})\frac{1-h}{3+h}$ instead of $(5\tan\frac{\pi}{5})\frac{1-h}{3+h}$ as in the case of $k \leq 5$. Note that

$$\begin{aligned} 5\tan\frac{\pi}{5} - 2\sqrt{3} &= 0.168611\cdots, \\ 2\sqrt{3}\Big/\left(5\tan\frac{\pi}{5}\right) &= 0.953585\cdots. \end{aligned} \tag{38}$$

Therefore, the last summand of the above individual lower bound estimates is equal to

$$\frac{2\sqrt{3}}{5\tan\pi/5}L(h) = 0.953585L(h)\,. \tag{39}$$

Thus, it suffices to show that the total sum of that lower bound of Sublemma 4.3.2 for $1 \leq j \leq 6$ is always larger than 0.0465 times $L(h)$.

The analysis of the lower bound of θ_j, namely, the function $\hat{\theta}_j(h, h_j, h_{j+1}, h'_j, h'_{j+1})$ defined by the equations of (12), shows that

$$\hat{\theta}_j(h, h_j, h_{j+1}, 0, 0) > \pi/3 \tag{40}$$

and $\hat{\theta}(h, h_j, h_{j+1}, 0, 0)$ *cannot* even get close to $\pi/3$ except when h is small and at least one of h_j and h_{j+1} are already getting close to the upper limit of 0.3. Therefore, $\{h_j, h'_j; 1 \leq j \leq 6\}$ must contain several h_j quite close to the upper limit of 0.3 while the total sum of $\{h'_j\}$, $\sum_{j=1}^{6} h'_j$, cannot be too small. However, each non-zero h'_j will add a layer of average thickness roughly equal to $\frac{4}{3\sqrt{3}}h'_j$ to the jth trapezoidal-side-face of the frustum (cf. (2) of Sublemma 4.3.1). Therefore, just a total sum of $\sum_{j=1}^{6} h'_j \geq 0.12$ will already produce more than enough additional volume, i.e. more than $(0.0465)L(h)$. Therefore, it suffices to consider the remaining subcase that $\sum_{j=1}^{6} h'_j < 0.12$. In such a subcase all the six central angles $\{\theta_j, 1 \leq j \leq 6\}$ must be very close to $\pi/3$, h must be small and there are at least three of $\{h_j\}$ very close to 0.3. With the above self-imposed constraints, it is quite simple to apply Sublemmas 4.3.1 and 4.3.2 to show that the contribution of those lower bounds of (28), at least three of them, is already more than $0.0465L(h)$, even without the additional contribution coming from $\sum_{j=1}^{6} h'_j$. See Example 4.3.1 for the value of (28) (i.e. $hF(h)$) for the case of $h \leq 0.1$, $h_2 = 0.3$, $h'_2 = 0$ and $\theta_1 = \theta_2 = \frac{\pi}{3}$.

Finally, let us discuss the only remaining case of $k \geq 7$, i.e. at least seven crucial neighbors. Note that a central angle of at most $2\pi/7$ can possibly occur only with rather large h'_j while the difference between the areas of the regular circumscribing pentagon and septagon is equal to

$$r^2\left(5\tan\frac{\pi}{5}-7\tan\frac{\pi}{7}\right)=(0.26169\cdots)r^2\,. \tag{41}$$

Therefore, the last summand for the case of $k=7$ is equal to

$$\frac{7\tan\pi/7}{5\tan\pi/5}L(h)=0.927963L(h)\,. \tag{42}$$

Hence, the rather large amount of $\sum h'_j$ already brings in a large enough summand that makes the lower bound estimate exceed $L(h)$. This completes the proof. □

Remark The lower bound estimate of $L(h)$ for the volume of the h-slab is no longer optimal as h exceed 0.05146. In fact, it is always possible to improve the above estimate by a factor slightly larger than 1 for h considerably larger than 0.05146. We include here such an improvement for a rather specific geometric setting, which will facilitate the proof of the Dodecahedron Conjecture in Chap. 6.

Lemma 4.3.1′ *If h is at least equal to* 0.145 *and $\sum_{j=1}^{k} h_j$ is at most equal to* 0.015, *then the volume of such an h-slab is at least equal to* $1.1L(h)$.

Proof Let us first study the special case of $h=0.145$. Set

$$\begin{aligned}
a&=\cos^{-1}\frac{1+h}{2}=\cos^{-1}0.5725=0.96124457\cdots,\\
\theta&=2\sin^{-1}\frac{1}{2\sin a}=1.31168446\cdots,\\
\varphi&=2\pi-3\theta=2.34813192\cdots.
\end{aligned} \tag{43}$$

In the case that $k\le 4$, it is easy to show that the volume of such an h-slab is at least equal to

$$\frac{4}{5\tan\frac{\pi}{5}}L(h)=(1.101105\cdots)L(h)\,. \tag{44}$$

Thus, it suffices to study the remaining case of $k = 5$ whose associated $St(A)$ is a 5Δ-star as indicated in Figure 4.16.

From the assumption that $\sum_{j=1}^{k} h_j$ is at most equal to 0.015, all $\ell(h, h_j)$ are very close to a while all $\ell(h_j, h_{j+1})$ are very close to $\frac{\pi}{3}$. Let us first consider the special case that all h_j are equal to zero. The bottom face of the h-slab is a pentagon on T_A as indicated in Figure 4.17.

The distance between A to its side on $T_A \cap T_{A_j}$ is given by $d_j = \tan \frac{1}{2} r_j$. Suppose that r_j, $1 \leq j \leq 4$, are actually equal to their lower bound $a = \cos^{-1} \frac{1+h}{2}$. Then the pentagon of Figure 4.17 can be regarded as a truncated circumscribing quadrilateral (i.e. of the circle of radius $\tan \frac{a}{2}$ centered at A), while the truncation line is $T_A \cap T_{A_5}$. In such a geometric situation, it is easy to see that the area of the

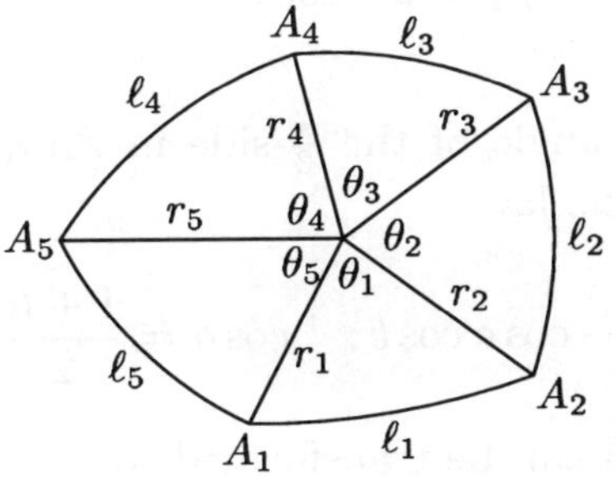

$$r_j \geq \ell(h, h_j)$$
$$\ell_j \geq \ell(h_j, h_{j+1})$$

Figure 4.16

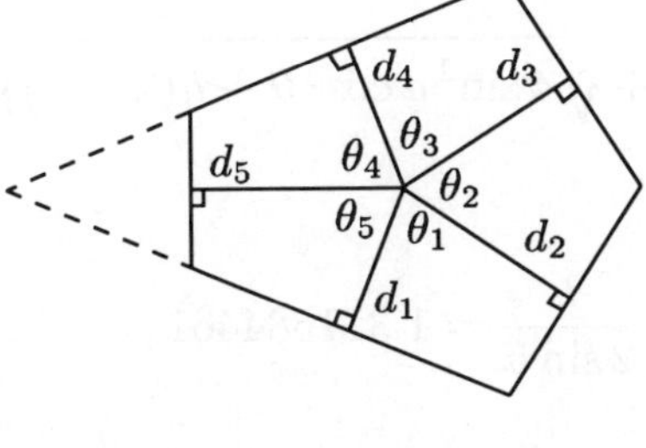

$$d_j = \tan \tfrac{1}{2} r_j$$

Figure 4.17

above pentagon for the case $\ell_4 = \ell_5 = \frac{\pi}{3}$ will always be smaller than otherwise. Moreover, the area of such a truncated circumscribing quadrilateral is bounded below by that of another one with $\theta_1' = \theta_2' = \theta_3' = \frac{1}{3}(\theta_1 + \theta_2 + \theta_3)$. Therefore, it is quite simple to check that the area of the bottom face for such a special case (i.e. $\sum h_j = 0$, $d_j = a$, $1 \le j \le 4$) always exceeds $4\tan^2 \frac{a}{2} = 4\left(\frac{1-h}{3+h}\right)$, and hence the volume of such an h-slab always exceeds $(1.1011055)L(h)$.

In the slightly more general case that $\sum h_j = 0$ but with at least two r_j longer than a, we claim that the *total amount of the radial edge-excess* of $St(A)$ of Figure 4.16, i.e. $\sum_{j=1}^{5}(r_j - a)$, has its minimum when all of its boundary edges are equal to $\frac{\pi}{3}$ and all of its radial edge-excess are concentrated in a single $(r_j - a)$, namely

$$\sum_{j=1}^{5}(r_j - a) \ge 0.245258\,. \tag{45}$$

Set θ to be the opposite angle of the $\frac{\pi}{3}$-side in $\triangle(a, b, \frac{\pi}{3})$, $a = \cos^{-1}\frac{1+h}{2}$. Then, by the cosine law

$$\sin a \sin b \cos\theta = \frac{1}{2} - \cos a \cos b\,, \quad \cos a = \frac{1+h}{2}\,. \tag{46}$$

Set $t = \tan\frac{b}{2}$. Then, Eq. (46) can be transformed to

$$(2+h)t^2 - (4\sin a\cos\theta)t - h = 0\,. \tag{47}$$

Hence, one has

$$\tan\frac{b}{2} = \frac{1}{2+h}\{2\sin a\cos\theta + \sqrt{4\sin^2 a\cos^2\theta + h(2+h)}\}\,. \tag{48}$$

Set

$$\begin{aligned} \theta_0 &= 2\sin^{-1}\frac{1}{2\sin a} = 1.311684461\,, \\ \theta_1 &= \pi - \frac{3}{2}\theta_0 = 1.174065962\,. \end{aligned} \tag{49}$$

Then, it is easy to use the differentiation of (48), namely

$$\frac{1}{2}\sec^2\frac{b\,db}{2\,d\theta} = -\frac{2\sin a\sin\theta}{2.145}\left\{1+\frac{2\sin a\cos\theta}{\sqrt{4\sin^2 a\cos^2\theta + 0.311025}}\right\} \tag{50}$$

to verify that $b(\theta)$ is a concave function for the range of $\theta_1 \leq \theta \leq \theta_0$. Using the above concavity of $b(\theta)$, it is not difficult to show that the total number of radial edge-excess of $St(A)$ of Figure 4.16 reaches its minimum when all its boundary edges are equal to $\frac{\pi}{3}$ and it has four radial edges equal to $a = \cos^{-1} 0.5725 = b(\theta_0)$ while its only other edge is equal to $b(\theta_1) = a + 0.245258$. From the above lower bound estimate of the total radial edge-excess and the fact that smaller radial edge-excess will contribute a strip-area with larger proportionate factors, one shows that the area of the bottom face for such an h-slab always exceeds $4\tan^2\frac{a}{2}$.

Finally, let us consider the general setting of Lemma 4.3.1′, namely, $0 < \sum_{j=1}^{5} h_j \leq 0.015$. Notice that the total amount of $\{h_j\}$ is assumed to be at most 0.015, thus all h_j are very small and hence all the boundary (resp. radial) edge-length bounds $\{\ell(h_j, h_{j+1})\}$ (resp. $\{\ell(0.145, h_j)\}$) are very close to $\frac{\pi}{3}$ (resp. $\cos^{-1} 0.5725$). Moreover, the same method as in the proof of Lemma 4.3.1 enables us to replace the h-slab with a smaller h-slab where the bottom face with some d_j less than $\tan\frac{a}{2}$ is replaced by a corresponding one with $d'_j = \tan\frac{a}{2}$. Therefore, we shall discount those negative radial edge-excess, if any, thus only counting the *total number of positive radial edge-excess.* For a given set of $\{h_j\}$ (with $\sum_{j=1}^{5} h_j \leq 0.015$), the same reason will again show that the total number of positive radial edge-excess reaches its minimal when all its boundary edges are equal to their individual lower bounds and it has only one positive radial edge-excess. Straightforward numerical computation will show that the minimal area of their corresponding bottom face is given by the geometric arrangement with $\ell_1 = \ell_2 = \ell_3 = \frac{\pi}{3}$, $\ell_4 = \ell_5 = \ell(0.015, 0)$ and $r_j = a$, $1 \leq j \leq 4$. However, at such a minimal setting, the area still exceeds $(1.1)5\tan\frac{\pi}{5}\tan^2\frac{a}{2}$ by a tiny amount. Such a lower bound estimate on the areas of the corresponding bottom faces clearly proves that the volume of the h-slab is at least equal to $(1.1)L(h)$ for the beginning case of $h = 0.145$. However, the above method of proof clearly also works without modification for $h > 0.145$, because all the lower estimates clearly still hold for $h > 0.145$. This completes the proof. □

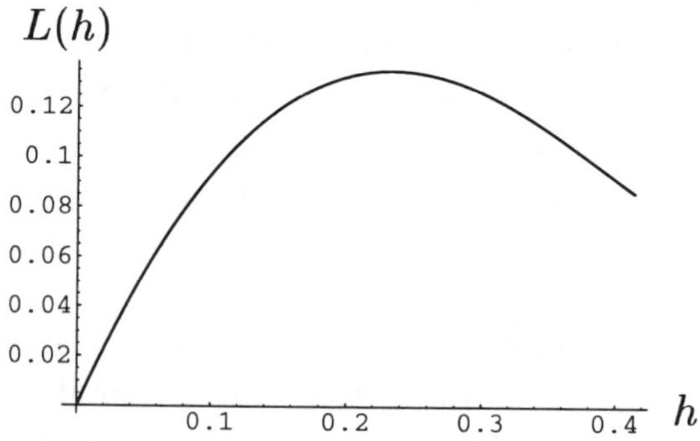

(a)

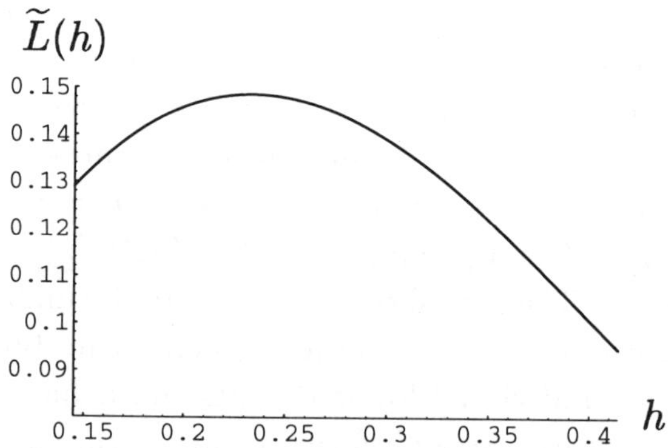

(b)

Figure 4.18

Remarks (i) Figure 4.18(a) is a plotting of $L(h)$ for $0 \leq h \leq (\sqrt{2}-1)$.

(ii) Figure 4.18(b) is a plotting of $\widetilde{L}(h) = 1.1L(h)$ for $0.145 \leq h \leq (\sqrt{2}-1)$.

4.4 Volume Estimation of Local Cells

The volume estimation techniques of Secs. 4.2 and 4.3 are developed in [Hsi-2] [Hsi-3] mainly for the purpose of achieving the *optimal estimates* of the *local density* (resp. the *locally averaged density*) of sphere packings. As has already been discussed in Chap. 3, the former amounts to the optimal lower bound estimate of the volume of a single local cell while the latter is essentially the optimal estimate of a *specific weighted average* of the volumes of the *cluster of local cells of a core packing.* At first glance, the latter seems to be much more demanding than the former. However, with the big help of the pertinent geometric understanding on type I configurations as well as on local extensions (cf. Chap. 7), the *collective* volume estimation of the cluster of local cells *for each given type of core packing* turns out to be much simpler than that of a single local cell, namely, the proof of the Dodecahedron Conjecture.

Both as a test and as a demonstration of the volume estimation techniques of Secs. 4.2 and 4.3, we shall apply them to one of the most critical cases in the proof of the Dodecahedron Conjecture, namely, the case that $\mathcal{L}(S)$ contains twelve touching neighbors. Let Σ be the set of twelve touching points of those touching neighbors of a given $\mathcal{L}(S)$ and $\mathcal{S}(\Sigma)$ be the spherical configuration with Σ as its set of vertices. All the edge-lengths of $\mathcal{S}(\Sigma)$ are at least $\frac{\pi}{3}$. We shall call such spherical configurations type I configurations, and denote the circumradius of a face $\sigma \in \mathcal{S}(\Sigma)$ by $r(\sigma)$. If all $r(\sigma)$ are at most equal to $\frac{\pi}{4}$, then $\mathcal{L}(S)$ cannot have any other (non-touching) neighbor and all $\sigma \in \mathcal{S}(\Sigma)$ contain their own circumcenter (cf. Example 4.2.1). Therefore, direct application of Lemmas 4.2.2

and 4.2.3 (cf. (67) of Sec. 4.2) already proves the lower bound of L_{12} for such a restrictive case. However, in the general case, on the one hand $\mathcal{L}(S)$ may contain additional neighbors thus making the local cell $C(\mathcal{L})$ a *truncated tangent polyhedron* instead of $T(\Sigma)$ itself; and on the other hand, some of the faces of $\mathcal{S}(\Sigma)$ may *not* contain their own circumcenter while the lower bound estimate of $f(\sigma) \geq f(\bar{\sigma}_\triangle)$ has, so far, only been proved under the condition that σ contains its own circumcenter. Anyhow, we need to modify the way of applying the volume estimation techniques of Secs. 4.2 and 4.3 in order to extend their applicability to the above two kinds of more general situations.

Let us begin with some basic geometry of isosceles triangles. Set $\sigma(a,\theta)$ to be the triangle with $\{a,a,c\}$ as its side-lengths and $\theta = C$ as its top angle. Then,

$$\cos c = \sin^2 a \cos\theta + \cos^2 a\,,$$

$$D = \sin^2 a \sin\theta = (1+\cos a)(1-\cos a)\sin\theta\,,$$

$$\begin{aligned} u &= 1 + 2\cos a + \cos^2 a + \sin^2 a\cos\theta \\ &= (1+\cos a)\left(2\cos^2\frac{\theta}{2} + 2\cos a\sin^2\frac{\theta}{2}\right), \end{aligned}$$

$$\tan\frac{\triangle}{2} = \frac{D}{u} = \frac{(1-\cos a)\sin\frac{\theta}{2}\cos\frac{\theta}{2}}{\cos^2\frac{\theta}{2} + \cos a\sin^2\frac{\theta}{2}} = \frac{(1-\cos a)\tan\frac{\theta}{2}}{1+\cos a\tan^2\frac{\theta}{2}}, \tag{1}$$

$$\begin{aligned} \tan^2 R &= \frac{2}{D^2}(1-\cos a)^2(1-\cos c) = \frac{2(1-\cos^2 a)(1-\cos\theta)}{(1+\cos a)^2\sin^2\theta} \\ &= \frac{2(1-\cos a)}{(1+\cos a)(1+\cos\theta)} = \tan^2\frac{a}{2}\sec^2\frac{\theta}{2}, \end{aligned}$$

$$R = \tan^{-1}\left\{\tan\frac{a}{2}\sec\frac{\theta}{2}\right\}.$$

Sublemma 4.4.1 *Set* $t = \tan\frac{\theta}{2}$ *and* $f(\sigma(a,\theta))$ *to be the volume function of* $\sigma(a,\theta)$ *(cf.* (4) *and* (9) *of Sec.* 4.2*). Then,*

$$f(\sigma(a,\theta)) = \frac{t}{3}\tan^2\frac{a}{2}\left\{1 + \frac{2+\cos a - t^2\cos a}{1+t^2\cos^2 a}\right\}, \tag{2}$$

$$\frac{\partial f}{\partial t} = \frac{\tan^2\frac{a}{2}}{3(1+t^2\cos^2 a)^2} \times (1-t^2\cos a)[3+\cos a + (\cos^2 a - \cos^3 a)t^2]. \tag{3}$$

Proof Recall that

$$f(\sigma) = \frac{1}{6}\sin\triangle\left\{\frac{8}{u} - \tan^2 R\right\}. \tag{4}$$

In the case of $\sigma = \sigma(a,\theta)$,

$$\sin\triangle = \frac{2\tan\frac{\triangle}{2}}{1+\tan^2\frac{\triangle}{2}} = \frac{2t(1-\cos a)(1+t^2\cos a)}{(1-\cos a)^2t^2 + (1+t^2\cos a)^2} = (1-\cos a)\sin\theta\frac{1+t^2\cos a}{1+t^2\cos^2 a}. \tag{5}$$

Hence, direct substitution and algebraic simplification will give

$$f(\sigma(a,\theta)) = \frac{t}{3}\tan^2\frac{a}{2}\frac{4-(1-\cos a)(1+t^2\cos a)}{1+t^2\cos^2 a} = \frac{t}{3}\tan^2\frac{a}{2}\left\{1+\frac{2+\cos a - t^2\cos a}{1+t^2\cos^2 a}\right\}. \tag{6}$$

Straightforward differentiation of (2) with respect to t and algebraic simplification will show that

$$\frac{\partial f}{\partial t} = \frac{\tan^2\frac{a}{2}}{3(1+t^2\cos^2 a)^2}(1-t^2\cos a)[3+\cos a+(\cos^2 a - \cos^3 a)t^2].$$

□

Corollary 4.4.1 *For a given a, $f(\sigma(a,\theta))$ has a unique maximal at $\tan^2\frac{\theta}{2} = \sec a$, namely*

$$\cos\theta = \frac{\cos a - 1}{\cos a + 1}, \tag{7}$$

which means $\sigma(a,\theta)$ is exactly the half of an a-square.

Notice that this is also the place that the area of $\sigma(a,\theta)$ reaches its unique maximal.

Set

$$\begin{aligned} \varphi_a(t) &= \left(1 - 2\cos\frac{\pi+\triangle}{3}\right)\tan\frac{\pi+\triangle}{6} \quad (= f(\bar\sigma_\triangle)), \\ \triangle &= 2\tan^{-1}\left\{\frac{(1-\cos a)t}{1+t^2\cos a}\right\}, \\ \psi_a(t) &= \frac{t}{3}\tan^2\frac{a}{2}\left\{1 + \frac{2+\cos a - t^2\cos a}{1+t^2\cos^2 a}\right\} \quad (= f(\sigma(a,\theta)). \end{aligned} \tag{8}$$

For each given value of a, e.g. $a = \frac{\pi}{3}$, it is simple to use a computer to plot the graph of $\psi_a(t) - \varphi_a(t)$ and to determine the unique value of $t_0 > \sqrt{\sec a}$ which makes $\psi_a(t_0) = \varphi_a(t_0)$. For example, in the important special case of $a = \frac{\pi}{3}$, one has

$$t_0 = 1.8226598 \tag{9}$$

and the graph of $\psi_a - \varphi_a$ as indicated in Figure 4.19.

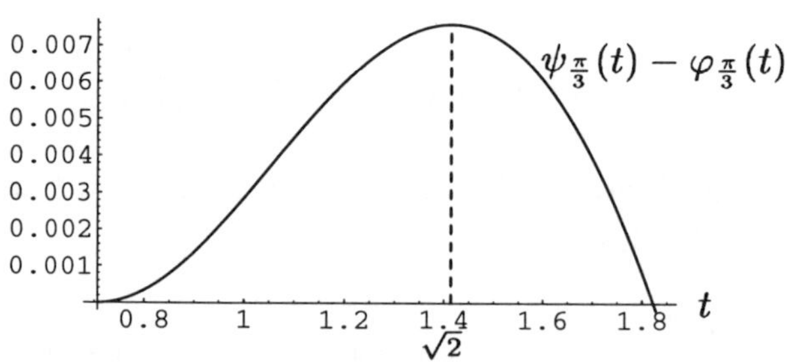

Figure 4.19

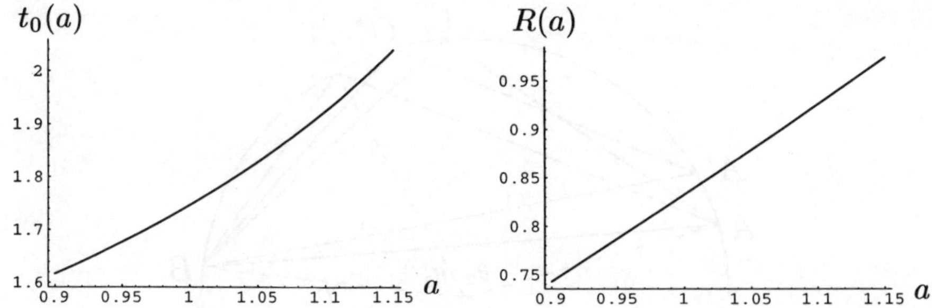

Figure 4.19′

The above computation shows that $f(\sigma(\frac{\pi}{3}, \theta)) \geq f(\bar{\sigma}_\triangle)$ still holds beyond $\cos^{-1}(-\frac{1}{3})$ up until

$$\theta = \theta_0\left(\frac{\pi}{3}\right) = 2\tan^{-1} t_0 = 2.137982289\,. \tag{10}$$

In terms of its circumradius, the above upper limit on its top angle corresponds to the upper limit of

$$R \leq \tan^{-1}\left\{\tan\frac{a}{2}\sec\frac{\theta_0}{2}\right\} = \tan^{-1}\sqrt{\frac{1+t_0^2}{3}} = 0.8761769\,. \tag{11}$$

Remark It is not difficult to show that $t_0(a)$ becomes larger as a increases. Figure 4.19′ is the computer plot of the graphs of $t_0(a)$ and $R(a) = \tan^{-1}\{\tan\frac{a}{2}\sqrt{t_0(a)^2+1}\}$.

Sublemma 4.4.2 *Let σ be a spherical triangle with side-lengths of at least $\frac{\pi}{3}$. If $r(\sigma)$ is at most equal to* 0.8761769, *then $f(\sigma) \geq f(\bar{\sigma}_\triangle)$ still holds.*

Proof If σ contains its own circumcenter, then $f(\sigma) \geq f(\bar{\sigma}_\triangle)$ has already been proved in Lemma 4.2.2. Thus, it suffices to prove the remaining case that the circumcenter of σ is situated outside, and its circumradius is at most equal to 0.8761769.

As indicated in Figure 4.20, $\sigma = \widehat{\triangle} ABC$, $\sigma' = \widehat{\triangle} ABC'$, $\sigma^* = \widehat{\triangle} A^*BC^*$ have the same circumcircle, $\theta_1 + \theta_2 < \pi$, both σ' and σ^*

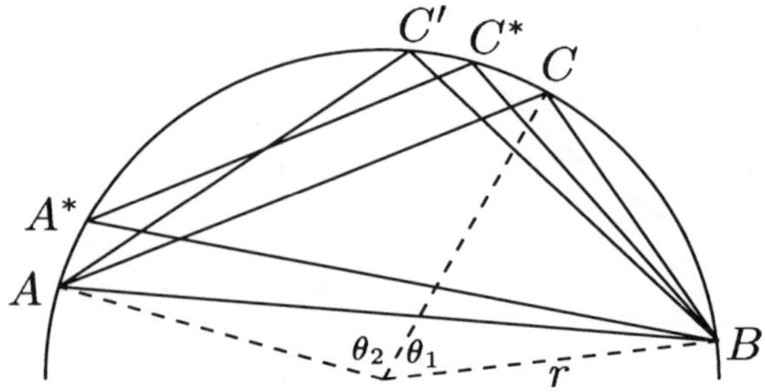

Figure 4.20

are isosceles and σ, σ^* have the same area, i.e. $\triangle = \triangle^*$.

$$\begin{aligned}\cos a + \cos b &= \sin^2 r(\cos\theta_1 + \cos\theta_2) + 2\cos^2 r\\ &= 2\left(\sin^2 r\cos\frac{\theta_1+\theta_2}{2}\cos\frac{\theta_1-\theta_2}{2} + \cos^2 r\right)\\ &\le 2\left(\sin^2 r\cos\frac{\theta_1+\theta_2}{2} + \cos^2 r\right) = 2\cos\widehat{AC'}\,. \qquad (12)\end{aligned}$$

Hence,

$$u(\sigma) \le u(\sigma') \le u(\sigma^*) \qquad (13)$$

and the equality holds when and only when $\sigma \cong \sigma^*$, namely, σ is itself isosceles. Therefore,

$$f(\sigma) = \frac{1}{6}\sin\triangle\left\{\frac{8}{u} - \tan^2 r\right\} \ge \frac{1}{6}\sin\triangle\left\{\frac{8}{u^*} - \tan^2 r\right\} = f(\sigma^*)\,, \qquad (14)$$

where σ^* is an isosceles triangle with its equal side-length $a \ge \frac{\pi}{3}$. Hence, it follows from the upper bound on circumradius that

$$f(\sigma) \ge f(\sigma^*) \ge f(\bar{\sigma}_\triangle)\,. \qquad (15)$$

□

Geometry of truncation Suppose that σ is a face of $\mathcal{S}(\Sigma)$ with $r(\sigma)$ exceeding $\frac{\pi}{4}$. Let $\{S_i, 1 \leq i \leq 3 \text{ (or 4)}\}$ be those touching neighbors in $\mathcal{L}(S)$ with the vertices of σ as their touching points, say $\{A_i\}$. Set V_σ to be the common intersection of the tangent planes $\{T_{A_i}\}$. Then, $|\overrightarrow{OV_\sigma}| = \sec r(\sigma)$ while the distance between the origin and the plane, $\Pi(\sigma)$, spanned by the centers of $\{S_i\}$ is equal to $2\cos r(\sigma)$. Set O' to be the *symmetric* point of O with respect to $\Pi(\sigma)$ and S' to be the unit sphere centered at O'. Then S' touches $\{S_i\}$ from the other side and $\Pi(\sigma)$ is exactly the perpendicular bisector of $\overline{OO'}$. Moreover,

$$\overrightarrow{OO'} = \frac{4}{|\overrightarrow{OV_\sigma}|^2}\overrightarrow{OV_\sigma}\,, \quad |\overrightarrow{OO'}| = 4\cos r(\sigma) \tag{16}$$

and hence, such a sphere S' is a neighbor with its *buckling height* $h = (2\cos r(\sigma) - 1)$, while the height of the corresponding truncated tip of $T(\Sigma)$ is given by $(\sec r(\sigma) - 2\cos r(\sigma))$.

Remark Using the technique of *area estimates* that we shall discuss in Chap. 5, it is not difficult to show that the circumradii of faces of a type I configuration are at most equal to $\cos^{-1} 0.58$ which corresponds to the lower bound of buckling height of 0.16 for such an additional neighbor on the top of such a big-hole.

In the general situation that those spheres in the directions of the vertices of a given face σ are not all touching neighbors, one has the following basic geometry on the truncation associated to such a big-hole.

Sublemma 4.4.3 *Let the position vectors of those corner spheres of a given big-hole be* $2(1+h_1)\mathbf{a}$, $2(1+h_2)\mathbf{b}$ *and* $2(1+h_3)\mathbf{c}$ *and* S' *be the unique unit sphere which touches the above triple from the outside. Set*

$$\frac{1}{2}\overrightarrow{OO'} = \frac{1}{\mathbf{a}\cdot\mathbf{b}\times\mathbf{c}}\{k_1\mathbf{b}\times\mathbf{c} + k_2\mathbf{c}\times\mathbf{a} + k_3\mathbf{a}\times\mathbf{b}\}\,. \tag{17}$$

Then, $\{k_1, k_2, k_3\}$ *is the unique solution of the following system, with* k_i *close to* $2\cos^2 r(\sigma)$

$$\begin{cases} 2(1+h_1)k_1 - 2(1+h_2)k_2 = (h_1-h_2)(2+h_1+h_2)\,, \\ 2(1+h_1)k_1 - 2(1+h_3)k_3 = (h_1-h_3)(2+h_1+h_3)\,, \\ Q(k_1,k_2,k_3) = 2(1+h_1)k_1 - 2h_1 - h_1^2\,, \end{cases} \tag{18}$$

where

$$Q(k_1,k_2,k_3) = \frac{1}{D^2}|k_1\mathbf{b}\times\mathbf{c} + k_2\mathbf{c}\times\mathbf{a} + k_3\mathbf{a}\times\mathbf{b}|^2\,. \tag{18'}$$

Proof Set $\mathbf{w}$ to be the vector given by (17). Then the geometric condition that $|\overrightarrow{O'O_i}| = 2$, $1 \le i \le 3$, simply means that

$$\begin{cases} |\mathbf{w} - (1+h_1)\mathbf{a}|^2 = |\mathbf{w}|^2 - 2(1+h_1)k_1 + (1+h_1)^2 = 1\,, \\ |\mathbf{w} - (1+h_2)\mathbf{b}|^2 = |\mathbf{w}|^2 - 2(1+h_2)k_2 + (1+h_2)^2 = 1\,, \\ |\mathbf{w} - (1+h_3)\mathbf{c}|^2 = |\mathbf{w}|^2 - 2(1+h_3)k_3 + (1+h_3)^2 = 1\,. \end{cases} \tag{19}$$

It is easy to see that (18) and (19) are algebraically equivalent to each other. Moreover, it is straightforward to use the pair of linear equations in (18) to solve k_2 and k_3 in terms of k_1, and then to substitute such linear expressions into the third equation to obtain a quadratic equation of k_1 only. It has two real roots with exactly one of them close to $2\cos^2 r(\sigma)$. □

Remarks (i) In the special case that $h_1 = h_2 = h_3 = \bar{h}$, the first two equations imply $k_1 = k_2 = k_3$, while the third equation implies that

$$k_1 = k_2 = k_3 = \cos^2 r\{(1+\bar{h}) + (1 - (2\bar{h} + \bar{h}^2)\tan^2 r(\sigma))^{\frac{1}{2}}\}\,. \tag{20}$$

(ii) In the general case, set $\bar{h} = \frac{1}{3}(h_1 + h_2 + h_3)$. Then the *averaged value* of $\{k_i\}$ is approximated by the right-hand side of (20) up to the first order of $\bar{h}$, which is, itself, approximated by

$\cos^2 r\{2-(\tan^2 r-1)\bar{h}\}$. Actually, one may use the following *approximate* linear equation, namely

$$\frac{1}{3}(k_1+k_2+k_3) \equiv \cos^2 r\left\{2-(\tan^2 r-1)\frac{1}{3}(h_1+h_2+h_3)\right\} \tag{21}$$

together with the pair of linear equations of (18) to solve for the *first order approximation* of $\{k_1, k_2, k_3\}$, which is already highly accurate in the case of small $\{h_i\}$, or rather, small $\{h_i - \bar{h}\}$.

Lemma 4.4.1 *If the core packing of a given local packing* $\mathcal{L}(S)$ *consists of twelve touching neighbors, then* $\operatorname{vol} C(\mathcal{L})$ *is at least equal to* $L_{12} = 5.550291\cdots$.

Proof Let Σ be the set of twelve touching points of S and its touching neighbors and $\mathcal{S}(\Sigma)$ be its associated configuration. If all the circumradii $\{r(\sigma), \sigma \in \mathcal{S}(\Sigma)\}$ are at most equal to $\frac{\pi}{4}$, then $\mathcal{L}(S) = \widehat{\mathcal{L}}(S)$, $C(\mathcal{L}) = T(\Sigma)$ and all σ containing their own circumcenter. In such a special case, the lower bound of L_{12} has already been proved in Example 4.2.1 as a direct application of Lemmas 4.2.2 and 4.2.3. □

In general, $\mathcal{L}(S) \neq \widehat{\mathcal{L}}(S)$ and $C(\mathcal{L})$ is a proper subset of $T(\Sigma)$ with some of its tips truncated by the perpendicular bisectors of those intervals, $\overrightarrow{OO'}_j$, linking the center of S and the centers of those *non-touching neighbors* $\{S'_j\}$. Set B_j to be the intersection point of $\overline{OO'}_j$ and S, and h'_j to be the *buckling height* of S'_j, namely

$$\overrightarrow{OO'_j} = 2(1+h'_j)\overrightarrow{OB_j}\,. \tag{22}$$

We shall assume that $h'_1 \leq h'_2 \leq \cdots$ etc. Notice that each h'_j is at least equal to 0.16 and by Lemma 4.3.1′, the volume of the h'_j-slab is at least equal to $\widetilde{L}(h'_j)$. Set $\Sigma_1 = \Sigma \cup \{B_1\}$, $\Sigma_2 = \Sigma_1 \cup \{B_2\}$, etc. If $\widetilde{L}(h'_1)$ (resp. $\widetilde{L}(h'_1) + \widetilde{L}(h'_2)$, etc.) are at least equal to $L_{12} - L_{13} = 0.13194$ (resp. $L_{12} - L_{14} = 0.25059$, etc.) then the proof of

Lemma 4.4.1 can be reduced to the proof of $\mathrm{vol}(C(\mathcal{L}) \cap T(\Sigma_1)) \geq L_{13}$ (resp. $\mathrm{vol}(C(\mathcal{L}) \cap T(\Sigma_2)) \geq L_{14}$, etc.) Let V_j be the vertex of the truncated tip corresponding to S'_j and σ_j be the face of $\mathcal{S}(\Sigma)$ *dual* to V_j. Geometrically speaking, σ_j is the big-hole that S'_j sits upon. It is natural to use $r(\sigma_j)$ as a measurement of the size of such a big-hole. For the purpose of lower bound volume estimation, one may assume without loss of generality that

$$h'_j = 2\cos r(\sigma_j) - 1\,. \tag{23}$$

For otherwise, the volume of the truncated tip is always smaller than that of the truncation corresponding to the fitting of S'_j touching all the corner spheres of σ_j from the outside. Thus, we shall assume (23) in the following discussion, and hence, $r(\sigma_1) \geq r(\sigma_2) \geq \cdots$, etc.

If $r(\sigma_2)$ is at least equal to $\cos^{-1} 0.67$, then

$$\widetilde{L}(h'_1) + \widetilde{L}(h'_2) \geq 2\widetilde{L}(0.34) > L_{12} - L_{14} \tag{24}$$

while $\widetilde{L}(\sqrt{2} - 1)$ already exceeds $L_k - L_{k+1}$, $k \geq 14$. Therefore, it follows readily from such a stepwise reduction that

$$\mathrm{vol}\, C(\mathcal{L}) \geq \widetilde{L}(h'_1) + \widetilde{L}(h'_2) + L_{14} \geq L_{12} \tag{25}$$

provided that $r(\sigma_2) \geq \cos^{-1} 0.67$.

Thus, what remains to prove are the following two subcases.

Subcase 1: S'_1 is the only non-touching neighbor.

Subcase 2: There are additional non-touching neighbors with $r(\sigma_j) < \cos^{-1} 0.67$, $j = 2, \ldots$, etc.

Proof of Subcase 1 If $r(\sigma_1) \geq \cos^{-1} 0.66$, then by Sublemma 4.4.2 and Lemmas 4.2.2, 4.2.3 the volume of $T(\Sigma_1)$ is at least L_{13}, and hence,

$$\mathrm{vol}\, C(\mathcal{L}) \geq \widetilde{L}(h'_j) + \mathrm{vol}\, T(\Sigma_1) \geq \tilde{L}(0.32) + L_{13} = L_{12} + 0.00109\,. \tag{26}$$

Thus, it suffices to consider the remaining one that $r(\sigma_1) < \cos^{-1} 0.66$. If σ_1 is a (cocircular) *quadrilateral*, then σ_1 splits into a quadruple of isosceles triangles $\{\sigma(r,\theta_i), 1 \leq i \leq 4\}$ in $\mathcal{S}(\Sigma_1)$. It is not difficult to use the restrictions $r < \cos^{-1} 0.66$ and $\theta_i \geq 2\sin^{-1}\frac{1}{2\sin r}$ to check that $f(\sigma) \geq f(\bar{\sigma}_\triangle)$ still holds for all $\sigma(r,\theta_i)$, $1 \leq i \leq 4$ (cf. Sublemma 4.4.1). Therefore, one has the following estimate, namely

$$\operatorname{vol} C(\mathcal{L}) \geq \widetilde{L}(h_1') + \operatorname{vol} T(\Sigma_1) \geq \widetilde{L}(h_1') + 4f(\bar{\sigma}_A) + 18 f(\bar{\sigma}_{\bar{\triangle}}), \tag{27}$$

where

$$\begin{aligned} A &= 2\tan^{-1}\frac{1-\cos r}{1+\cos r} = \frac{1}{4}\square(r) \geq \frac{1}{4}\sum_{i=1}^{4}\triangle(\sigma(r,\theta_i)), \\ \bar{\triangle} &= \frac{1}{18}(4\pi - 4A). \end{aligned} \tag{27'}$$

It is a simple accounting to check that the lower bound estimate of (27) already exceeds L_{12} by a comfortable margin.

Next let us consider the only remaining situation that σ_1 is a triangle with $r(\sigma_1) < \cos^{-1} 0.66$. The local geometry surrounding such a big-hole is represented by the quadruple of triangular faces consisting of σ_1 and its adjacent triple $\{\sigma', \sigma'', \sigma'''\}$ as indicated in Figure 4.21.

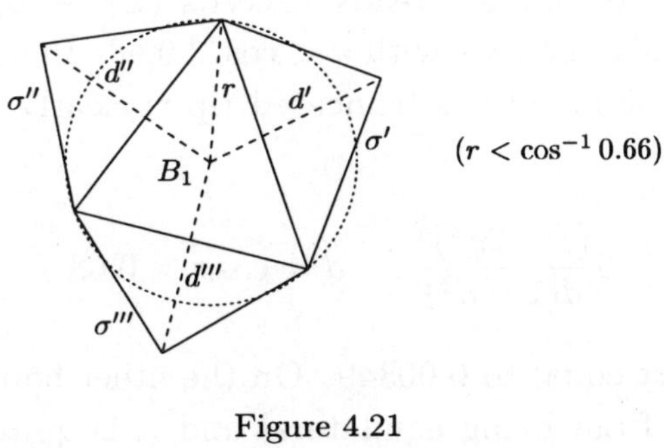

Figure 4.21

Let d', d'' and d''' be the distances between the circumcenter of σ_1 and the other vertices of σ', σ'' and σ''', respectively. Roughly speaking, the big-hole is said to be of *buckled quadrilateral* (resp. *lopsided triangular*, or *balanced triangular*) *type* if the shortest distance, say d', is only slightly longer than r (resp. considerably longer but still close to r, or much longer than r). For the sake of definitiveness of the following proof, we introduce here the following quantitative division for the above three types big-holes, namely

(i) buckled quadrilateral type: $\tan\frac{d'}{2} < \frac{5}{4}\tan\frac{r}{2}$,
(ii) lopsided triangular type: $\frac{5}{4}\tan\frac{r}{2} \le \tan\frac{d'}{2} \le \frac{3}{2}\tan\frac{r}{2}$,
(iii) balanced triangular type: $\tan\frac{d'}{2} > \frac{3}{2}\tan\frac{r}{2}$.

If the big-hole is of *buckled quadrilateral type*, it is quite simple to use Lemma 2.1.1′ to show that the sum of $\triangle(\sigma_1)$ and $\triangle(\sigma')$ can, at the most, exceed $\square(r) = 8\tan^{-1}\frac{1-\cos r}{1+\cos r}$ by a small factor of $\square(r)$. Hence, the same kind of estimate shows that

$$\operatorname{vol} C(\mathcal{L}) \ge \widetilde{L}(h_1') + 4f(\bar{\sigma}_A) + 18f(\bar{\sigma}_{\bar{\triangle}}) > L_{12}\,, \tag{28}$$

where $A = \frac{1}{4}(\triangle(\sigma_1) + \triangle(\sigma'))$ can be, at most, slightly larger than $\frac{1}{4}\square(r)$.

If the big-hole is of *lopsided triangular type*, then the volume of the h_1'-slab is at least equal to (1.12)-times of $\widetilde{L}(h_1')$, while $(1.12)\widetilde{L}(0.36) = 0.132118$ already exceeds $(L_{12} - L_{13})$. Therefore, it suffices to study the case with $r < \cos^{-1} 0.68$. In such a geometric situation, the volume of the truncated tip is clearly bounded above by

$$3\frac{(1-2d^2)^3}{d(1-d^2)}\,, \quad d = \cos r > 0.68\,, \tag{29}$$

which is at most equal to 0.00349. On the other hand, the shape of σ' must be far from being equilateral and it is quite simple to use

Sublemma 4.4.1 to show that

$$f(\sigma') - f(\bar{\sigma}_{\triangle'}) > 0.00349\,. \tag{30}$$

Hence,

$$\begin{aligned}\operatorname{vol} C(\mathcal{L}) &= \operatorname{vol} T(\Sigma) - \operatorname{vol}(\text{the truncated tip})\\ &\geq \sum_{i=1}^{20} f(\sigma_i) - 0.00349 \geq L_{12}\,.\end{aligned} \tag{31}$$

If the big-hole is of *balanced triangular type*, then the volume of h_1'-slab is at least (1.21)-times of $\widetilde{L}(h_1')$, while $(1.21)\widetilde{L}(0.38) = 0.13247$ already exceeds $(L_{12} - L_{13})$. Therefore, it suffices to study this type of situation with $r < \cos^{-1} 0.69$. In such a geometric situation, the volume of the truncated tip is bounded above by

$$\frac{5(1-2d^2)^3}{2d(1-d^2)}\,, \qquad d = \cos r > 0.69\,, \tag{32}$$

which is at most equal to 0.000756. On the other hand, the area of such a non-lopsided σ_1 is clearly bounded below by $(0.57)\square(r)$. Hence, one has the following simple estimate, namely

$$\operatorname{vol} C(\mathcal{L}) \geq f(\bar{\sigma}_A) + 19 f(\bar{\sigma}_{\bar{\triangle}}) - 0.000756 > L_{12}\,, \tag{33}$$

where $A \geq (0.57)\square(r)$, $\bar{\triangle} = \frac{1}{19}(4\pi - A)$, $r < \cos^{-1} 0.69$.

This completes the proof of Lemma 4.4.1 for Subcase 1. $\square$

Proof of Subcase 2 The proof of Subcase 2 is essentially the same as the above proof of Subcase 1. The same way of volume estimation applying to both of the big-holes that S_1' and S_2' sit upon works just as well while the accounting, in fact, becomes more favorable than that of a single big-hole. For example, in the geometric situation that both of them are quadrilateral-big-holes, one has the lower bound estimate of

$$\widetilde{L}(h_1') + \widetilde{L}(h_2') + 4f(\bar{\sigma}_{A_1}) + 4f(\bar{\sigma}_{A_2}) + 16f(\bar{\sigma}_{\bar{\triangle}}), \tag{34}$$

where $A_1 = \frac{1}{4}(\triangle(\sigma_1) + \triangle(\sigma_1'))$, $A_2 = \frac{1}{4}(\triangle(\sigma_2) + \triangle(\sigma_2'))$ and $\bar{\triangle} = \frac{1}{16}(4\pi - 4A_1 - 4A_2)$. It is not difficult to check that the estimate of (34) exceeds L_{12} with a bigger margin than that of (28). We shall leave such a checking by straightforward accounting to the reader. □

Chapter 5

Estimates of Total Buckling Height

Among the basic geometric features of single-layer local packings (or rather, reduced local packings), the most important are the following, namely

(i) the *overcrowdedness* of reduced local packings with thirteen or more reduced (or close) neighbors,
(ii) the *non-tightness* of reduced local packings with twelve or fewer reduced (or close) neighbors.

As has already been pointed out in the discussion in Chap. 3, some kind of refined, quantitative understanding of the above two basic geometric features will always be necessary in order to achieve the *optimal estimates* of both the *local density* and the *locally averaged density* of sphere packings. We shall study the quantitative analysis on the *geometry of overcrowdedness* in this chapter and that of *non-tightness* in Chap. 7.

The problem of thirteen spheres, discussed by D. Gregory and I. Newton three centuries ago, is a qualitative one which asks whether it is possible to surround a sphere by thirteen *touching* neighbors. The *negative answer* to the above problem, first proved in 1953 by

Schütte and van der Waerden [SW-2], provides a qualitative understanding of the overcrowdedness of thirteen or more neighbors. However, such a qualitative result, just by itself, will be of little use in the study of densities of sphere packings. For example, in the volume estimation of the peripheral part of a local cell, the total sum of the lower bound estimates of individual slabs will, of course, depend on the collection of buckling height. In case that all $\{h_j\}$ are at most 0.09, the total sum $\sum_j L(h_j)$ is to a large extent determined by the *total buckling height* $\sum_j h_j$. Therefore, the *total buckling height* of a reduced (resp. core) local packing with thirteen or more neighbors is a kind of quantitative measurement of the overcrowdedness of such local packings, that will be useful for the study of densities. Hence, some "good enough" *lower bound estimates* on the total buckling height of various types of *overcrowded* local packing will provide the needed crucial geometric understanding, for the purpose of achieving the optimal estimates of both the local density and the locally averaged density of sphere packing.

In this chapter, we shall give the proofs of the following set of lower bound estimates of [Hsi-2] on the total buckling height of various types of overcrowded local packings, which are already *good enough* for the purpose of achieving the optimal estimates of the two kinds of local densities.

Lemma 5.3.1 *Let $\hat{\mathcal{L}}(S_0)$ be a core packing with thirteen close neighbors. Then its total buckling height is at least* 0.325.

Lemma 5.3.2. *Let $\hat{\mathcal{L}}(S_0)$ be a core packing with fourteen close neighbors. Then, its total buckling height is at least* 0.56.

Lemma 5.3.3. *A core packing has at most fifteen close neighbors and the total buckling height of a core packing with fifteen close neighbors is at least* 1.

The proofs given here are essentially a simplification of that of [Hsi-2] and [Hsi-3], where the analysis of the functional relationships between the area estimates and the buckling height has been systematically replaced by simple geometric estimates (cf. Sec. 5.1), instead of the more algebraic method of Taylor's expansion and quadratic form analysis. The basic idea of the proofs of the above estimates on total buckling height is roughly as follows.

The geometry of a given core packing $\hat{\mathcal{L}}(S_0)$ can be concisely represented by its associated spherical configuration $\mathcal{S}(\Sigma)$ together with the set of buckling height $\{h_j\}$ assigned to its vertices. Let $\widehat{A_iA_j}$ be a given edge of $\mathcal{S}(\Sigma)$. Then, its length is bounded below by

$$\ell(h_i, h_j) = \cos^{-1}\left\{\frac{(1+h_i)^2 + (1+h_j)^2 - 1}{2(1+h_i)(1+h_j)}\right\}. \tag{1}$$

Thus, the associated spherical configuration $\mathcal{S}(\Sigma)$ is the kind of spherical configuration whose edge-lengths are constrained by the above *individualized* lower bound of (1) in terms of the assigned buckling height. The geometric task of proving the above lemmas amounts to proving the *non-existence* of such a spherical configuration with thirteen (resp. fourteen, fifteen) vertices together with the edge-length bounds of (1) where $0 \leq h_j \leq 0.09$ and Σh_j is less than 0.325 (resp. 0.56, 1).

If one compares the geometric complexity that one encounters in such a proof to that of the non-existence of thirteen touching neighbors (cf. Subsec. 2.2.3), the following two factors clearly makes the former much more complicated than the latter, namely

(i) the individualized lower bounds on the edge-lengths depend on the *distribution* of buckling height instead of the uniform lower bound of $\pi/3$ for the latter,

(ii) furthermore, the distribution of $\{h_j\}$ are free and arbitrary except the upper limit of at most 0.09 and that of the total amount.

Therefore, we need to find a good way of handling such a geometric complexity in order to prove the non-existence of such spherical configurations. Intuitively speaking, the necessity of having a certain amount of total buckling height of a given type of *overcrowded* local packing is a kind of *quantitative manifestation of the lack of room* for accommodating that many touching neighbors while the total area of 4π is the basic measurement of the available room for local packing. This geometric intuition naturally suggests the following method of area estimates, namely, to prove the *non-existence* of such spherical configurations by proving that the total area of such a *hypothetical spherical configuration* necessarily exceeds 4π ! This type of proof, of course, relies heavily on the geometric techniques of area estimates together with some kind of geometric correlation between buckling heights and area estimates. However, this method, at least, has the advantage of having a *total accounting of area estimates at the end*, thus providing a natural way of circumventing the above two types of geometric complexities.

5.1 The Correlation Between Buckling Heights and Area Estimates

Let σ be a triangular (resp. quadrilateral) face of $\mathcal{S}(\Sigma)$ with $\{h_i\}$, $1 \leq i \leq 3$ (resp. 4) as the assigned buckling heights at its vertices in cyclic order. Then, its edge-lengths (in cyclic order) are bounded below by $\ell(h_i, h_{i+1})$ (cf. (1)), $1 \leq i \leq 3$ (resp. 4). Suppose that its edge-lengths are, indeed, equal to their respective lower bounds, i.e. $\ell(h_i, h_{i+1})$. Then, its area is a function of $\{h_i\}$ that we shall denote by $\triangle(h_1, h_2, h_3)$ (resp. $\square(h_1, h_2, h_3, h_4)$).

Sublemma 5.1.1 *Let $\ell(h, h_1)$ be the function given by* (1), $0 \leq h, h_1 \leq 0.3$. *Then, $\frac{\partial}{\partial h_1}\ell(h, h_1)$ is monotonically decreasing with respect to h_1.*

Proof Let $\triangle\, OAB$ be the triangle with side-lengths $\{1+h_1, 1, 1+h\}$ and $\{\ell, \psi, \theta\}$ be its three angles. Then, the same computation as that of Sublemma 4.3.1 will show that

$$\frac{\partial \ell}{\partial h_1} = -\frac{1}{1+h_1}\cot\psi, \qquad \left.\frac{\partial \ell}{\partial h_1}\right|_{h=0} = -\frac{1}{2\sin\psi}\,. \tag{2}$$

Therefore, $\left.\frac{\partial \ell}{\partial h_1}\right|_{h=0}$ is a decreasing function of h_1. Next let us consider the general case of $h > 0$. As indicated in Figure 5.1, the perpendicular line of $\overline{OA}$ intersects AB at C. Set $\overline{OC} = L$, $\angle ABA' = d\theta$, $\angle AOA' = -d\ell$, $\angle COC' = d\alpha$. Then,

$$\cot\psi = \frac{1+h_1}{L}, \quad \frac{\partial \ell}{\partial h_1} = -\frac{1}{1+h_1}\cot\psi = -\frac{1}{L},$$

$$\cos\psi = -\frac{(1+h_1)d\ell}{d\theta}, \quad \sin\psi = \cos\left(\frac{\pi}{2}-\psi\right) = \frac{Ld\alpha}{\overline{BC}d\theta}, \tag{3}$$

$$L = (1+h_1)\tan\psi\,, \quad \overline{BC} = (1+h_1)\sec\psi - 1\,.$$

Therefore,

$$\frac{d\alpha}{d\theta} + \frac{d\ell}{d\theta} = \frac{1+h_1-2\cos\psi}{1+h_1} > 0\,, \tag{4}$$

which implies that $\angle A'OC'$ becomes larger than $\pi/2$. Hence, L is a decreasing function of h_1 and hence, $\frac{\partial \ell}{\partial h_1} = -\frac{1}{L}$ is also a decreasing function of h_1. □

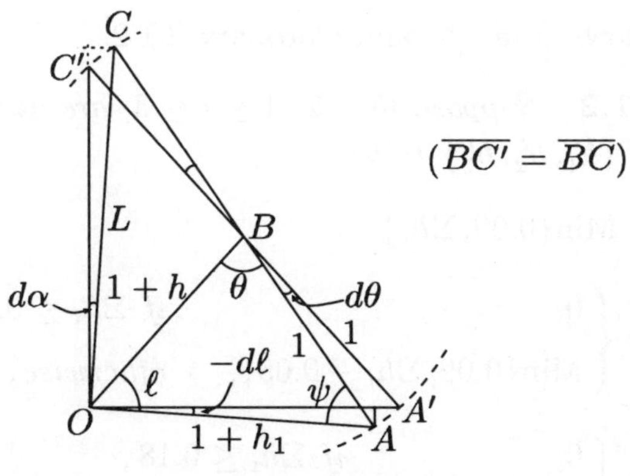

Figure 5.1

Corollary 5.1.1 *If* $h_1 + h_2 = h_1' + h_2'$ *and* $|h_1 - h_2| < |h_1' - h_2'|$, *then*

$$\ell(h, h_1) + \ell(h, h_2) > \ell(h, h_1') + \ell(h, h_2'). \tag{5}$$

Proof For fixed $h \geq 0$, the function $\ell(h, x)$, $0 \leq x \leq 0.3$, is a concave function of x. □

Sublemma 5.1.2 *Suppose that* $h_2 + h_3 = h_2' + h_3' \leq 0.3$ *and* $|h_2 - h_3| < |h_2' - h_3'|$. *Then,*

$$\triangle(h_1, h_2, h_3) > \triangle(h_1, h_2', h_3'). \tag{6}$$

Proof Set $h_1' = h_1$ and $\widehat{\triangle}ABC$ (resp. $\widehat{\triangle}A'B'C'$) to be the spherical triangle with $\{\ell(h_i, h_{i+1})\}$ (resp. $\{\ell(h_i', h_{i+1}')\}$), $1 \leq i \leq 3$, as the side-lengths. Then, by the above corollary,

$$\ell(h_1, h_2) + \ell(h_1, h_3) > \ell(h_1', h_2') + \ell(h_1', h_3'). \tag{7}$$

and it follows easily from the assumption that

$$\ell(h_2, h_3) > \ell(h_2', h_3'),$$

$$|\ell(h_1, h_2) - \ell(h_1, h_3)| < |\ell(h_1', h_2') - \ell(h_1', h_3')|. \tag{8}$$

Hence, (6) follows from (8) and Corollary 2.1.2. □

Corollary 5.1.2 *Suppose that* $h_i, 1 \leq i \leq 3$, *are at most* 0.09. *Set* $\bar{h} = \frac{1}{3}\Sigma h_i$ *and* $\{h_1', h_2', h_3'\}$ *to be*

$$\begin{aligned} h_1' &= \mathrm{Min}\{0.09, \Sigma h_i\}, \\ h_2' &= \begin{cases} 0, & \textit{if } \Sigma h_i \leq 0.09, \\ \mathrm{Min}\{0.09, \Sigma h_i - 0.09\}, & \textit{otherwise}, \end{cases} \\ h_3' &= \begin{cases} 0, & \textit{if } \Sigma h_i \leq 0.18, \\ \Sigma h_i - 0.18, & \textit{otherwise}. \end{cases} \end{aligned} \tag{9}$$

Then one has the following bounds on $\triangle(h_1, h_2, h_3)$, namely

$$\triangle(\bar{h}, \bar{h}, \bar{h}) \geq \triangle(h_1, h_2, h_3) \geq \triangle(h'_1, h'_2, h'_3)\,. \tag{10}$$

Proof It follows directly from Sublemma 5.1.2 and the above definition of $\{h'_1, h'_2, h'_3\}$ by (9). □

Remark Similar bounds of (10)-type also hold if one replaces 0.09 by the maximum of $\{h_1, h_2, h_3\}$.

Sublemma 5.1.3 *Let $\widehat{\triangle} A_1A_2A_3$ be a spherical triangle with $\ell_1 \leq \ell_2 \leq \ell_3$ as the* individual lower bounds *on its side-lengths. Set $c_i = \cos \ell_i$, and $\{\alpha_i\}, \triangle(\ell_1, \ell_2, \ell_3)$ to be the angles and the area of the spherical triangle with $\{\ell_i\}$ as its side-lengths. If the area of $\widehat{\triangle} A_1A_2A_3$ is less than $\triangle(\ell_1, \ell_2, \ell_3)$, then the largest angle (resp. longest side) of $\widehat{\triangle} A_1A_2A_3$ exceeds*

$$\alpha_1 + \alpha_2 = \cos^{-1}\left\{(1-c_1^2)^{-\frac{1}{2}}(1-c_2^2)^{-\frac{1}{2}}\left[\frac{(c_1+c_2)^2}{1+c_3} - 1 - c_1c_2\right]\right\},$$

$$\left(\textit{resp.}\ \cos^{-1}\left\{\frac{(c_1+c_2)^2}{1+c_3} - 1\right\}\right) \tag{11}$$

and moreover, its circumradius exceeds

$$\tan^{-1}\left[\frac{2}{D^2}(1-c_1)(1-c_2)\left(2 - \frac{(c_1+c_2)^2}{1+c_3}\right)\right]^{1/2}. \tag{12}$$

Proof The spherical "parallelogram" as indicated in Figure 5.2 consists of a pair of triangles with side-lengths $\{\ell_i\}$ and with ℓ_3 as one of its diagonal lengths.

Set d to be the length of the other diagonal. Then,

$$\begin{aligned} (1+\cos d)(1+c_3) &= (c_1+c_2)^2\,, \\ (1-c_1^2)^{\frac{1}{2}}(1-c_2^2)^{\frac{1}{2}}\cos(\alpha_1+\alpha_2) &= \cos d - c_1c_2 \end{aligned} \tag{13}$$

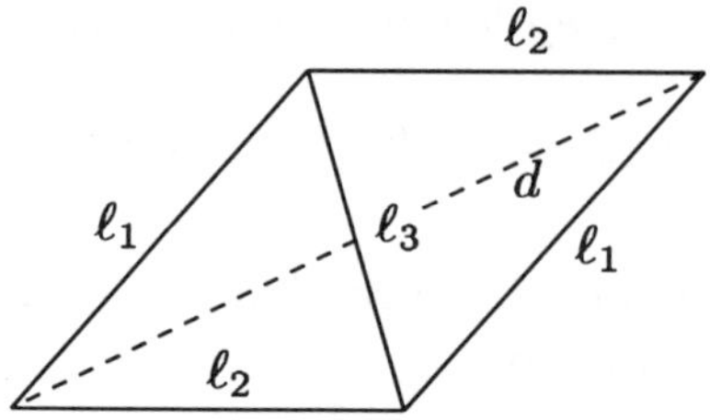

Figure 5.2

while the circumradius r of the spherical triangle with ℓ_1, ℓ_2 and d as the side-lengths is given by

$$r = \tan^{-1}\left[\frac{2}{D^2}(1-c_1)(1-c_2)(1-\cos d)\right]^{1/2}, \tag{14}$$

where D is the determinant associated to the above triangle. On the other hand, the area of $\widehat{\triangle} A_1A_2A_3$ can be smaller than $\triangle(\ell_1, \ell_2, \ell_3)$ only when the largest angle (resp. longest side and circumradius) all exceed that of the above triangle (cf. Corollary 2.1.1). □

Sublemma 5.1.4 *Let $\square A_1A_2A_3A_4$ be a cocircular quadrilateral with $\{\ell_i,\ 1 \le i \le 4\}$ as the set of* individual lower bounds *on its side-lengths in cyclic order, $\pi/3 \ge \ell_i \ge 2\sin^{-1}\frac{1}{2.6}$. If its circumradius is less than $\pi/3$, then its area is at least equal to the area of the cocircular quadrilateral with the above lower bounds as its edge-lengths.*

Proof Without loss of generality, let us assume that $\ell_1 \le \ell_2 \le \ell_3 \le \ell_4$. Set $\square(r; \ell_1, \ell_2, \ell_3)$ to be the area of cocircular quadrilateral with r as its circumradius and with $\{\ell_i,\ 1 \le i \le 3\}$, as three of its edge-lengths. Moreover, set r_0 to be the circumradius of the cocircular quadrilateral with $\{\ell_i, 1 \le i \le 4\}$ as its edge-lengths. Then, the same proof as that of Sublemma 2.1.3 shows that

$$\frac{\partial \square}{\partial r} > 0,\ r_0 \le r < \pi/3\,. \tag{15}$$

Hence, the area of $\square\, A_1A_2A_3A_4$ is bounded below by the area of the cocircular quadrilateral with $\{\ell_i, 1 \le i \le 4\}$ as its edge-lengths, i.e. $\square\,(r_0; \ell_1, \ell_2, \ell_3)$. $\square$

Notation We shall denote the area of the cocircular quadrilateral with $\{\ell_i, 1 \le i \le 4\}$ as its edge-lengths by $A(\ell_1, \ell_2, \ell_3, \ell_4)$. Moreover, in the special case that $\ell_i = \ell(h_i, h_{i+1})$, we shall denote it simply by $\square\,(h_1, h_2, h_3, h_4)$. Notice that A is a symmetric function of the $\{\ell_i\}$ while $\square$ is, in general, *not* a symmetric function of the $\{h_i\}$. In fact, $\square$ is only cyclically symmetric in $\{h_i\}$.

Sublemma 5.1.5 (i) *Suppose that* $h_2 + h_4 = h'_2 + h'_4 \le 0.3$ *and* $|h_2 - h_4| < |h'_2 - h'_4|$. *Then,*

$$\square\,(h_1, h_2, h_3, h_4) > \square\,(h_1, h'_2, h_3, h'_4)\,. \tag{16}$$

(ii) *Suppose that* $h_3 + h_4 = h'_3 + h'_4 \le 0.3$ *and* $|h_3 - h_4| < |h'_3 - h'_4|$. *Then,*

$$\square\,(h, h, h_3, h_4) > \square\,(h, h, h'_3, h'_4)\,. \tag{17}$$

Proof Let us first prove (i). Set $h'_1 = h_1$, $h'_3 = h_3$ and $\square\, ABCD$ (resp. $\square\, A'B'C'D'$) to be the *cocircular* quadrilateral with $\{\ell(h_i, h_{i+1})\}$ (resp. $\{\ell(h'_i, h'_{i+1})\}$), $1 \le i \le 4$, as the edge-lengths in cyclic order. Then, it follows from the assumption of (i) and the corollary of Sublemma 5.1.1 that

$$\begin{aligned} \ell(h_1, h_2) + \ell(h_1, h_4) &> \ell(h'_1, h'_2) + \ell(h'_1, h'_4)\,, \\ \ell(h_3, h_2) + \ell(h_3, h_4) &> \ell(h'_3, h'_2) + \ell(h'_3, h'_4)\,, \\ |\ell(h_1, h_2) - \ell(h_1, h_4)| &< |\ell(h'_1, h'_2) - \ell(h'_1, h'_4)|\,, \\ |\ell(h_3, h_2) - \ell(h_3, h_4)| &< |\ell(h'_3, h'_2) - \ell(h'_3, h'_4)|\,. \end{aligned} \tag{18}$$

Applying shearing deformation to $\square\, ABCD$ one obtains another quadrilateral $\square\, A^*B^*C^*D^*$ with $\widehat{B^*D^*} = \widehat{B'D'}$. Then, on the one hand the area of $\square\, A^*B^*C^*D^*$ is less than that of $\square\, ABCD$; and on the other hand, it follows from (18) and Corollary 2.1.2 that

$$\triangle A^*B^*D^* > \triangle A'B'D', \quad \triangle B^*C^*D^* > \triangle B'C'D'. \tag{19}$$

Hence,

$$\square(h_1, h_2, h_3, h_4) > \square(h_1', h_2', h_3', h_4'). \tag{20}$$

Next let us prove (ii). Again, it follows from the assumption of (ii) and the Corollary 5.1.1 that

$$\begin{aligned} \ell(h_3, h_4) &> \ell(h_3', h_4'), \\ \ell(h, h_3) + \ell(h, h_4) &> \ell(h, h_3') + \ell(h, h_4'), \\ |\ell(h, h_3) - \ell(h, h_4)| &< |\ell(h, h_3') - \ell(h, h_4')|. \end{aligned} \tag{21}$$

Therefore, by (21) and Corollary 2.1.2

$$\begin{aligned} \square(h, h, h_3, h_4) &> A(\ell(h, h), \ell(h, h_3), \ell(h_3', h_4'), \ell(h_4, h)) \\ &= A(\ell(h, h), \ell(h_3', h_4'), \ell(h, h_3), \ell(h, h_4)) \\ &> A(\ell(h, h), \ell(h_3', h_4'), \ell(h, h_3'), \ell(h, h_4')) \\ &= \square(h, h, h_3', h_4'). \end{aligned} \tag{22}$$

□

Corollary 5.1.3 *Suppose that all* h_i, $1 \le i \le 4$, *are at most* 0.09. *Set* $\bar{h} = \frac{1}{4}\sum_{i=1}^{4} h_i$, $h = \frac{1}{2}(h_1 + h_3)$ *and* $h' = \frac{1}{2}(h_2 + h_4)$. *Then,*

$$\begin{aligned} \square(\bar{h}, \bar{h}, \bar{h}, \bar{h}) &\ge \square(h, h', h, h') \ge \square(h_1, h_2, h_3, h_4) \\ &\ge \square(h_1', h_2', h_3', h_4'), \end{aligned} \tag{23}$$

where

$$h_1' = \mathrm{Min}\{0.09, h_1 + h_3\}, \qquad h_2' = \mathrm{Min}\{0.09, h_2 + h_4\},$$

$$h_3' = \begin{cases} 0, & \text{if } h_1 + h_3 \le 0.09, \\ h_1 + h_3 - 0.09, & \text{otherwise}, \end{cases}$$

$$h_4' = \begin{cases} 0, & \textit{if } h_2 + h_4 \leq 0.09\,, \\ h_2 + h_4 - 0.09, & \textit{otherwise}\,. \end{cases} \tag{24}$$

Moreover, in case $h_1' = h_2' = 0.09$, *one has*

$$\square(h_1, h_2, h_3, h_4) \geq \square(0.09, 0.09, h_3', h_4') \geq \square(0.09, 0.09, h_3^*, h_4^*)\,, \tag{25}$$

where

$$h_3^* = \mathrm{Min}\{0.09, 4\bar{h} - 0.18\}\,,$$
$$h_4^* = \begin{cases} 0, & \textit{if } 4\bar{h} \leq 0.27\,, \\ 4\bar{h} - 0.27, & \textit{otherwise}\,. \end{cases} \tag{26}$$

Proof Note that $\square(h, h', h, h')$ is, in fact, the area of the cocircular quadrilateral of uniform side-length of $\ell(h, h')$. Hence, it follows from $\ell(h, h') \leq \ell(\bar{h}, \bar{h})$ that $\square(\bar{h}, \bar{h}, \bar{h}, \bar{h}) \geq \square(h, h', h, h')$. The remaining inequalities of the above corollary are direct consequences of (16), (17) and (24), (26). □

Example 5.1.1

$$\triangle(3\bar{h}, 0, 0) = 2\cos^{-1}\left\{\frac{5 + 6\bar{h}}{\sqrt{3}(3 + 3\bar{h})}\right\}.$$

Proof $\triangle(3\bar{h}, 0, 0)$ is equal to twice the area of the right-angle triangle with

$$\cos c = \frac{1 + 3\bar{h}}{2}, \quad \cos b = \frac{\sqrt{3}}{2}, \quad \cos a = \frac{1 + 3\bar{h}}{\sqrt{3}}\,. \tag{27}$$

Hence, by Lemma 2.1.1′

$$\triangle(3\bar{h}, 0, 0) = 2\cos^{-1}\left\{\frac{5 + 6\bar{h}}{\sqrt{3}(3 + 3\bar{h})}\right\}. \tag{28}$$

□

Example 5.1.2

$$\triangle(\bar{h},\bar{h},\bar{h}) = 3\cos^{-1}\left\{\frac{\cos\ell(\bar{h},\bar{h})}{1+\cos\ell(\bar{h},\bar{h})}\right\} - \pi$$
$$= 3\cos^{-1}\left\{\frac{2(1+\bar{h})^2-1}{4(1+\bar{h})^2-1}\right\} - \pi\,.$$

Example 5.1.3

$$A(a,b,a,b) = 4\tan^{-1}\sqrt{\frac{(1-\cos a)(1-\cos b)}{2(\cos a+\cos b)}}\,.$$

Proof Set d to be the equal length of the pair of diagonals of the cocircular quadrilateral with $\{a,b,a,b\}$ as side-lengths. Then, $1+\cos d = \cos a + \cos b$. Therefore,

$$u = \cos a + \cos b + \cos d + 1 = 2(\cos a + \cos b)\,,$$
$$D^2 = 2(1+\cos a)(1+\cos b)(1+\cos d) - u^2 \qquad (29)$$
$$= 2(\cos a+\cos b)(1-\cos a)(1-\cos b)\,.$$

Hence,

$$A(a,b,a,b) = 4\tan^{-1}\frac{D}{u} = 4\tan^{-1}\sqrt{\frac{(1-\cos a)(1-\cos b)}{2(\cos a+\cos b)}}\,. \qquad (30)$$

□

Example 5.1.4 Applying (30) above to compute $\square(4\bar{h},0,0,0)$ (resp. $\square(\lambda\bar{h},0,(4-\lambda)\bar{h},0)$, $\square(h,h',h,h')$, $\square(\bar{h},\bar{h},\bar{h},\bar{h})$), one has the following explicit formulas, namely

(i) $\square(4\bar{h},0,0,0) = 4\tan^{-1}\sqrt{\dfrac{1-4\bar{h}}{8(1+2\bar{h})}}$,

(ii) $\square(\lambda\bar{h},0,(4-\lambda)\bar{h},0) = 4\tan^{-1}\sqrt{\dfrac{(1-\lambda\bar{h})(1-(4-\lambda)\bar{h})}{8(1+2\bar{h})}}$,

(iii) $\square(h, h', h, h')$

$$= 4\tan^{-1}\left\{\frac{1}{2\sqrt{2}}\frac{1-(h-h')^2}{\sqrt{(1+h)(1+h')(1+2h+2h'+h^2+h'^2)}}\right\},$$

(iv) $\square(\bar{h}, \bar{h}, \bar{h}, \bar{h}) = 4\tan^{-1}\left\{\frac{1}{2\sqrt{2}}\frac{1}{(1+\bar{h})\sqrt{1+4\bar{h}+2\bar{h}^2}}\right\}.$

$$(31)$$

Lemma 5.1.1 *If all* $h_i, 1 \le i \le 3$, *are at most equal to* 0.09, *then*

$$\triangle(0,0,0) - \triangle(h_1,h_2,h_3) \le (0.4757122)\sum_{i=1}^{3} h_i\,,$$

$$\triangle(0,0,0) - \triangle(h_1,h_2,h_3) \le (1.0724)(\triangle(0,0,0) - \triangle(\bar{h},\bar{h},\bar{h}))\,, \tag{32}$$

where $\bar{h} = \frac{1}{3}\sum_{i=1}^{3} h_i$.

Proof By Sublemma 5.1.2 and Corollary 5.1.2, it is easy to check that the ratio between $(\triangle(0,0,0) - \triangle(h_1,h_2,h_3))$ and $\sum_{i=1}^{3} h_i$ (resp. $(\triangle(0,0,0) - \triangle(\bar{h},\bar{h},\bar{h}))$) reaches its maximum at (0.09, 0, 0). Thus, both estimates of (32) can easily be checked by straightforward computations. □

Remark Using the formula of $\triangle = 2\tan^{-1}\frac{D}{u}$, it is straightforward to compute the Taylor's expansion of $\triangle(h_1,h_2,h_3)$ up to the quadratic terms, namely

$$\triangle(h_1,h_2,h_3) = \triangle(0,0,0) - \frac{\sqrt{2}}{3}\sum_{i=1}^{3} h_i + \frac{7\sqrt{2}}{9}\sum_{i<j} h_i h_j - \frac{\sqrt{2}}{36}\left(\sum h_i\right)^2 + \cdots\,. \tag{33}$$

Note that the constant factor of the linear estimate of (32) is equal to $\frac{\sqrt{2}}{3} + 0.0043$, which is only slightly larger than that of (33). This is due to the small negative contribution of the quadratic term in the most lopsided distribution of $(h_1, 0, 0)$.

Next let us study the linear estimate of $\square(0,0,0,0) - \square(h_1, h_2, h_3, h_4)$, $0 \le h_i \le 0.09$, where $\square(0,0,0,0) = \square_{\pi/3} = 4\tan^{-1}\frac{1}{2\sqrt{2}}$.

Lemma 5.1.2 *If all h_i, $1 \le i \le 4$, are at most equal to* 0.09, *then*

$$4\tan^{-1}\frac{1}{2\sqrt{2}} - \square(h_1, h_2, h_3, h_4) \le \frac{2\sqrt{2}}{3}\sum_{i=1}^{4} h_i\,. \tag{34}$$

Proof Set $\sum_{i=1}^{4} h_i = 4\bar{h}$. If $4\bar{h} \le 0.3$, then it follows easily from Sublemma 5.1.5 that

$$\begin{aligned}\square(h_1, h_2, h_3, h_4) &\ge \square(4\bar{h}, 0, 0, 0)\\ &= 4\tan^{-1}\sqrt{\frac{1-4\bar{h}}{8(1+2\bar{h})}} \ge 4\tan^{-1}\frac{1}{2\sqrt{2}} - \frac{2\sqrt{2}}{3}\sum_{i=1}^{4} h_i\,.\end{aligned} \tag{35}$$

Hence, (34) holds for the case $\sum_{i=1}^{4} h_i \le 0.3$. In the remaining case of $4\bar{h} > 0.3$, one has

$$\square(h_1, h_2, h_3, h_4) \ge \square(0.09, 0.09, 0.09, 4\bar{h} - 0.27) \tag{36}$$

and it is easy to see that (34), in fact, holds for a constant factor considerably smaller than $\frac{2\sqrt{2}}{3}$. $\square$

Example 5.1.5 $\square(0.09, 0.09, 0, 0) = A(a, b, a, c)$, $\cos a = 0.545$, $\sin a = 0.838436$, $\cos b = \frac{1}{2}$, $\cos c = 0.57916$. Hence, from (92) and (93) of Sec. 2.1,

$$\square(0.09, 0.09, 0, 0) = 1.203113 = 4\tan^{-1}\frac{1}{2\sqrt{2}} - (0.86797)(0.18)\,. \tag{37}$$

Example 5.1.6 $\square(0.09, 0.09, 0.09, 0) = A(a, a, b, b) = A(a, b, a, b)$, $\cos a = 0.57916$, $\cos b = 0.545$. Hence, from (30),

$$\begin{aligned}\square(0.09, 0.09, 0.09, 0) &= 1.13579142\\ &= 4\tan^{-1}\frac{1}{2\sqrt{2}} - (0.827986)(0.27)\,.\end{aligned} \tag{38}$$

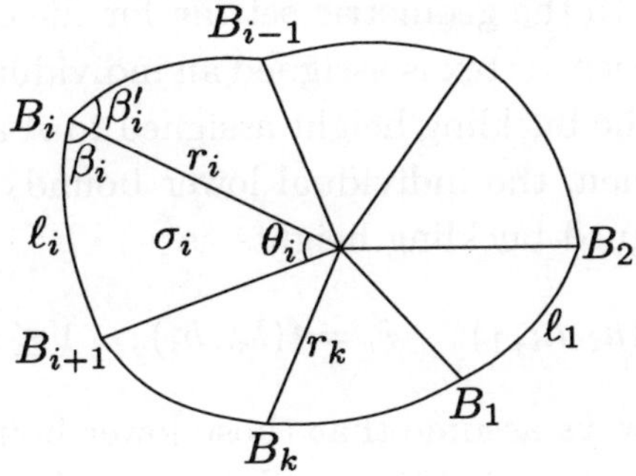

Figure 5.3

5.2 Area Estimates of Star Configurations

Let A be a vertex in a given spherical configuration $\mathcal{S}(\Sigma)$. The star configuration of A in $\mathcal{S}(\Sigma)$ is the sub-configuration consisting of those faces of $\mathcal{S}(\Sigma)$ with A as their common vertex. We shall denote it simply by $St(A)$ if there is no danger of confusion. Moreover, we shall always cut those non-triangular faces in $St(A)$, if any, into triangles by adding their diagonals radiating from A. As indicated in Figure 5.3, such a star configuration has the same number of triangles, radial edges and boundary edges. We shall call it a $k\triangle$-star if that same number is equal to k.

Notations As indicated in Figure 5.3, we shall denote the central vertex by A and its boundary vertices in cyclic order by $\{B_i; 1 \leq i \leq k\}$ and the triangles $\widehat{\triangle AB_iB_{i+1}}$ by σ_i. Moreover, we shall denote the lengths of the boundary (resp. radial) edges by $\{\ell_i\}$ (resp. $\{r_i\}$), while the *assigned* individual lower bounds of the boundary edges (resp. radial edges) will be denoted by $\{\hat{\ell}_i\}$ (resp. $\{\hat{r}_i\}$), namely

$$\widehat{B_iB_{i+1}} = \ell_i \geq \hat{\ell}_i\,, \quad \widehat{AB_i} = r_i \geq \hat{r}_i\,, \quad 1 \leq i \leq k \tag{1}$$

and the central angles (resp. the pair of angles at B_i) will be denoted by $\{\theta_i\}$ (resp. β_i, β_i').

In application to the geometric setting for the estimation of total buckling height, each vertex is assigned an individual buckling height. We shall denote the buckling height assigned to A and B_i, $1 \leq i \leq k$, by h_0 and h_i. Then, the individual lower bounds of (1) is given in terms of the assigned buckling heights as

$$\hat{\ell}_i = \ell(h_i, h_{i+1}) , \quad \hat{r}_i = \ell(h_0, h_i) , \quad 1 \leq i \leq k . \tag{2}$$

Thus, we shall always assume that those lower bounds $\{\hat{\ell}_i\}$ and $\{\hat{r}_i\}$ of (1) are either $\pi/3$ or slightly smaller than $\pi/3$, say at least 0.95.

Suppose that the three edge-lengths of σ_i (i.e. $\widehat{\triangle AB_iB_{i+1}}$) are actually equal to their individual lower bounds, namely, $\{\hat{r}_i, \hat{\ell}_i, \hat{r}_{i+1}\}$. Then, its central angle θ_i is quite close to $\cos^{-1}\frac{1}{3}$, namely, the angle of $\pi/3$-equilateral triangle. Notice that $6\cos^{-1}\frac{1}{3}$ *exceeds* 2π *by an amount of* $1.10257\ldots$. Therefore, in the case of $k \geq 6$, some of the radial edges of such a $k\triangle$-star must be considerably longer than their assigned lower bounds. For otherwise, the total sum of the k central angles would exceed 2π by a large amount. Roughly speaking, whenever the central angles of a certain cluster of triangles are in such a "squeezed situation" which forces some radial edges to be considerably longer than their individual bounds, the total area of such a cluster must be *considerably larger* than the *sum* of their *individual triangular area estimates.* In the special and geometrically simpler case of uniform bound of $\pi/3$, the total area of a $6\triangle$-star is at least equal to $2\cos^{-1}(-\frac{1}{3})$, which is 0.51355288 in excess of $6\triangle_{\pi/3}$ (cf. Lemma 2.2.2).

In this section, we shall extend the above type of *collective area estimate* of $6\triangle$-stars to the more general setting of non-uniform lower bounds on edge-lengths. The estimates of this section provide a lower bound of such *extra amounts of area estimates* in terms of the geometric data of such clusters. The basic techniques for obtaining this type of collective area estimates are the following four types of *area-decreasing deformations* of star configurations, which have already been used in the proof of Lemma 2.2.2.

5.2.1 *Area decreasing deformations of star configurations*

Recall that a star configuration is conceptually just the *localized version* of a spherical configuration. Thus, by definition, a star configuration also satisfies the same kind of *convexity property* as that of a spherical configuration, namely, every face of $St(A)$ is cocircular and the interior of each circumcircle contains *no* vertices of $St(A)$. For the purpose of lower bound area estimates of star configurations, the following simple types of *area-decreasing* deformations of star configurations naturally constitute a set of powerful tools.

The cycle of boundary edges $\{\widehat{B_iB_{i+1}};\ 1 \leq i \leq k\}$ of a given star configuration $St(A)$ looks like a "belt" around $St(A)$. If some of its segments are longer than their individual lower bounds and moreover, if there exists a certain simple type of *area-decreasing* deformation which shortens such segments, then such a deformation can be used to reduce the task of a lower bound area estimate to that of the shortened ones, such a deformation will be called a "*belt-tightening.*" In the proof of Lemma 2.2.2, we use two simple types of belt-tightening which can be reformulated for the more general setting of non-uniform edge-length-bounds as follows.

1. First kind of belt-tightening If

$$r_i \geq r_{i-1},\ r_i \geq r_{i+1} \quad \text{and} \quad r_i > \hat{r}_i\,, \quad \ell_{i-1} + \ell_i > \hat{\ell}_{i-1} + \hat{\ell}_i \tag{3}$$

then it is easy to see that $\frac{\pi}{2} < \beta_i + \beta_i' < \pi$. Hence, the deformation of shortening $\widehat{B_{i-1}B_i} + \widehat{B_iB_{i+1}}$ simply by moving B_i inward of $\widehat{\Box AB_{i-1}B_iB_{i+1}}$ is certainly area-decreasing and within the given family of star configurations, unless $\widehat{AB_i}$ is the side of a cocircular quadrilateral.

2. Second kind of belt-tightening Suppose that

$$\hat{\ell}_i < \ell_i \leq \cos^{-1}(\cos r_i + \cos r_{i+1} - 1)\,. \tag{4}$$

Then, the shortening of $B_i\overset{\frown}{B_{i+1}}$ will decrease both the area of σ_i and its central angle θ_i. Moreover, the above decrement of θ_i can always be accompanied by the shearing of a suitable quadrilateral $\overset{\frown}{\square}AB_{j-1}B_jB_{j+1}$ with r_j considerably larger than $\hat{r}_j$, so as to match the decrement of θ_i by the same amount of increment of $(\theta_{j-1}+\theta_j)$. It follows from the convexity property of the star configuration that the above accompanied shearing deformation is also area-decreasing. Thus, this kind of belt-tightening is, in fact, doubly area-decreasing, up until it is blocked by the lower bound $\hat{r}_j$ or the convexity condition of $St(A)$ as a whole.

3. Shearing of a quadrilateral along the boundary As indicated in Figure 2.25 of Chap. 2, the shearing deformation of $\overset{\frown}{\square}B_1B_2B_3B_4$ is applicable when both r_2 and r_3 are larger than their lower bounds $\hat{r}_2$ and $\hat{r}_3$, while *none* of $\{\overset{\frown}{AB_i}, 1 \leq i \leq 4\}$ is the diagonal of a cocircular quadrilateral. Such an area-decreasing shearing of the star configuration, whenever applicable, shortens one of the adjacent pairs of radial edges $\{\overset{\frown}{AB_2}, \overset{\frown}{AB_3}\}$ and lengthens the other, while keeping all the other edge-lengths fixed. Moreover, such an area-decreasing shearing remains applicable *up until* either the shortening one has already reached its lower bound or it is blocked by the convexity condition, namely, one of the pair of quadrilaterals with the shortening radial edge as their common side becomes *cocircular.*

4. Pairs of central-angle compensating shearings The fourth type of area-decreasing deformation is a coupling of a pair of shearings of two non-overlapping regions of the following two special types, so that their changes of central angles compensate each other:

(i) the shearing of $\overset{\frown}{\square}AB_{i-1}B_iB_{i+1}$ which varies the length of $\overset{\frown}{AB_i}$ *within* its *lower bound* and the diagonal length that makes it *cocircular*,

(ii) the shearing of a pentagon which consists of an adjacent pair of a *cocircular* quadrilateral and a triangle by varying the length of their common side between its lower bound and the length that makes the pentagon cocircular.

To each coupled pair of such shearings, there are two directions to deform. As it turns out, there is always a direction whose total effect is area-decreasing up until such a deformation is again *blocked* either by the *lower bound* of the shortening radial edge, or by the *convexity condition*, i.e. one of the coupled pairs of quadrilaterals or pentagons becomes *cocircular*. Anyhow, this type of area-decreasing deformation works just as well for the general setting of non-uniform lower bounds as for the special case of uniform lower bound of $\pi/3$ in Sec. 2.2.

5.2.2 *Area estimates of star configurations*

Let us first consider the important case of 6$\triangle$-stars with individual lower bounds of $\{\hat{\ell}_i, \hat{r}_i, 1 \leq i \leq 6\}$ on edge-lengths. In the special case of *uniform* lower bound of $\pi/3$ on edge-lengths, the areas of such a family of 6$\triangle$-stars have a global minimum of $2\cos^{-1}(-\frac{1}{3})$ while the minimal stars consist of a pair of cocircular quadrilaterals and a pair of triangles with $\pi/3$ side-length (cf. Lemma 2.2.2). In the general setting of non-uniform lower bounds on edge-lengths, the global minimum of the areas of such a family of 6$\triangle$-stars will certainly depend on the given set of individual bounds, and it is, in general, difficult to write down a formula of the global minimum in terms of the given set of individual bounds. However, the four kinds of area-decreasing deformations that we use to prove Lemma 2.2.2 are still applicable to the general setting of non-uniform bounds. Such area-decreasing deformations enable us to reduce the problem of lower bound area estimates of a given family of star configurations (such as the family of 6$\triangle$-stars with the edge-length-bounds of $\{\hat{\ell}_i, \hat{r}_i\}$) to that of the subfamily of *non-deformable* 6$\triangle$-stars with respect to the above four kinds of area-decreasing deformations. Therefore, let us begin with some specific examples of non-deformable $6\triangle$-stars.

5.2.2.1 *Examples of non-deformable $6\triangle$-stars with a given set of edge-length bounds*

Set $\hat{\sigma}_i$ to be the spherical triangle with side-lengths $\{\hat{r}_i, \hat{\ell}_i, \hat{r}_{i+1}\}$ and $\hat{\theta}_i$ to be its "central angle" (i.e. the angle between the $\hat{r}_i$-side and the $\hat{r}_{i+1}$-side). Set $\widehat{\square}_i$ to be the cocircular spherical quadrilateral with side-lengths $\{\hat{r}_i, \hat{\ell}_i, \hat{\ell}_{i+1}, \hat{r}_{i+2}\}$ in cyclic order, and $\hat{\varphi}_i$ to be its central angle (i.e. the angle between the $\hat{r}_i$-side and the $\hat{r}_{i+2}$-side). Notice that the six indices are cyclically ordered and there are no essential differences if one changes the labelling of $\{\hat{\ell}_i, \hat{r}_i, 1 \leq i \leq 6\}$ by a cyclic permutation of indices.

Example 5.2.1 Suppose that $\hat{\theta}_1 + \hat{\theta}_4 + \hat{\varphi}_2 + \hat{\varphi}_5 = 2\pi$. Then, the pair of triangles $\{\hat{\sigma}_1, \hat{\sigma}_4\}$ and the pair of cocircular quadrilaterals $\{\widehat{\square}_2, \widehat{\square}_5\}$ can be pieced together to form a $6\triangle$-star which is *non-deformable.*

Example 5.2.1′ Suppose that $\hat{\theta}_1 + \hat{\theta}_4 + \hat{\varphi}_2 + \hat{\varphi}_5 < 2\pi$. Then, there are two non-deformable $6\triangle$-stars which consist of $\{\hat{\sigma}_1, \hat{\sigma}_4, \widehat{\square}_2, \widehat{\boxslash}_5\}$ and $\{\hat{\sigma}_1, \hat{\sigma}_4, \widehat{\boxslash}_2, \widehat{\square}_5\}$, respectively, where $\widehat{\boxslash}_2$ (resp. $\widehat{\boxslash}_5$) is the shearing of $\widehat{\square}_2$ (resp. $\widehat{\square}_5$) with its central angle equal to $2\pi - \hat{\theta}_1 - \hat{\theta}_4 - \hat{\varphi}_5$ (resp. $2\pi - \hat{\theta}_1 - \hat{\theta}_4 - \hat{\varphi}_2$).

Example 5.2.2 Suppose that $\hat{\theta}_1 + \hat{\theta}_2 + \hat{\varphi}_3 + \hat{\varphi}_5 = 2\pi$. Then, there is a non-deformable $6\triangle$-star which consists of $\{\hat{\sigma}_1, \hat{\sigma}_2, \widehat{\square}_3, \widehat{\square}_5\}$ as its faces.

Example 5.2.2′ Suppose that $\hat{\theta}_1 + \hat{\theta}_2 + \hat{\varphi}_3 + \hat{\varphi}_5 < 2\pi$. Then, there are two non-deformable $6\triangle$-stars which consist of $\{\hat{\sigma}_1, \hat{\sigma}_2, \widehat{\square}_3, \widehat{\boxslash}_5\}$ and $\{\hat{\sigma}_1, \hat{\sigma}_2, \widehat{\boxslash}_3, \widehat{\square}_5\}$ as their faces, respectively.

Lemma 5.2.1 *Let $St(A)$ be a $6\triangle$-star with individual edge-length-bounds $\{\hat{\ell}_i, \hat{r}_i, 1 \leq i \leq 6\}$. Then, its total area is bounded below by the minimum of the following set of total areas, namely*

(i) *the total areas of those* $6\triangle$*-stars of Examples* 5.2.1$'$ *and* 5.2.2$'$ *with respect to suitable cyclical re-indexing,*

(ii) *the total areas of* $\{\hat{\sigma}_1, \hat{\sigma}_2, \widehat{\Box}_3, \widehat{\Box}_5\}$ *(resp.* $\{\hat{\sigma}_1, \hat{\sigma}_4, \widehat{\Box}_2, \widehat{\Box}_5\}$*) with respect to that cylical reindexing having* $\hat{\theta}_1 + \hat{\theta}_2 + \hat{\varphi}_3 + \hat{\varphi}_5 \geq 2\pi$ *(resp.* $\hat{\theta}_1 + \hat{\theta}_4 + \widehat{\varphi}_2 + \hat{\varphi}_5 \geq 2\pi$*).*

Proof It suffices to prove that the total area of a *non-deformable* $6\triangle$-star is at least equal to one of the above two types of total area. We shall divide the proof into two cases according to the number of non-minimal radial edges of the given *non-deformable* $6\triangle$*-star* being at most two or at least three.

Case 1: At most two non-minimal radial edges It is straightforward to check that

$$\hat{\varphi}_i + \sum_{j \neq i, i+1} \hat{\theta}_j > 2\pi, \quad 1 \leq i \leq 6\,. \tag{5}$$

Hence, it is impossible to have such a $6\triangle$-star with only one non-minimal radial edge. Suppose that $St(A)$ is a $6\triangle$-star with $\{\widehat{AB_2}, \widehat{AB_3}\}$ as its set of non-minimal radial edges. Then, the area-decreasing shearing of $\widehat{\Box} B_1B_2B_3B_4$ is always applicable unless one of $\{\widehat{AB_2}, \widehat{AB_3}\}$ is the diagonal of a cocircular quadrilateral. However, even in such a case, the fourth kind of area-decreasing deformation is always applicable. Therefore, the pair of non-minimal radial edges of such a non-deformable $6\triangle$-star *cannot* be adjacent. Thus, they are respectively the diagonals of a pair of quadrilaterals with minimal side-lengths. Moreover, at least one of the pair of quadrilaterals must be cocircular. For otherwise, the fourth kind of area-decreasing deformation would again be applicable. This proves that those $6\triangle$-stars of the type of Examples 5.2.1, 5.2.1$'$, 5.2.2 and 5.2.2$'$ are the only *non-deformable* $6\triangle$-stars with *at most two non-minimal radial edges.*

Case 2: At least three non-minimal radial edges Let $St(A)$ be a *non-deformable* $6\triangle$-star with at least three non-minimal radial edges. We may assume that the total area of $St(A)$ is at most equal to

$2\cos^{-1}(-\frac{1}{3})$. For otherwise, the lower bound of Lemma 5.2.1 clearly holds for $St(A)$.

Let us first consider the case that $St(A)$ has exactly a triple of non-minimal radial edges. If their relative locations are cosecutive such as $\{\overset{\frown}{AB_1}, \overset{\frown}{AB_2}, \overset{\frown}{AB_3}\}$, then the *non-applicability* of the third kind of area-decreasing deformation implies that either $\overset{\frown}{\square}AB_1B_2B_3$ is cocircular or both $\overset{\frown}{\square}AB_6B_1B_2$ and $\overset{\frown}{\square}AB_2B_3B_4$ are cocircular. In the former case, the fourth kind of area-decreasing deformation would be applicable. Therefore, it must be of the latter case for the non-deformable $St(A)$. Hence, the total area of such a $St(A)$ is necessarily larger than the lower bound of Lemma 5.2.1. In the case that the non-minimal triple are separated by the minimal edges either into isolated ones or an adjacent pair and an isolated one, it again follows from the non-applicability of the four kinds of area-decreasing deformations that $St(A)$ must contain a pair of cocircular quadrilaterals. Hence, its total area already exceeds the lower bound of Lemma 5.2.1.

The proof of the remaining case of non-deformable $6\triangle$-stars with more than three non-minimal edges is essentially the same as that of the above. $\square$

Remarks (i) The underlying geometric reasons of the above area estimate are the convexity property of star configurations and the central angle squeezed situation of such $6\triangle$-stars which forces some radial edges to be much longer than $\frac{\pi}{3}$.

(ii) Roughly speaking, the two kinds of belt-tightening area-decreasing deformations show that any amount of boundary edge-excesses is strongly area-increasing. The underlying geometric reason of the above fact is that each increment of a boundary edge in a $6\triangle$-star not only produces an increment of the area of the triangle, but also produces an increment of its central angle which always makes the area of the remaining part of the $6\triangle$-star increase by a corresponding amount.

(iii) The two kinds of area-decreasing shearings always shorten one radial edge while lengthening another radial edge up until such a deformation has already reached the limitation of either the edge-length-bound or the convexity condition, namely, one of the deforming quadrilaterals has already become cocircular. Roughly speaking, the two kinds of area-decreasing shearings actually show that star configurations with more a *lopsided distribution* of *radial edge-excess* are, in general, of smaller area than otherwise, while the most lopsided distributions are the ones which concentrate most of the radial edge-excess into the diagonals of cocircular quadrilaterals.

5.2.2.2 *$7\triangle$-star with individual lower bounds on edge-lengths close to $\frac{\pi}{3}$*

The central angle of a minimal triangle (resp. cocircular quadrilateral) with individual bounds close to $\pi/3$ is close to $\cos^{-1}\frac{1}{3}$ (resp. $\cos^{-1}(-\frac{1}{3})$). Note that

$$3\cos^{-1}\left(-\frac{1}{3}\right) + \cos^{-1}\frac{1}{3} = 2\pi + 0.67967\cdots. \tag{6}$$

Therefore, there are roughly 0.5 in the "*shortage of room*" for the central angles if one tries to fit together a triple of minimal cocircular quadrilaterals and a minimal triangle in order to build a $7\triangle$-star. Such a severe shortage of room for accommodating the central angles of a $7\triangle$-star forces the *minimal* $7\triangle$*-stars* to contain a triple of cocircular quadrilaterals, and at least one of them must be non-minimal.

Lemma 5.2.2 *The area of a $7\triangle$-star with individual lower bounds on edge-lengths of at least $a = 2\sin^{-1}\frac{1}{2.18}$ exceeds the sum of areas of a triple of minimal quadrilaterals and a minimal triangle by at least* 0.3.

Proof Due to the extreme shortage of room for accommodating the central angles of a $7\triangle$-star, it is quite straightforward to apply the four kinds of area-decreasing deformations of Subsec. 5.2.1 to deform such a 7△-star to a smaller one containing a triple of cocircular quadrilaterals. Moreover, even though all the lower bounds are equal to $a = 2\sin^{-1}\frac{1}{2.18}$, the sum of central angles of a triple of minimal cocircular quadrilaterals and a minimal triangle still exceeds 2π by at least 0.433752, namely

$$3\cos^{-1}\frac{\cos a - 1}{\cos a + 1} + \cos^{-1}\frac{\cos a}{\cos a + 1} = 2\pi + 0.433752\cdots. \tag{7}$$

Therefore, the central angles of the triple of cocircular quadrilaterals and the triangle in such a 7△-star must be considerably *smaller* than that of the minimal cocircular quadrilaterals (resp. triangles). Hence, their areas must be considerably *larger* than that of the minimal ones. Moreover, it is quite straightforward to estimate the area increments in terms of the angular decrements (of the central angles). For example, in the case of cocircular quadrilaterals, the area-increment is at least 1.35 times the corresponding angular decrement. Therefore, a total amount of 0.43 angular decrement will certainly produce more than 0.3 of area-increment. □

Remark The technique of area estimate by means of area-decreasing shearing can also be applied to a cluster of triangles which only constitutes a partial star. For example, as indicated in Figure 5.4, $AB_1B_2B_3B_4$ is the union of a triple of triangular faces

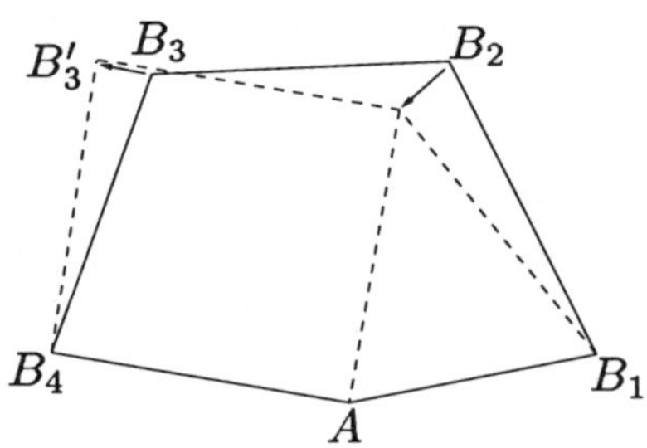

Figure 5.4

of a given spherical configuration. One can again apply area-decreasing shearing to $\Box B_1B_2B_3B_4$ to obtain a lower bound area estimate of such a triple. In particular, if both $\overline{AB_1}$ and $\overline{AB_4}$ are close to $\pi/3$ and the total central angle is at most slightly larger than π, then the total area of such a triple is at least quite close to the sum of the area of a minimal cocircular quadrilateral and a minimal triangle.

5.3 Estimation of Total Buckling Height

In this section, we shall present the proof of the following lower bound estimates on the total buckling height of core packings with thirteen or more close neighbors.

Lemma 5.3.1 *The total buckling height is at least* 0.325 *for a core packing with thirteen close neighbors.*

Lemma 5.3.2 *The total buckling height is at least* 0.56 *for a core packing with fourteen close neighbors.*

Lemma 5.3.3 *The total buckling height is at least* 1 *for a core packing with fifteen close neighbors, and a core packing has at most fifteen neighbors.*

The proof given here is essentially a simplification and a clarification of that of [Hsi-2] and [Hsi-3]. The basic techniques for such proofs are the following:

(1) the correlation between buckling heights and area estimates (cf. Sec. 5.1),
(2) area estimates on individual triangles, 6$\triangle$-stars, 7$\triangle$-stars and a cluster of triangles with squeezed central angles,
(3) geometric combinatorial analysis on the possible combinatorial structures of the associated spherical configurations.

The basic strategy of the proofs of Lemmas 5.3.1–5.3.3 is to prove the *non-existence* of a core packing with thirteen (resp. fourteen, fifteen) close neighbors whose total buckling height is less than the given bound of 0.325 (resp. 0.56, 1) by proving the associated spherical configuration of such a *hypothetical* core packing necessarily having a total area *exceeding* 4π (which is of course a contradiction to the most basic fact that the area of unit sphere is equal to 4π). We shall give the proofs of Lemmas 5.3.1–5.3.3 in reverse order because the latter are simpler than the former, and moreover, the simpler proofs of the latter will also provide a preview of the method of proof by area estimates.

5.3.1 *The proof of Lemma* 5.3.3

Let us first use the method of area estimate to prove the non-existence of a core packing with sixteen close neighbors. Suppose on the contrary $\hat{\mathcal{L}}(S_0)$ is a core packing with sixteen close neighbors and set $\mathcal{S}'(\Sigma)$ to be its associated (triangulated) configuration. By Euler's formula, $\mathcal{S}'(\Sigma)$ has twenty-eight triangles, forty-two edges and sixteen vertices. Set $a = 2\sin^{-1}\frac{1}{2.18}$. We may assume that Σ is already *a-saturated*. For otherwise, Σ can be extended to a set Σ^* of seventeen points with a separation of at least a, which implies that

$$\begin{aligned} 4\pi \geq 30 \cdot \triangle_a &= 30 \cdot \left\{ 3\cos^{-1}\frac{\cos a}{1+\cos a} - \pi \right\} \\ &= 30 \cdot (0.4442505) = 4\pi + 0.761\cdots, \end{aligned} \tag{1}$$

which is obviously a contradiction.

Therefore, the area of each triangle in $\mathcal{S}'(\Sigma)$ is at least equal to $\triangle_a$, while the total area of a $6\triangle$-star in $\mathcal{S}'(\Sigma)$ is quite *close* to $2\triangle_a + 2\square_a$; that of a $7\triangle$-star, if any, will exceed $\triangle_a + 3\square_a + 0.3$. On the other hand, either $\mathcal{S}'(\Sigma)$ has at least one $7\triangle$-star or $\mathcal{S}'(\Sigma)$ has at least four 6-fork vertices. Since $\square_a = 1.079023$,

$$28 \cdot \triangle_a + 4(\square_a - 2\triangle_a)$$
$$= 28 \cdot (0.4442505) + 4 \cdot (0.19052235) = 4\pi + 0.63473\cdots , \quad (2)$$

which is again an obvious contradiction, thus proving the *impossibility* of having sixteen close neighbors.

Next let us proceed to prove the estimate of at least 1 for the total buckling height of a core packing with fifteen close neighbors. Let $\hat{\mathcal{L}}(S_0)$ be such a core packing, $\mathcal{S}'(\Sigma)$ be its associated (triangulated) configuration and $\{h_j\}$ be the buckling heights of its close neighbors. Suppose the contrary that $\Sigma h_j < 1$. We shall use the method of area estimate to show the *non-existence* of such a configuration $\mathcal{S}'(\Sigma)$ and hence also the *non-existence* of such a core packing.

First of all, the area of each individual triangular face is at least equal to $\triangle(h_i, h_j, h_k)$ where $\{h_i, h_j, h_k\}$ are the assigned buckling heights at its vertices, and moreover, its area is close to the above lower bound *only when* its three side-lengths are close to $\ell(h_i, h_j), \ell(h_j, h_k)$ and $\ell(h_k, h_i)$, respectively. On the other hand, it follows from Euler's formula that $\mathcal{S}'(\Sigma)$ has twenty-six triangles, thirty-nine edges and fifteen vertices. Therefore, either $\mathcal{S}'(\Sigma)$ has at least three 6-fork vertices, or one 7-fork and at least one 6-fork vertices, or at least two 7-fork vertices.

In the case where $\mathcal{S}'(\Sigma)$ has exactly three 6-fork vertices and suppose that they are connected by edges, it follows directly from Lemmas 5.1.1, 5.1.2 and 5.2.1 that the total sum of the triangular and the 6$\triangle$-star *area estimates* is very closely approximated by

$$18\triangle_{\pi/3} + 4\square_{\pi/3} - 2\sqrt{2}\Sigma h_j > 4\pi - 0.034267\,. \quad (3)$$

Although the above inequality, just by itself, is not a contradiction, it implies that the twelve triangular faces outside of the union of the three 6$\triangle$-stars must be all very, very close to be with *minimal* side-edges, namely, the total sum of their edge-lengths in excess of their individual lower bounds of $\ell(h_i, h_j)$ is at most 0.05. However, such a collection of almost minimal triangles clearly cannot be fitted around the union of the three $6\triangle$-stars. (The total angles at quite a few vertices of such a hypothetical fitting together are much bigger than 2π!) Hence, such a spherical configuration is geometrically *impossible*.

The proof of other cases is simpler because the area estimates of Lemmas 5.1.1, 5.1.2, 5.2.1 and 5.2.2 already add up to an amount *exceeding* 4π, if $\Sigma h_j < 1$.

This completes the proof. $\square$

5.3.2 *The proof of Lemma* 5.3.2

Suppose on the contrary that there exists a core packing $\hat{\mathcal{L}}(S_0)$ with fourteen close neighbors whose total buckling height is less than 0.56. Let $\mathcal{S}'(\Sigma)$ be its associated configuration and $\{h_j\}$ be its buckling heights. Again, by Euler's formula, $\mathcal{S}'(\Sigma)$ has twenty-four triangles, thirty-six edges and fourteen vertices. Hence, either $\mathcal{S}'(\Sigma)$ has at least two 6-fork vertices, or $\mathcal{S}'(\Sigma)$ has at least one 7-fork vertex. Moreover, straightforward geometric combinatorial analysis shows that a configuration does *not* exist with one 7-fork vertex and thirteen 5-fork vertices, and moreover, the combinatorial type as indicated in Figure 5.5 is the only possible spherical configuration with twelve 5-fork vertices and a pair of 6-fork vertices. In the case that $\mathcal{S}'(\Sigma)$ is of the combinatorial type as indicated in Figure 5.5, it follows from direct application of Lemmas 5.1.1, 5.1.2 and 5.2.1 that the total sum of the triangular and the $6\triangle$-star *area estimates* is already very close to

$$16\triangle_{\pi/3} + 4\square_{\pi/3} - 2\sqrt{2}\Sigma h_j\,, \tag{4}$$

which is approximately $4\pi - 0.10767\cdots$. Therefore, the total amount of the edge-excesses contained in the twenty-four side-edges of the twelve triangles fitting between the pair of $6\triangle$-stars is at most 0.18. It is again quite simple to show such a geometric fitting is far from geometrically feasible. Hence, such a configuration $\mathcal{S}'(\Sigma)$ is geometrically impossible.

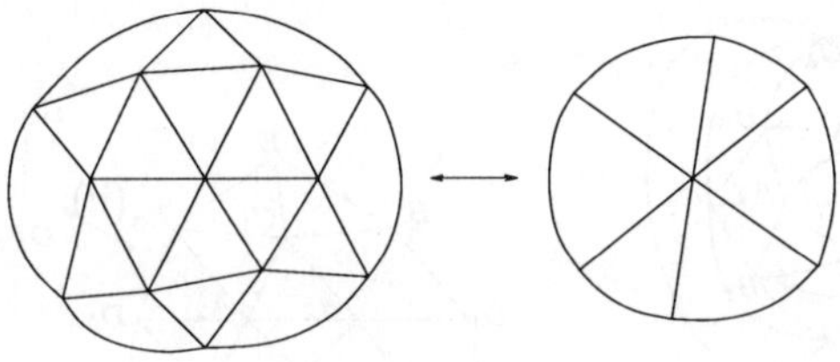

Figure 5.5

The proof of the remaining cases of Lemma 5.3.2 is simpler because the area estimates of Lemmas 5.1.1, 5.1.2, 5.2.1 and 5.2.2 already add up to an amount exceeding 4π, if $\Sigma h_j < 0.56$.

This completes the proof. □

5.3.3 *The proof of Lemma* 5.3.1

Suppose on the contrary that there exists a core packing $\hat{\mathcal{L}}(S_0)$ with thirteen close neighbors whose total buckling heights is less than 0.325. Set $\mathcal{S}'(\Sigma)$ to be its associated configuration and $\{h_j\}$ to be its buckling heights. By Euler's formula, $\mathcal{S}'(\Sigma)$ has twenty-two triangles, thirty-three edges and thirteen vertices. A simple combinatorial analysis shows that it is impossible to have such a spherical configuration with *only one* vertex of the type of either 6-fork or 7-fork. A slightly more detailed, but quite straightforward combinatorial analysis, shows that the combinatorial types as indicated in Figure 5.6 are the only possible ones with at most three 6-fork vertices and the other vertices are either 5-fork ones or 4-fork ones, namely, only one combinatorial type with two 6-fork vertices and three combinatorial types with three 6-fork vertices. In other words, all the other combinatorial types of such spherical configurations either have at least one 7-fork vertex or have at least four 6-fork vertices.

In the case that $\mathcal{S}'(\Sigma)$ is of the combinatorial type of either Figure 6 (ii) or Figure 6 (iv), $\mathcal{S}'(\Sigma)$ contains a pair of overlapping $6\triangle$-stars, $St(A_1)$ and $St(A_2)$, and another $6\triangle$-star, $St(M)$, outside of the above pair. Hence, the total sum of the triangular and the

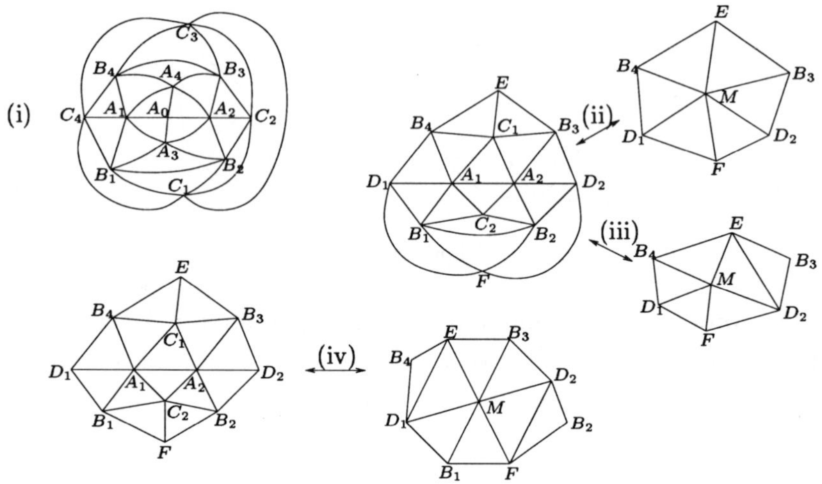

Figure 5.6

6$\triangle$-star *area estimates* by direct applications of Lemmas 5.1.1, 5.1.2 and 5.2.1 is already very close to

$$12\triangle_{\pi/3} + 5\square_{\pi/3} - 2\sqrt{2}\Sigma h_j > 4\pi - 0.07345\,. \tag{5}$$

The above lower bound area estimate implies that the total amount of "edge-excesses" contained in the side-edges of the six triangular faces outside of the union of $\{St(A_1), St(A_2), St(M)\}$ is at most 0.12. Hence, all of the six triangles are quite close to being minimal, which implies that the central angles at C_1 will add up to a sum considerably larger than 2π. Therefore, such a spherical configuration is, in fact, geometrically *impossible.*

The proof of *non-existence* of such a spherical configuration with a combinatorial type other than the four indicated in Figure 5.6 is simpler, because the same area estimates by direct applications of Lemmas 5.1.1, 5.1.2, 5.2.1 and 5.2.2 already add up to an amount exceeding 4π, if $\Sigma h_j < 0.325$. Therefore, it suffices to give the proof of non-existence of such spherical configurations whose combinatorial type is as indicated in Figure 5.6(i) or Figure 5.6(iii).

Case 1: Suppose that $\mathcal{S}'(\Sigma)$ is of the combinatorial type of Figure 5.6(i) As indicated in Figure 5.6(i), each of the stars of the vertices $\{B_i, 1 \leq i \leq 4, C_2, C_4\}$ contain a cluster of triple triangles outside of $St(A_1) \cup St(A_2)$. On the other hand, it follows from the proof of Lemma 5.2.1 that a $6\triangle$-star with an adjacent pair of radial edges considerably longer than $\pi/3$ will have an area considerably larger than the sum of the areas of a pair of *minimal* cocircular quadrilaterals and a pair of *minimal* triangles, namely, the approximate lower bound area estimate for a given $6\triangle$-star. Therefore, we may assume that either $\{\overline{A_1B_1}, \overline{A_1B_4}\}$ (resp. $\{\overline{A_2B_2}, \overline{A_2B_3}\}$) are both quite close to $\pi/3$, or $\overline{A_1C_4}$ (resp. $\overline{A_2C_2}$) is quite close to $\pi/3$. Hence, either both outer angles at $\{B_1, B_4\}$ (resp. $\{B_2, B_3\}$) or the outer angle at C_4 (resp. C_2) of $St(A_1)$ (resp. $St(A_2)$) at most slightly exceed π. Therefore, by the simple area estimate of such a triple of triangles with total central angle close to π (cf. the remark at the end of Sec. 5.2.2), the total area of such a triple of triangles is at least quite close to the sum of the areas of a *minimal cocircular quadrilateral* and a minimal triangle. Thus, the total sum of the area estimates of the non-overlapping pair of $6\triangle$-stars and such triples of triangles as well as the simple triangular area estimates for the remaining triangles is approximately equal to

$$10\triangle_{\pi/3} + 6\square_{\pi/3} - 2\sqrt{2}\Sigma h_j > 4\pi + 0.1833\cdots, \tag{6}$$

which is an obvious contradiction. This proves the non-existence of such a core packing whose associated configuration is of the combinatorial type of Figure 5.6(i).

Case 2: Suppose that $\mathcal{S}'(\Sigma)$ is of the combinatorial type of Figure 5.6(iii) The combinatorial type of Figure 5.6(iii) can also be represented as in Figure 5.7. Note that both $\{\overline{A_2C_1}, \overline{A_2C_2}\}$ (resp. $\{\overline{A_2B_2}, \overline{A_2B_3}\}$) are boundary edges of $St(A_1)$ (resp. $St(D_2)$). Hence, the above quadruple of radial edges of $St(A_2)$ must be all quite close to $\pi/3$. For otherwise, the areas of $St(A_1)$ (resp. $St(D_2)$) would be considerably larger than the lower bound estimates of

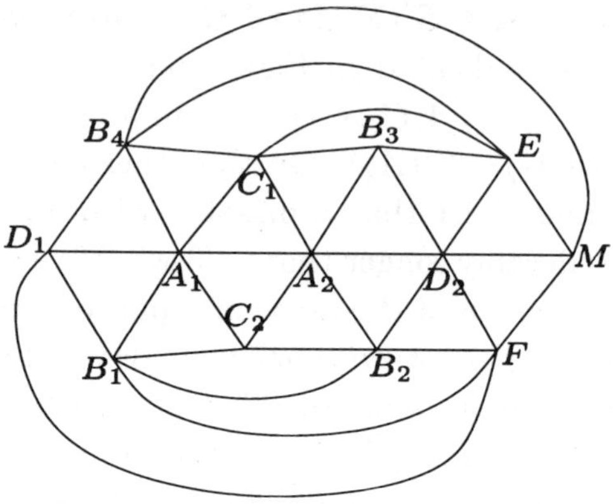

Figure 5.7

Lemma 5.2.1. Therefore, both $\square A_2B_2D_2B_3$ and $\square A_2C_1A_1C_2$ must be close to cocircular. Hence, the outer angles of the union of the triple of 6$\triangle$-stars at both C_1 and B_2 at most slightly exceed $\cos^{-1}(-\frac{1}{3})$, which implies that the areas of both $\square C_1B_3EB_4$ and $\square B_2C_2B_1F$ are close to that of a minimal cocircular quadrilateral. Thus, the above area estimate of the quadruple of triangles plus the area estimates of $St(A_1)$ and $St(D_2)$ and the simple triangular area estimates of the remaining triangles already add up to an amount very close to (6), which is again an obvious contradication. This proves the non-existence of such a core packing whose associated configuration is of the combinatorial type of Figure 5.6(iii), and hence also completes the proof of Lemma 5.3.1. □

Remarks (i) The lower bound estimates of Lemmas 5.3.1, 5.3.2 and 5.3.3 are not optimal for core packings with thirteen, fourteen and fifteen close neighbors, respectively. However, they are already much stronger than what are actually needed for the purpose of proving Main Theorem I in Chap. 8.

(ii) It is quite easy to use the same proof to obtain a somewhat smaller lower bound estimate of total buckling height for the setting of reduced local packing with only one non-close neighbor. In particular, it is quite simple to use the area estimate of Sec. 5.2 to prove that the total buckling height of a local packing with thirteen neighbors is at least 0.16.

(iii) One of the fundamental geometric features of a local packing with *more than twelve* neighbors is the *overcrowdedness* of such a local packing, while that of a local packing with *at most twelve* neighbors is the *non-tightness* of such a surrounding. The *lower bound estimates* on *the total buckling heights* of various types of overcrowded local packings, such as that of Lemmas 5.3.1, 5.3.2 and 5.3.3, can be regarded as a kind of *quantitative* understanding of the *geometry of overcrowdedness.* It should also be pointed out here that such *overcrowded* local packings are still very much *untight.* Geometrically, a star consisting of five small triangles can be regarded as the kind of locally tight arrangement. However, it is not difficult to see from the *combinatorial structure* of the *associated spherical configurations* of local packings with more than twelve neighbors that such a locally tight arrangement cannot occur *unless* with much higher total buckling height than that of the estimates.

Chapter 6

The Proof of the Dodecahedron Conjecture

The *Dodecahedron Conjecture* [Fej-2] on sphere packings asserts that the volume of any local cell of packings of unit spheres is at least equal to the volume of the *regular circumscribing dodecahedron*, namely

$$L_{12} = 20\left(1 - 2\cos\frac{2\pi}{5}\right)\tan\frac{\pi}{5} = 5.550291\cdots \tag{1}$$

and it is equal to the above lower bound when and only when the local cell is congruent to it. Both as a test and as an important application of the techniques of volume estimation of local cells of Chaps. 4 and 5, we shall combine them to give a rather clean-cut proof of the above conjecture.

Theorem *The volume of a local cell, $C(\mathcal{L})$, is at least equal to L_{12} of* (1) *and it is equal to L_{12} when and only when $C(\mathcal{L})$ is a regular circumscribing dodecahedron* (*of the unit sphere*) (*i.e. the local packing $\mathcal{L}$ consists of twelve symmetrically arranged touching neighbors*).

Proof Let $\mathcal{L}(S)$ be a given local packing of unit spheres and $\mathcal{L}'(S)$ (resp. $\widehat{\mathcal{L}}(S)$) be its *reduced* (resp. *core*) subpacking. Set n (resp. m) to be the number of reduced (resp. close) neighbors. We shall label the reduced neighbors in such a way that

$$h_j \le 0.09 \text{ for } j \le m, \text{ and } h_{m+1} \le h_{m+2} \le \cdots \le h_n\,. \tag{2}$$

We shall divide the proof of the above theorem into three cases according to $m = 12$, $m \le 11$ and $m \ge 13$. □

6.1 The Proof of Case 1: $m = 12$

First of all, the special but also the most critical subcase of twelve touching neighbors has already been proved as Lemma 4.4.1. Moreover, it is quite easy to check that the proof of Lemma 4.4.1 still works without any modification provided that $\sum_{j=1}^{12} h_j$ is at most equal to 0.015 (cf. Lemma 4.3.1′, the lower bound estimate of $\widetilde{L}(h'_i)$ that we use in the proof of Lemma 4.4.1 still holds). Therefore, it suffices to consider the case that $\sum_{j=1}^{12} h_j$ exceeds 0.015.

Set $\mathcal{S}(\Sigma)$ to be the associated configuration of $\mathcal{L}'(S)$. We shall assume, without loss of generality for the proof, that all the circumradii $\{r(\sigma), \sigma \in \mathcal{S}(\Sigma)\}$ are less than $\cos^{-1} 0.65$. For otherwise, one can always extend $\mathcal{L}'(S)$ by fitting an additional reduced neighbor on top of a big-hole of such a size. If the twelve close neighbors are the only reduce neighbors (i.e. $n = m = 12$) and S'_1 is a non-reduced neighbor with buckling height h'_1, then

$$\sum_{j=1}^{12} L(h_j) + L(h'_1) \ge L(0.015) + L(h'_1)\,, \tag{3}$$

which is less than 0.13194 only when the corresponding $r(\sigma_1)$ is less than $\cos^{-1} 0.669$. Therefore, the same proof as that of Lemma 4.4.1 will again provide the same kinds of lower bound estimates according to the types of σ_1-hole. In fact, due to the extra contribution of $\sum_{j=1}^{12} L(h_j) \ge 0.017528$, such estimates are slightly larger than the respective estimates given in the proof of Lemma 4.4.1.

Next let us consider the subcase that $n = 13$ and $\sum_{j=1}^{12} h_j > 0.015$. As was already pointed out at the end of Sec. 5.3, $\sum_{j=1}^{13} h_j$ is at least 0.16, while h_{13} cannot be very close to 0.09, say at least 0.11. Therefore, by the concavity of $L(h)$, one has

$$\sum_{j=1}^{13} L(h_j) \geq L(0.015) + L(0.145) = 0.134721\,, \tag{4}$$

which already exceeds $(L_{12} - L_{13}) = 0.13194$ with a small margin. Hence, the same kind of estimates as that of the proof of Lemma 4.4.1 again provide the proof for such a subcase.

In the remaining case of $n \geq 14$ and $\sum_{j=1}^{12} \geq 0.015$, the big-holes of the other reduced neighbors bring in additional area estimates. Thus, it is quite simple to use the method of area estimate to show that

$$\sum_{j=1}^{12} h_j + h_{13},\ \sum_{j=1}^{12} h_j + h_{14},\ \text{etc.} \tag{5}$$

all exceed 0.18. Therefore, it is easy to check that

$$\sum_{j=1}^{14} L(h_j) > L_{12} - L_{14},\quad \text{etc.} \tag{6}$$

Hence, the proof of the remaining case of $n \geq 14$ again follows from the same kind of estimates as that of Lemma 4.4.1.

6.2 The Proof of Case 2: $m \leq 11$

Again, we may assume that all the circumradii of $\mathcal{S}(\Sigma)$ are less than $\cos^{-1} 0.65$. Moreover, we may also assume that $n \geq 13$. For otherwise, the lower bound estimate of L_{12} becomes very simple. The same kind of lower bound estimates as that of Case 1 will show that

$$\operatorname{vol} C(\mathcal{L}) \geq L_n + (n-m)L(0.09)\,, \quad m \leq 11,\ n \geq 13\,, \tag{7}$$

which always exceeds L_{12} by a comfortable margin.

6.3 The Proof of Case 3: $13 \leq m \leq 15$

In this case, Lemmas 5.3.1–5.3.3 prove that

$$\sum_{j=1}^{m} h_j \geq \begin{cases} 0.325 \\ 0.56 \\ 1 \end{cases} \quad \text{for} \quad m = \begin{cases} 13\,, \\ 14\,, \\ 15\,. \end{cases} \tag{7}$$

Again, the same kind of volume estimates will show that

$$\operatorname{vol} C(\mathcal{L}) \geq L_n + (n-m)L(0.09) + \sum_{j=1}^{m} L(h_j)\,, \tag{8}$$

which always exceeds L_{12} with a rather big margin because the lower bounds of (7) are much larger than what is needed for the lower bound estimate of L_{12}.

This completes the proof of Case 3 and hence, also the proof of the Dodecahedron Conjecture.

Chapter 7

Geometry of Type I Configurations and Local Extensions

After the proof of the Dodecahedron Conjecture, the next major tasks in the study of sphere packings are naturally the proof of Kepler's Conjecture and the study of the least action principle of crystal formations. As has already been pointed out in Chap. 3, both the proof of Kepler's Conjecture and that of the geometric theorem, which provides the least action principle of crystal formation of dense packing type, *can be reduced* to the proof of the Main Theorem of the 1993 paper [Hsi-2] which asserts that $\pi/\sqrt{18}$ *is the optimal upper bound* of the *locally averaged density* and moreover, the locally averaged density of a given double layer local packing, $\mathcal{L}^2(S)$, is equal to $\pi/\sqrt{18}$ when and only when it is one of the six hexagonal double-layer local packings (cf. Chap. 3). Conceptually, the above locally averaged density is a geometric invariant defined in terms of a *double layer local packing*, which consists of a central sphere, S, its neighbors and the neighbors of its neighbors. Technically, it is quite natural to consider such a double layer local packing, $\mathcal{L}^2(S)$, as an *extension* of its single layer local packing of S, $\mathcal{L}(S)$. In the definition of the above locally averaged density, the role of those additional spheres of the second layer is just to make the *local cells* of those close neighbors (of the central sphere) geometrically defined. Let S_j be such a close neighbor and $\mathcal{L}(S_j)$ be the local packing of S_j contained in $\mathcal{L}^2(S)$. Then, $\mathcal{L}(S_j) \cap \mathcal{L}(S)$ already constitutes a *partial* local packing of S_j, while the others of $\mathcal{L}(S_j)$ are exactly those additional neighbors belonging to the second layer of $\mathcal{L}^2(S)$. Thus, the *restriction* of the extension, $\mathcal{L}(S) \subset \mathcal{L}^2(S)$, to that of $\mathcal{L}(S_j)$ is an extension from the partial one, $\mathcal{L}(S_j) \cap \mathcal{L}(S)$, to a full-fledged local packing. Geometrically, this is exactly the *localization* of the extension, $\mathcal{L}(S) \subset \mathcal{L}^2(S)$, to the vicinity of S_j which shall be called the *local extension at* S_j of the given double layer extension. Notice that the local extensions of a given double layer extension at a pair of neighboring spheres, say S_j and S_k, often share a few of those *additional neighbors*. Therefore, the local extension at a given S_j will have some strong effects on those neighboring local extensions (cf. Subsec. 7.2.2).

The technical advantages of taking such a two-stage approach in managing the geometry of double layer local packings are two-fold. On the one hand, the geometries of single layer and double layer local packings are both very complicated, however, the complexity of the former is relatively much simpler than that of the latter; on the other hand, the geometry of the former plays a much more important role in the proof of Main Theorem I than that of the latter. Therefore, as one shall see in the next chapter, the proof of Main Theorem I is divided into cases according to the geometric types of the single layer local packings. Moreover, among the various cases, the proof for the case of local packings with twelve touching neighbors turns out to be the most crucial and also the most demanding one. We shall call this type of (single-layer) local packing *type* I *local packings* and their corresponding spherical configurations *type* I *configurations.* Geometrically speaking, type I configurations are those spherical configurations with *twelve vertices* and with *edge-lengths of at least* $\pi/3$.

In this chapter, we shall study the geometry of type I configurations as well as the geometry of local extensions, thus providing the geometrical understanding of these two important aspects needed in the proof of Main Theorem I (cf. Chap. 8).

7.1 Geometry of Type I Local Packings and Type I Configurations

Among all type I configurations, the most symmetric one consists of twenty equilateral triangles of the same size, namely, with edge-lengths of $\tan^{-1} 2 = (\pi/3 + 0.05995\cdots)$. Therefore, each of its twelve vertices has two degrees of freedom for the local deformations of such a configuration within type I configurations. This shows that the *moduli space* of all *congruence classes of type* I *configurations* constitutes a space of 21 dimensions.

The faces of a type I configuration are either triangular or quadrilateral, namely, there can be *neither pentagons nor hexagons* among the faces of a type I configuration. It is obvious that there can be *no* hexagons because the only cociruclar hexagon with side-lengths of at least $\pi/3$ is a *hemi-sphere*. The *non-existence* of pentagonal faces in a type I configuration can be proved as follows.

Note that the circumradius of the cocircular $\frac{\pi}{3}$-equilateral pentagon is given by

$$r_0 = \sin^{-1}\left(\frac{1}{2\sin\frac{\pi}{5}}\right) = \pi/3 - 0.02997558\cdots. \tag{1}$$

Suppose on the contrary that there exists a type I configuration, $\mathcal{S}(\Sigma)$, with a pentagonal face. Then, the *circumradius* of such a cocircular pentagon is at least equal to r_0. Therefore, an additional sphere of buckling height of at most $h_0 = 2\cos r_0 - 1 = 0.051462\cdots$ can be added (in the direction of its circumcenter) to the type I local packing with Σ as the set of twelve touching points. This, of course, grossly contradicts the estimate of total buckling heights for thirteen close neighbors (cf. Chap. 5). [The non-existence of a pentagonal face in a type I configuration can also be proved directly by a much simpler area estimate than that of Lemma 5.3.1.]

Let F_3 and F_4 be respectively the number of triangles and quadrilaterals in a given type I configuration, $\mathcal{S}(\Sigma)$. Then, it follows from Euler's formula that

$$F_3 + 2F_4 = 2n - 4 = 20\,. \tag{2}$$

Moreover, it follows from Lemma 2.2.1,

$$\begin{gathered} F_3\triangle_{\frac{\pi}{3}} + F_4\square_{\frac{\pi}{3}} \le 4\pi\,, \\ \triangle_{\frac{\pi}{3}} = 3\cos^{-1}\frac{1}{3} - \pi\,, \quad \square_{\frac{\pi}{3}} = 4\cos^{-1}\left(-\frac{1}{3}\right) - 2\pi\,. \end{gathered} \tag{3}$$

Therefore, F_4 is at most equal to six, and in the case of $F_4 = 6$, all triangles (resp. quadrilaterals) of such a type I configuration are actually the $\pi/3$-equilateral ones (cf. Sublemma 3.3.1). It is easy to see that there are exactly two such configurations, namely, the configurations corresponding to f.c.c. h.c.p. local packings as indicated in Figure 7.1.

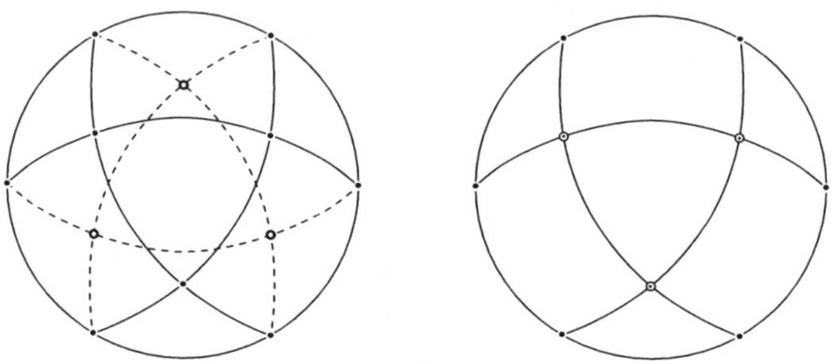

f.c.c. h.c.p.

Figure 7.1

7.1.1 *A strong local characterization of f.c.c. and h.c.p. configurations*

Set $\alpha = \cos^{-1}(\frac{1}{3})$ and $\beta = \cos^{-1}(-\frac{1}{3})$. The local geometric structure at each vertex of the above f.c.c. configuration consists of four edges

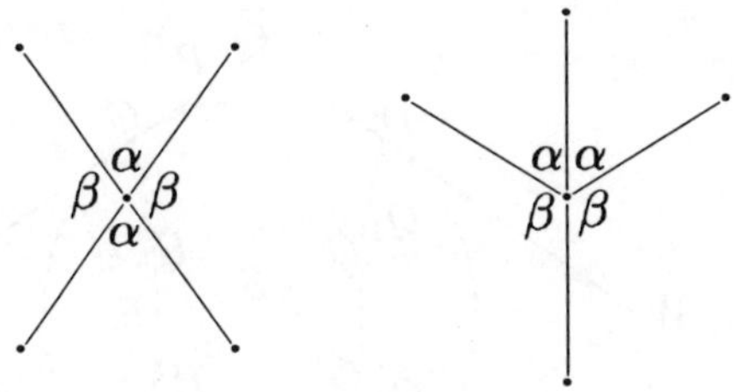

Figure 7.2

of uniform length of $\pi/3$ arranged in $(\alpha, \beta, \alpha, \beta)$-angular separation, while that of the h.c.p. configuration has two types of angular arrangements, namely, six of them in $(\alpha, \beta, \alpha, \beta)$-angular arrangement and the other six of them in $(\alpha, \alpha, \beta, \beta)$-angular arrangement (cf. Figure 7.2).

It is a remarkable result that just the existence of a single vertex with $(\alpha, \beta, \alpha, \beta)$-angular arrangement already suffices to characterize a *type* I *configuration* to be either the f.c.c. or the h.c.p. configuration.

Lemma 7.1.1 *Let $\mathcal{S}(\Sigma)$ be a type I configuration. If it has at least one vertex with $(\alpha, \beta, \alpha, \beta)$-angular arrangement, then it is congruent to either the f.c.c. or the h.c.p. configuration (cf. Figure* 7.1).

Proof Up to a congruence, one may assume that the north pole N is such a vertex of $\mathcal{S}(\Sigma)$, i.e. with $(\alpha, \beta, \alpha, \beta)$-angular arrangement of the edges at N. Set A, B, C and D to be its four neighboring vertices and let us first consider the case that the above four edges, namely, $\overline{NA}, \overline{NB}$, etc. are all of the minimal length of $\pi/3$. Set Ω to be the *complementary region* of the union of the five open $\pi/3$-disks centered at N, A, B, C and D, respectively. It is easy to see that Ω is a subregion of the southern hemi-sphere. Therefore, it can be conveniently represented via the *stereographic projection* by a subregion inside the unit disk. It is the parallelogram, as indicated in Figure 7.3, whose four sides represent the $\pi/3$-circular arcs centered at A, B, C and D, respectively.

Ω :

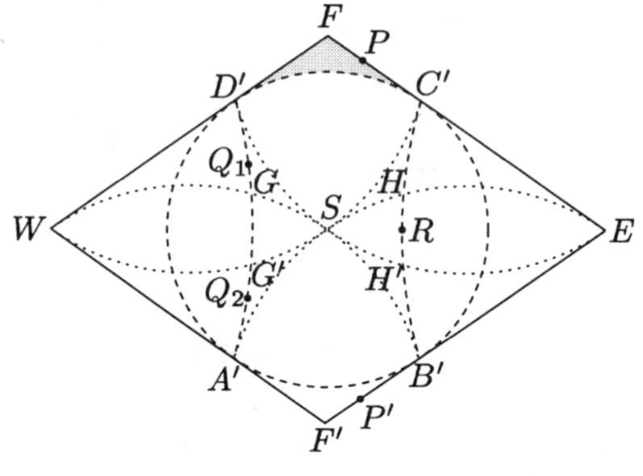

Figure 7.3

In the above stereographic representation of the complementary region Ω, S is the south pole; A', B', C' and D' are the antipodal points of A, B, C and D, respectively, while those dotted circular arcs represent the (spherical) $\pi/3$-circular arcs centered at S, E, W, A', B', C' and D', respectively. Since the remaining seven points of Σ must lie in Ω, we shall study all the geometric possibilities of accommodating seven $\frac{\pi}{3}$-separated points inside Ω. First of all, the f.c.c. and the h.c.p. configurations are, of course, such possibilities, namely, as indicated in Figure 7.3,

$$\begin{aligned} \text{f.c.c. config.} &\longleftrightarrow \{E, W, S, A', B', C', D'\}\,, \\ \text{h.c.p. config.} &\longleftrightarrow \{E, W, A', B', F, G, H\} \text{ or} \\ &\qquad \{E, W, D', C', F', G', H'\}\,. \end{aligned}$$

Therefore, we shall prove that the above three septuples are the *only* geometric possibilities of *accommodating a septuple of* $\frac{\pi}{3}$*-separated points inside* Ω.

Let us first consider the case that both E and W belong to Σ. Then, the remaining five points must be situated in the subregion Ω_1 bounded by the dotted circular arcs $A'F'B'C'FD'A'$.

Let us begin with the quantitative analysis of $(\mu - \gamma - \gamma')$ of the configuration as indicated in Figure 7.4, where $\overline{WS_u} = \frac{\pi}{2} + u$.

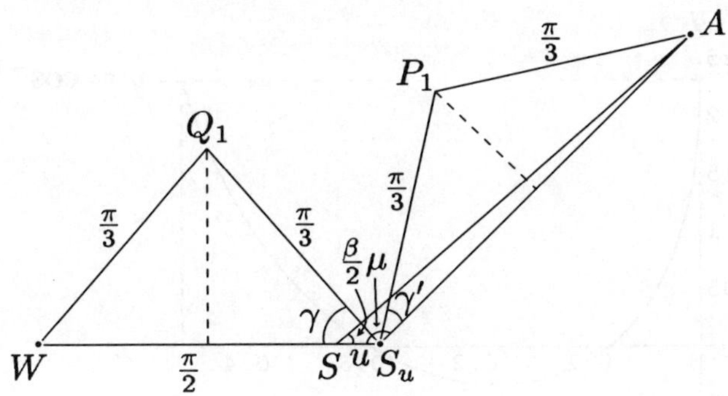

Figure 7.4

$$\begin{aligned}
\cos\gamma &= \frac{1}{\sqrt{3}}\tan\left(\frac{\pi}{4}+\frac{u}{2}\right) = \frac{1}{\sqrt{3}}\frac{\cos u}{1-\sin u}\,,\\
\cos\overline{AS}_u &= \frac{\sqrt{3}}{2}\sin u\cos\frac{\beta}{2} - \frac{1}{2}\cos u = \frac{1}{2}(\sin u - \cos u)\,,\\
\cos\gamma' &= \frac{1}{\sqrt{3}}\tan\left(\frac{1}{2}\overline{AS}_u\right) = \frac{1}{\sqrt{3}}\frac{\frac{1}{2}\sqrt{3+2\sin u\cos u}}{1+\frac{1}{2}(\sin u-\cos u)}\,,\\
\sin\mu &= \frac{\sqrt{3}}{2}\frac{\sin\frac{\beta}{2}}{\sin\overline{AS}_u} = \frac{1}{\sqrt{2}\sin\overline{AS}_u}\,.
\end{aligned} \tag{4}$$

Set $x = \sin u$. It is not difficult to use (4) to express $(\mu - \gamma - \gamma')$ as a function of x, namely

$$\begin{aligned}
\mu - \gamma - \gamma' = f(x) = \pi - \sin^{-1}&\left[\frac{1}{\sqrt{2}\sqrt{1-\frac{1}{4}(x-\sqrt{1-x^2})^2}}\right]\\
&- \cos^{-1}\frac{\sqrt{1-x^2}}{\sqrt{3}(1-x)} - \cos^{-1}\frac{\sqrt{3+2x\sqrt{1-x^2}}}{2\sqrt{3}(1+\frac{1}{2}(x-\sqrt{1-x^2}))}\,,
\end{aligned} \tag{4'}$$

while the following graph is the computer plot of $f(x)$, $0 \le x \le \sin 0.474$:

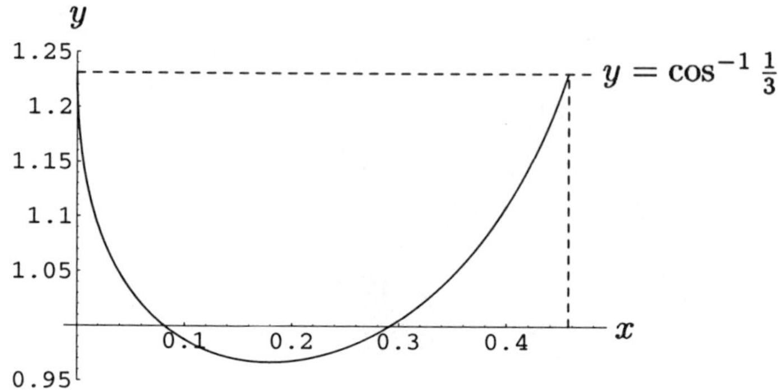

Set $r_0 = 0.474$ and $D_{r_0}(S)$ to be the spherical disc of radius r_0 centered at S. Using the above result on $(\mu - \gamma - \gamma')$, we shall prove that $\Sigma \cap D_{r_0}(S)$ is either $\{S\}$ or empty. Suppose on the contrary that $\Sigma \cap D_{r_0}(S) = \{S'\}$, $S' \neq S$. Set $u = \overline{SS'}$, and

$$\Omega_2 = \Omega_1 \setminus D^{\circ}_{\frac{\pi}{3}}(S') = \Omega_2^+ \cup \Omega_2^- ,$$

where $\Omega_2^{\pm}$ are the parts of Ω_2 situated in the upper and the lower half-planes. In the special case that $S' = S_u$, Ω_2^+ and Ω_2^- are reflectionally symmetric with respect to the x-axis and their diameters are *less than* $\frac{\pi}{3}$ (cf. Figure 7.5). In the general case that S' is situated in the upper (resp. lower) half-plane, the diameter of Ω_2^+ (resp. Ω_2^-) becomes smaller than the above symmetric case, while that of the other becomes larger. However, even the larger one can only accommodate at most a pair of $\frac{\pi}{3}$-separated points. Therefore, Ω_2 can, altogether, accommodate at most a triple of $\frac{\pi}{3}$-separated points which contradicts the assumption that $\#(\Sigma \cap \Omega_1) = 5$. Hence, $\Sigma \cap D_{r_0}(S)$ is either empty or $\{S\}$.

In the case that $S \in \Sigma$, it is quite obvious that the remaining quadruple of Σ must be $\{A', B', C', D'\}$. In the other case that all points of Σ are outside of $D_{r_0}(S)$, one may assume that there is a point above the circular arc of $D'C'$ as indicated by the shaded subregion in Figure 7.3. It is easy to see that there can be only one

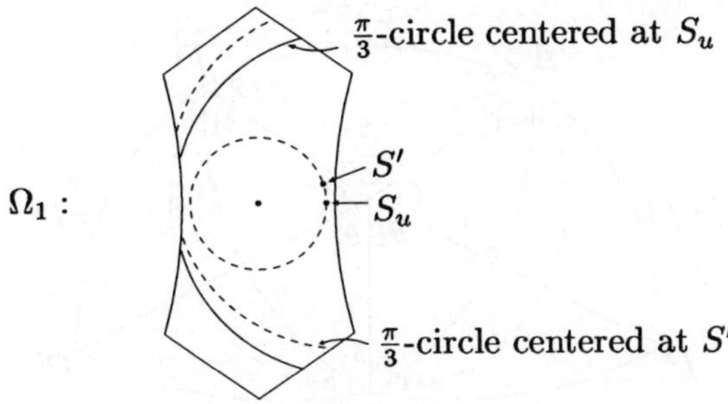

Figure 7.5

such point. We claim that such a point must be F and the remaining four points must be $\{A', B', G, H\}$. If either A' or B' is one of such five points in Ω_1, then one may assume without loss of generality that both of them belong to a set of five points in Ω_1 with $\pi/3$-separation, and moreover, it is then quite easy to show that such a set must be $\{F, A', B', G, H\}$. Therefore, it suffices to prove the *non-existence* of such a set with one point above $\widehat{D'C'}$ and another point below $\widehat{A'B'}$, say $P \neq F$ and $P^* \neq F'$. Suppose on the contrary that there exists such a set of five points. Again, one may assume without loss of generality for the proof of contradiction that both P and P^* are situated on the boundary of Ω_1. Set Q_1 (resp. R) to be the point on the circular arc $\widehat{D'A'}$ (resp. $\widehat{C'B'}$) with $\overline{PQ_1} = \pi/3$ (resp. $\overline{PR} = \pi/3$). Moreover, set P' (resp. Q_2) to be the point on the circular arc $\widehat{B'F'}$ (resp. $\widehat{A'D'}$) such that $\overline{P'R} = \pi/3$ (resp. $\overline{P'Q_2} = \pi/3$). Now, we shall analyze the spherical geometry of the above five points (so constructed).

The spherical geometry of the relative positions of $\{E, A, D, R, P, P'\}$ are as indicated in Figure 7.6. Set $\triangle$ to be the area of $\widehat{\triangle ARD}$.

$$\begin{aligned} &\triangle = 2\alpha_1 + 2\delta_1 + \beta - \pi\,, \quad 2\alpha_1 + 2\delta_1 = 2\pi - \varphi\,, \\ &\pi - \varphi = \triangle - \beta\,, \quad \theta_1 + \theta_1' = \pi - \frac{1}{2}\beta\,. \end{aligned} \tag{5}$$

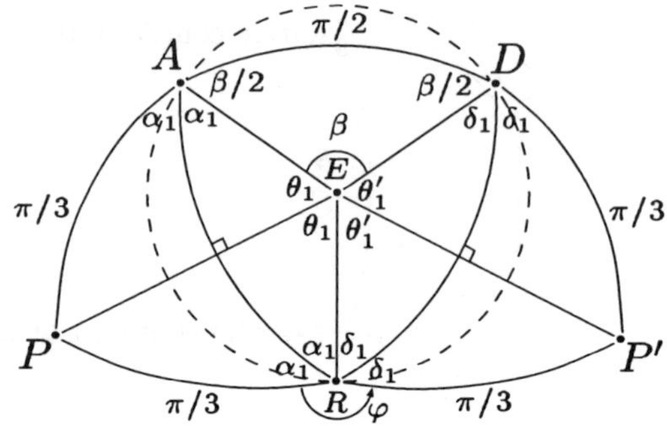

Figure 7.6

Set $\xi = \cos(\theta_1 - \theta_1')$ and u, D to be the ones associated to $\widehat{\triangle} ARD$. Then,

$$u = \frac{3}{2} + \frac{3}{4}(\cos 2\theta_1 + \cos 2\theta_1') = \frac{1}{2}(3 - \sqrt{3}\xi)\,,$$

$$D^2 = 1 - \frac{1}{16}\{(1 + 3\cos 2\theta_1)^2 + (1 + 3\cos 2\theta_1')^2\} = \frac{1}{8}(1 + \sqrt{3}\xi)^2\,, \quad (6)$$

$$\triangle = 2\tan^{-1}\frac{D}{u} = 2\tan^{-1}\frac{1}{\sqrt{6}}\left(\frac{1+\sqrt{3}\xi}{\sqrt{3}-\xi}\right).$$

Next let us analyze the geometry of the configuration as indicated in Figure 7.7. The following spherical triangles are all $\pi/3$-isosceles ones, namely, $\widehat{\triangle}\ PRE, \widehat{\triangle}\ PQ_1W, \widehat{\triangle}\ P'RE, \widehat{\triangle}\ P'Q_2W$, $\widehat{\triangle}\ RPR', \widehat{\triangle}\ RP'R'$, while their base-angles are respectively given as $\theta_1, \theta_2, \theta_1', \theta_2'$ and $\varphi/2$. Set $\overline{EP}$ (resp. $\overline{PW}$, $\overline{P'E}$, $\overline{P'W}$) to be $2b_1$ (resp. $2b_2$, $2b_1'$, $2b_2'$), and $\angle Q_1PR = \psi, \angle Q_2P'R = \psi', \angle R'PR = \overline{\varphi}, \angle Q_2WQ_1 = \lambda$.

Then, one has the following set of equations among them, namely

$$\begin{gathered}
b_1 + b_2 = \pi/2\,, \quad \tan b_1 = \sqrt{3}\cos\theta_1\,, \quad \tan b_2 = \sqrt{3}\cos\theta_2\,, \\
b_1' + b_2' = \pi/2\,, \quad \tan b_1' = \sqrt{3}\cos\theta_1'\,, \quad \tan b_2' = \sqrt{3}\cos\theta_2'\,, \\
\psi = \pi - \theta_1 - \theta_2\,, \quad \psi' = \pi - \theta_1' - \theta_2'\,, \\
\theta_1 + \theta_1' = \pi - \frac{\beta}{2}\,, \quad \theta_2 + \theta_2' + \lambda = \pi - \frac{\beta}{2}\,.
\end{gathered} \quad (7)$$

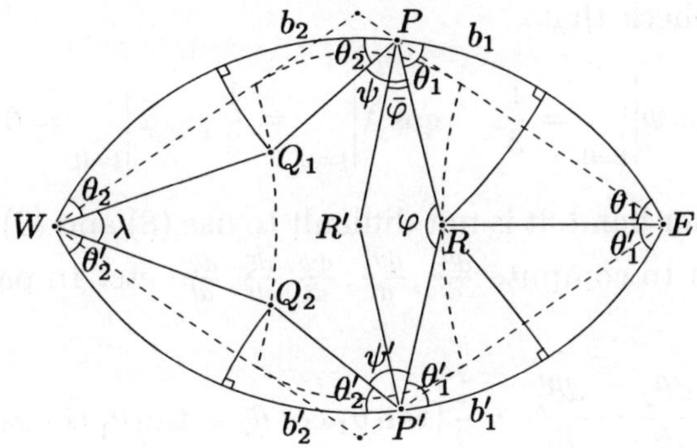

Figure 7.7

Therefore,

$$\cos\theta_1\cos\theta_2 = \frac{1}{3}\,, \quad \cos\theta_1'\cdot\cos\theta_2' = \frac{1}{3}\,,$$

$$\begin{aligned}\cos\psi &= -\cos\theta_1\cos\theta_2 + \sin\theta_1\sin\theta_2 \\ &= -\frac{1}{3} + \left[1 - \cos^2\theta_1 - \cos^2\theta_2 + \frac{1}{9}\right]^{1/2} \\ &= -\frac{1}{3} + \frac{1}{3\sqrt{2}}\left\{\frac{(1-\cos 2\theta_1)(7+9\cos 2\theta_1)}{1+\cos 2\theta_1}\right\}^{1/2}, \\ \cos\psi' &= -\frac{1}{3} + \frac{1}{3\sqrt{2}}\left\{\frac{(1-\cos 2\theta_1')(7+9\cos 2\theta_1')}{1+\cos 2\theta_1'}\right\}^{1/2}.\end{aligned} \tag{8}$$

Set $2\theta_1 = \beta + t$. Then, $2\theta_1' = 2\alpha - t$,

$$\begin{aligned}&\cos(\beta+t) + \cos(2\alpha - t) = \cos 2\theta_1 + \cos 2\theta_1' = -\frac{2}{\sqrt{3}}\xi\,, \\ &\cos\frac{\overline{\varphi}}{2} = \frac{1}{\sqrt{3}}\tan\sin^{-1}\left\{\frac{\sqrt{3}}{2}\cos\left[\tan^{-1}\frac{1}{\sqrt{6}}\right.\right. \\ &\qquad \left.\left.\times\left(\frac{1+\sqrt{3}\xi}{\sqrt{3}-\xi}\right) - \frac{1}{2}\cos^{-1}\left(-\frac{1}{3}\right)\right]\right\}.\end{aligned} \tag{9}$$

It is easy to check that

$$\cos\psi\Big|_{t=0} = \frac{1}{3}\,, \quad \cos\lambda\Big|_{t=0} = \frac{1}{3}\,, \quad \overline{\varphi}\Big|_{t=0} = 0\,. \tag{10}$$

On the other hand, it is not difficult to use (8) and (9) and implicit differentiation to compute $\frac{d\theta_2}{dt}, \frac{d\theta_2'}{dt}, \frac{d\psi}{dt}, \frac{d\xi}{dt}, \frac{d\overline{\varphi}}{dt}$, etc. In particular, one has

$$\frac{d\lambda}{dt} = -\frac{d\theta_2}{dt} - \frac{d\theta_2'}{dt} = \frac{1}{2}\{\tan\theta_1 \cot\theta_2 - \tan\theta_1' \cot\theta_2'\}\,,$$

$$\frac{d\psi}{dt}\Big|_{t=0} = 0\,, \quad \frac{d\overline{\varphi}}{dt}\Big|_{t=0} > 0\,, \quad (\psi - \overline{\varphi})\Big|_{t=(\alpha-\frac{\beta}{2})} = \alpha - 0.08058707\,. \tag{11}$$

The following are computer plots of the graphs of $\lambda(t)$ and $(\psi-\overline{\varphi})(t)$ as t varies from 0 to $(\alpha - \frac{\beta}{2})$:

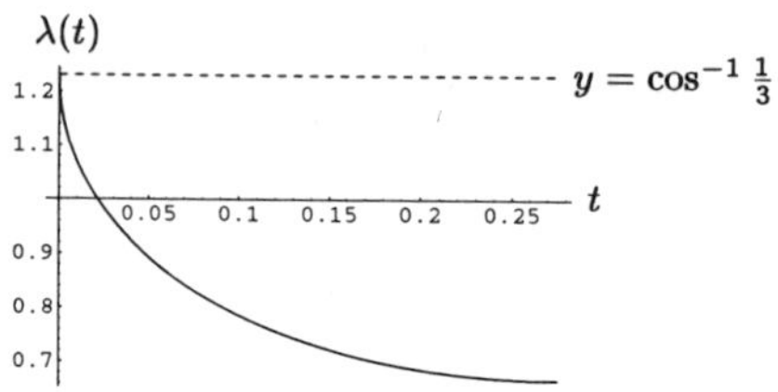

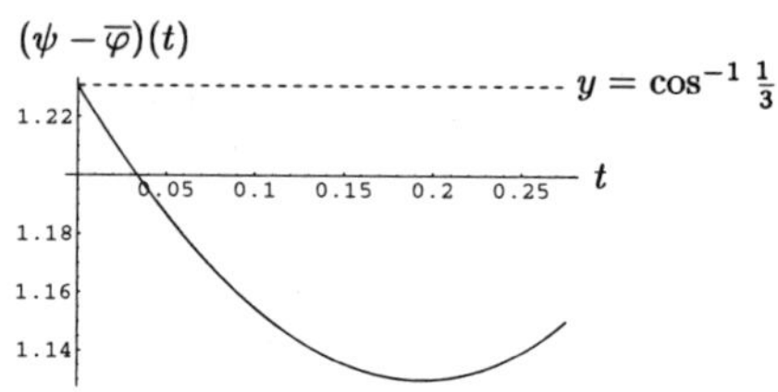

Set Ω_3 to be the complement of $D_{r_0}(S) \cup D^{\circ}_{\frac{\pi}{3}}(P)$ in Ω_1. We shall use the result of the above geometric analysis on the configuration of Figure 7.7 to prove that Ω_3 *cannot* accommodate a quadruple of $\frac{\pi}{3}$-separated points, thus proving the non-existence of such a Σ. Let us first introduce the following useful concept of *replacement set*.

Definition Let P be a point in a given region Ω. The *replacement set* of P in Ω with respect to a given separation δ, such as $\frac{\pi}{3}$, is defined to be

$$R_\delta(P,\Omega) = \{X \in \Omega;\ \Omega \setminus D^{\circ}_{\delta}(X) \subset \Omega \setminus D^{\circ}_{\delta}(P)\}\,. \tag{12}$$

In this section, we shall always assume that $\delta = \frac{\pi}{3}$ and hence, omit the specification of δ from the above notation.

For example, as indicated in Figure 7.8, the replacement set of R in Ω_3 covers almost entirely a rather large neighborhood of R in Ω_3 (i.e. except a tiny region to the west of R' in the case that R' is still outside of $D_{r_0}(S)$). Therefore, for the sake of the *proof of non-existence*, one may assume that either $R \in \Sigma$ or Σ contains a point in that tiny region to the west of R':

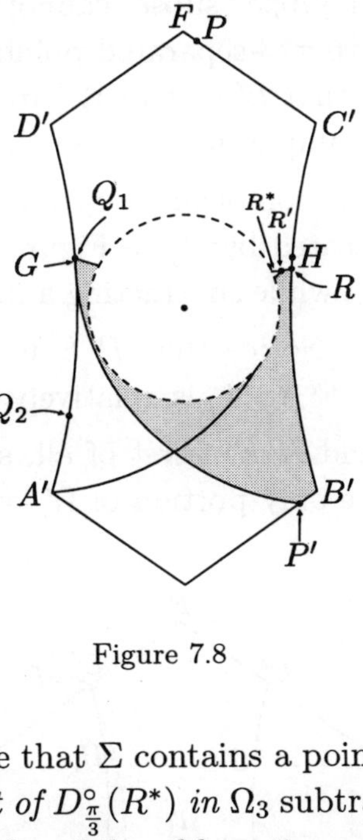

Figure 7.8

However, in the case that Σ contains a point, say R^*, in that tiny region, the *complement of* $D^\circ_{\frac{\pi}{3}}(R^*)$ *in* Ω_3 subtracts a substantial (but narrow) region below Q_1 while adds a tiny corner above P' to the *complement of* $D^\circ_{\frac{\pi}{3}}(R)$ *in* Ω_3. Therefore, the proof of non-existence of a triple of $\frac{\pi}{3}$-separated points in such a region is quite obvious. Thus, we shall only prove that of the latter as follows.

Set Ω_4 to be the complement of $D^\circ_{\frac{\pi}{3}}(R)$ in Ω_3. Again, by the method of replacement set of Q_1 in Ω_4, such a proof can be reduced to show that the complement of $D^\circ_{\frac{\pi}{3}}(Q_1)$ in Ω_4 cannot accommodate a pair of $\frac{\pi}{3}$-separated points. Geometrically, this is a direct consequence of the fact that $\overline{Q_1Q_2}$ is less than $\frac{\pi}{3}$. This proves the non-existence of such a quintuple of $\frac{\pi}{3}$-separated points inside Ω_1.

Finally, let us prove the non-existence of such a septuple without using both E and W. If E' (resp. W') lies inside the region bounded by the circular arcs $EHH'E$ (resp. $WGG'W$), as indicated by the shaded ones in Figure 7.9, then the complementary region of $D^\circ(E') \cup D^\circ(W')$ in Ω is a *proper subset* of Ω_1 and it follows directly from the above proof that such a *proper* subset cannot accommodate the only three possibilities of five $\pi/3$-separated points. Therefore, it suffices to consider those cases that E' and/or W' are situated *outside* of the shaded regions. For example, in the case that E' is situated below $\overset{\frown}{H'E}$ while $W' = W$, the complementary region of $D^\circ(E') \cup D^\circ(W)$ in Ω is the one as indicated by Ω_1' in Figure 7.9, which adds a tiny wedge to the east of C', while subtracting a much larger wedge along the west-side of $\overset{\frown}{B'C'}$. Notice that $\overset{\frown}{H'E}$ is tangent to $\overset{\frown}{B'E}$ at E. Thus, the distance of E' to E is relatively much larger than the distance between E' and $\overset{\frown}{H'E}$. First of all, since the tiny wedge in the vicinity of C' is the only portion of Ω_1' outside of Ω_1, it suffices

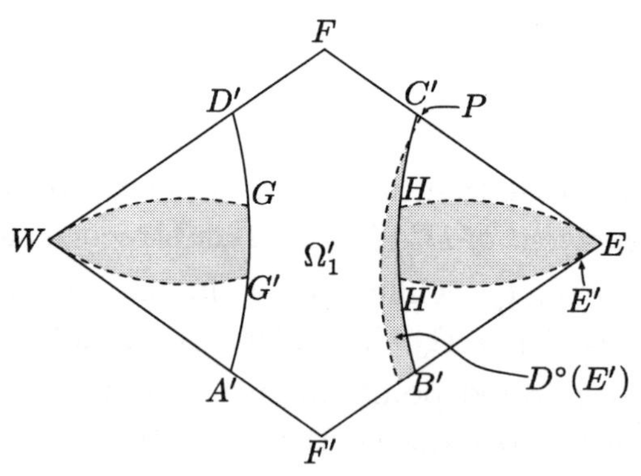

Figure 7.9

to consider the case that P actually belongs to that tiny wedge for the purpose of *the proof of non-existence.* However, such a set of $\pi/3$-separated five points inside of Ω_1' can be realized even inside the enlarged region of Ω_1 by adding that tiny wedge *only* by tiny deformation of $\{C', D', G', H', F'\}$. Therefore, it is impossible to accommodate such a tiny deformation inside of Ω_1' because a much wider strip is subtracted from Ω_1 along $\widehat{H'B'}$. This completes the proof of Lemma 7.1.1 for the case that the lengths of the four edges are all equal to $\pi/3$. However, the above proof clearly also shows that the lengths of the four edges must all be equal to $\pi/3$ lest its complementary region cannot possibly accommodate a septuple of $\pi/3$-separated points. □

Remarks (i) The above proof not only proves that $\{E, W, S, A', B', C', D'\}$, $\{E, W, A', B', F, G, H\}$ and $\{E, W, D', C', F', G', H'\}$ are the only arrangements of seven $\frac{\pi}{3}$-separated points inside the complementary region Ω of Figure 7.3, it also follows from the quantitative analysis of the above proof that any deformation of the above three arrangements of a septuple of points within Ω must be accompanied by a certain amount of *decrement in separation roughly of the size of the deformation.*

(ii) The characterization of the only two type I configurations with six quadrilaterals by the $(\alpha, \beta, \alpha, \beta)$-angular arrangement at a single vertex is, by itself, a remarkable geometric result. Moreover, such a remarkable characterization of both f.c.c. and h.c.p. local packing by a single local arrangement of $(\alpha, \beta, \alpha, \beta)$-type greatly facilitates the task of collective volume estimate for the proof of Main Theorem I in the above two most critical cases (cf. Lemma 7.2.2).

7.1.2 *Type* I *configuration with at least one 4-fork vertex*

Lemma 7.1.1 asserts that the mere existence of at least one 4-fork vertex with $(\alpha, \beta, \alpha, \beta)$-angular arrangement already suffices to characterize a type I configuration to be either f.c.c. or h.c.p. Such a result clearly suggests that the *existence of at least one 4-fork vertex*, in general, may already constitute a rather restrictive condition for type I configurations. Thus, it is natural to study this subfamily of type I configurations so as to understand the overall geometric implications of such a restrictive condition.

Let us begin with some examples of type I configurations with some 4-fork vertices, other than the f.c.c. and h.c.p. configurations.

A family of examples with five quadrilaterals Let (r, θ) be the spherical polar coordinate system with the north pole, N, as the base point and $\{\theta_j; 1 \le j \le 5\}$ be five angles with angular separation of at least α, namely

$$\{\delta_j = \theta_{j+1} - \theta_j,\ 1 \le j \le 5\}, \quad \delta_j \ge \alpha, \quad \Sigma\delta_j = 2\pi. \tag{13}$$

Set $\Sigma = \{N, S; A_j, B_j; 1 \le j \le 5\}$ to be the set of twelve points on $S^2(1)$ where the polar coordinates of A_j and B_j are respectively $(\pi/3, \theta_j)$ and $(2\pi/3, \theta_j)$. Then, it is easy to check that $\mathcal{S}(\Sigma)$ is a type I configuration while $\{A_j, B_j; 1 \le j \le 5\}$ are all vertices of 4-fork type. As indicated in Figure 7.10, such a type I configuration has two clusters of quintuple of $\pi/3$-isosceles triangles around the two

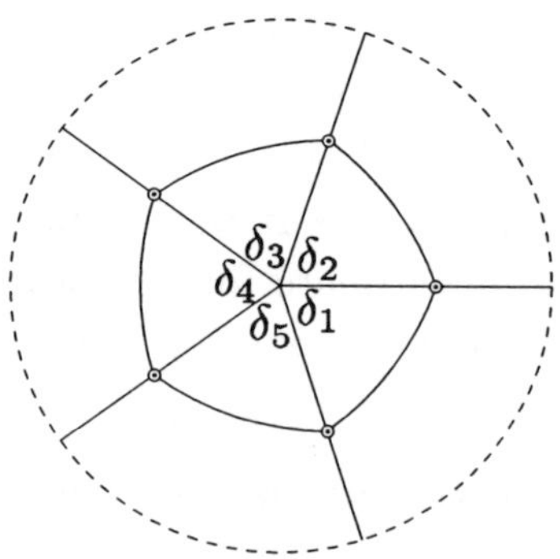

Figure 7.10

poles $\{N, S\}$ and five quadrilaterals arranged around the equator. Then, the congruence class of such a type I configuration is uniquely determined by its angular distribution, namely, (13). Notice that $2\pi - 5\alpha = 0.128388\cdots$. Thus, each one is just a small variation of the one with $\delta_j = \frac{2\pi}{5}$.

Small deformations of the above 5□-ones Let $\mathcal{S}(\Sigma)$ be one of the above family of type I configurations with five quadrilaterals, such as the one with $\delta_j = 2\pi/5, 1 \leq j \leq 5$. Then, there are always some small deformations of such a 5□-type I configuration which are still type I configurations and with at least one 4-fork vertex. For example, let $\mathcal{S}(\Sigma_1)$ be such a 5□-one with $\delta_1 = \delta_2 = \alpha$. Then, the local geometric structure of $\mathcal{S}(\Sigma_1)$ at A_2 (resp. B_2) is exactly the same as the second one of Figure 7.2. Therefore, one may deform such a 5□-one within type I configurations while fixing its local geometric structure at A_2, thus obtaining many examples of type I configurations, other than the h.c.p. configuration, which also have some 4-fork vertices of the $(\alpha, \alpha, \beta, \beta)$-angular arrangement.

Lemma 7.1.2 *Let $\mathcal{S}(\Sigma)$ be a type I configuration which has at least one vertex with four $\pi/3$-forks of the $(\alpha, \alpha, \beta, \beta)$-angular arrangement. Then, $\mathcal{S}(\Sigma)$ is either congruent to the h.c.p. configuration or congruent to a small deformation of the above 5□-ones.*

Proof Up to a congruence, one may assume that such a vertex of $\mathcal{S}(\Sigma)$ is situated at the north pole N. We shall again denote its four neighboring vertices by A, B, C and D, respectively. Set Ω to be the *complementary region* of the union of the five open $\pi/3$-disks centered at N, A, B, C and D, respectively. It is a region inside the southern hemisphere and we shall again represent it by the stereographic projection. It is now a trapezoid, as indicated in Figure 7.11, instead of the parallelogram in the case of $(\alpha, \beta, \alpha, \beta)$-angular arrangement.

The dotted circular arcs represent the spherical $\pi/3$-circular arcs centered at A', C', S, D', L and R, respectively, while the solid

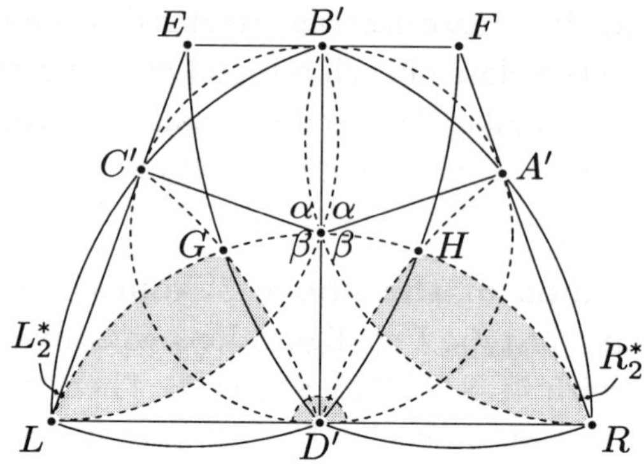

Figure 7.11

circular arcs represent the spherical geodesic arcs passing through $\{L, D'\}$, $\{D', R\}$, $\{R, A', B'\}$, $\{B', C', L\}$, $\{E, G, D'\}$ and $\{F, H, D'\}$, respectively.

Now, let us proceed to determine all the possible ways of arranging $\frac{\pi}{3}$-separated septuples inside the above complementary region Ω. First of all, the h.c.p. configuration and those 5□-ones provide the following corresponding arrangements of septuples of points, namely

(i) h.c.p. $\longleftrightarrow \{L, R, D', S, A', B', C'\}$,
(ii) 5□-ones $\longleftrightarrow \{L, R, D', E', F', G', H'\}$,

where E' (resp. F') is situated on the circular arc of $EB'F$ and G' (resp. H') is the intersection point of the geodesic arc of $D'E'$ (resp. $D'F'$) and the $\pi/3$-circle centered at D'.

The proof of Lemma 7.1.2 amounts to showing that the only ways of arranging such a septuple of points inside Ω are either exactly the ones corresponding to the h.c.p. configuration or *small deformations* of the second type of the above examples.

Notice that the angle between the pair of geodesic rays $D'GE$ and $D'HF$ is equal to $(2\pi - 4\alpha)$. Therefore, there is such a little angular room for H to swing westward so as to allow D' moving upward by a tiny distance. As indicated in Figure 7.12, the shortening of $\overline{D'F}$

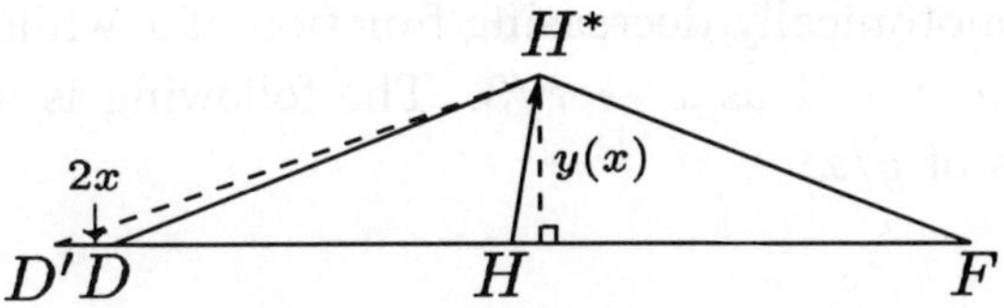

Figure 7.12

by an amount of $2x$ will produce a westward buckling of $H \to H^*$ by the amount of y while $y(x)$ is given by

$$\cos\left(\frac{\pi}{3} - x\right) \cdot \cos y = \frac{1}{2}. \tag{14}$$

The maximal room for such a buckling is approximately equal to 0.113, which means that D' can move toward F by at most 0.00743. Anyhow, the above brief analysis of some possible deformations of those 5□-ones actually suggests that all such type I deformation of the 5□-ones within Ω can only change the position of D' by a tiny amount. The above intuitive grasp of the *necessity of having a point in the vicinity of* D' for all possible arrangements of $\pi/3$-separated septuples inside Ω is actually the key insight of the following proof of Lemma 7.1.2.

Let us begin with some simple technical preparations that will be handy to use in the type of problem that we are going to study.

The magnified buckling phenomenon As indicated in Figure 7.12 and formula (14), a small shortening of $D'HF$ will produce a much larger buckling of H to H^*. The quantitative correlation between them is given by (14). Differentiate (14) with respect to x, one gets

$$y' = \cot y \tan\left(\frac{\pi}{3} - x\right), \tag{15}$$

which is a monotonically decreasing function of x while $y' \to \infty$ as $x \to 0_+$ and $y' \to 0$ as $x \to \pi/3$. Therefore, the ratio between y and

x is also a monotonically decreasing function of x while $y/x \to \infty$ as $x \to 0_+$ and $y/x \to 1$ as $x \to \pi/3$. The following is a list of a few specific values of y/x:

x	0.001	0.005	0.01	0.02	0.04	0.08
y/x	58.806	26.208	18.453	12.938	8.998	6.163

Note that the *replacement sets* of R (resp. L) in Ω are the pair of shaded regions as indicated in Figure 7.11. However, the replacement sets of R (resp. L) in the sub-region $\Omega' = \Omega \backslash D^\circ_\varepsilon(D')$, $\varepsilon = 0.15$, will be enlarged by adding a strip along $\overset{\frown}{HR}$ (resp. $\overset{\frown}{GL}$) as indicated by the pair of shaded regions in Figure 7.13.

$\Omega' = \Omega \setminus D^\circ_\varepsilon(D')$:

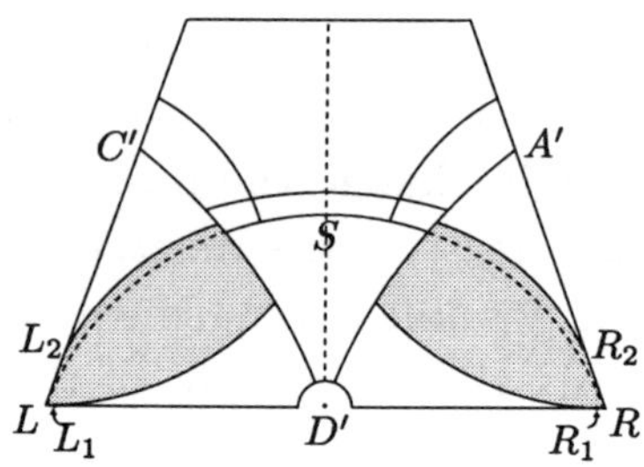

$\overline{RR_2} = \overline{LL_2}$
$= 2\tan^{-1} 2\sin 0.075$
$= 0.2975049$
(L_1, R_1 very close to L, R)

Figure 7.13

Sublemma 7.1.1 *A septuple of $\pi/3$-separated points in Ω of Figure* 7.11 *necessarily contains a point D^* with $\overline{D'D^*} < 0.15$.*

Proof It suffices to prove the *non-existence* of a $\frac{\pi}{3}$-separated septuple inside of $\Omega' = \Omega \setminus D^\circ_\varepsilon(D')$. By the method of replacement sets, it suffices to prove the non-existence of such a septuple in the unshaded subregion of Ω'. Let D^* (resp. L^*, R^*) be the point of such a *hypothetical* septuple which is closest to D' (resp. L, R). Again, by the method of replacement (i.e. for the unshaded subregion of Ω'), one may assume without loss of generality that L^* (resp. R^*) is either L (resp. R) or L_2 (resp. R_2), or L_1 (resp. R_1) very close to L (resp. R) as indicated in Figure 7.13.

In the case that $L^* = L$ and $R^* = R$, one may also assume that D^* belongs to the ε-circular arc outside of $D^\circ(L)$ and $D^\circ(R)$. Therefore, the remaining quadruple lies inside of $\Omega_3 = \Omega \backslash \{D^\circ(L) \cup D^\circ(R) \cup D^\circ(D^*)\}$ as indicated in Figure 7.14 by the region above the highlighted circular arcs.

Subdivide Ω_3 into a pair of subregions by $\overline{B'S}$. Each of them *cannot* accommodate a triple of $\pi/3$-separated points. Moreover, since a strip of roughly 0.15 in width has been cut off along $\widehat{GH}$, any pair of $\pi/3$-separated points inside one of the above subregions necessarily has at least 0.4 angular difference in their longitudes with $\{D, D'\}$ as the poles. Hence, it is impossible to put together two such pairs, one pair in each subregion, to form the remaining $\frac{\pi}{3}$-separated quadruple. This proves the *non-existence* of a $\pi/3$-separated septuple including L and R. In the remaining case that either $L^* = L_i$ or $R^* = R_i$, $i = 1$ or 2, one may again assume that D^* belongs to the ε-circular arc outside of $D^\circ(L^*)$ and $D^\circ(R^*)$ which is actually the same as that of the above case. Therefore, the remaining quadruple of the hypothetical $\frac{\pi}{3}$-separated septuple lies inside of a proper subset or a tiny variant (in the case of $i = 1$) of Ω_3 which is, again, impossible.

This completes the proof. □

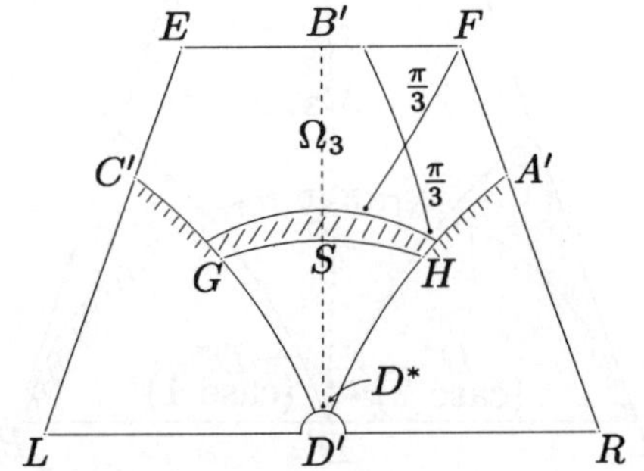

Figure 7.14

Sublemma 7.1.2 *If D' is one of a given $\frac{\pi}{3}$-separated septuple inside Ω, then such a septuple corresponds to either the h.c.p. or a small deformation of an 5□-one.*

Proof Let Σ_1 be such a septuple in Ω. Again, let us first consider the special case that both L and R also belong to Σ_1. Then, the remaining quadruple lies inside the region bounded by $\pi/3$-circular arcs of $GHA'FEC'G$. By the technique of replacement, we can assume that all four points are situated on the above chain of boundary arcs. In such a case, it is not difficult to use magnified buckling effect to show that the quadruple is either $\{A', B', C', S\}$ or a small deformation of $\{E, F, G, H\}$. Note that the former cannot be deformed while the latter can only be deformed toward the interior by tiny amount. In the general case that $\{L^*, R^*\} \neq \{L, R\}$, either L^* or R^* must be as indicated in Figure 7.11 by L_2^* or R_2^*. Hence, the quadruple can only be a small deformation of $\{E, F, G, H\}$. This completes the proof. □

Sublemma 7.1.3 *Let Σ_2 be a given $\frac{\pi}{3}$-separated septuple inside of Ω but $D' \notin \Sigma_2$. Then, Σ_2 is a small deformation of a 5□-one.*

Proof By Sublemma 7.1.1, Σ_2 must contain a point D^* inside of $D_\varepsilon^\circ(D')$, $\varepsilon = 0.15$. In view of the symmetry of Ω with respect to $\overline{B'D'}$,

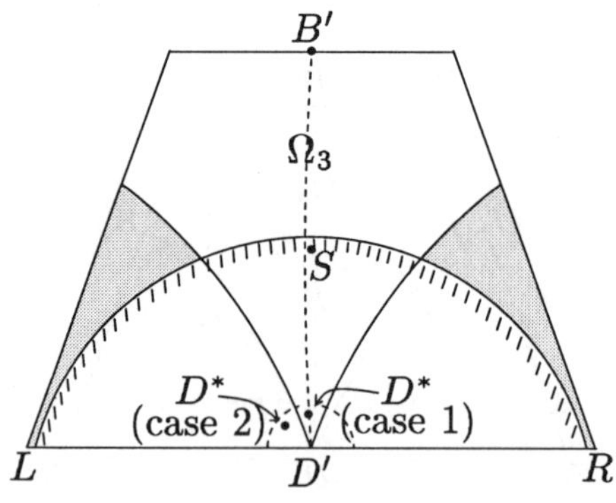

Figure 7.15

it suffices to consider those cases that D^* is situated in the left-half of $D^\circ_\varepsilon(D')$. We shall divide the proof into two cases according to $D^* \notin D^\circ(L)$ or $D^* \in D^\circ(L)$ (cf. Figure 7.15).

Case 1: $D^* \notin D^\circ(L)$. As indicated in Figure 7.15, the subregion $\Omega_1 = \Omega \backslash D^\circ(D^*)$ has two very, very narrow wedges in the vicinity of L and R, respectively. Therefore, again by the method of replacement, Σ_2 can always be deformed to Σ'_2 which contains L^* and R^*, very, very close to L and R, respectively. From here, the same reason as that of Sublemma 7.1.2 shows that Σ'_2 and hence also Σ_2 must be a small deformation of a septuple that corresponds to 5□-one.

Case 2: $D^* \in D^\circ(L)$. In this case, as indicated in Figure 7.16, the subregion $\Omega_1 = \Omega \backslash D^\circ(D^*)$ may still include S, but such a situation will occur only when D^* is outside of $D^\circ(S)$, thus the magnified buckling effect will push L^* upward by an amount much larger than $\overline{D'D^*}$. Therefore, the subregion $\Omega_3 = \Omega \backslash \{D^\circ(L^*) \cup D^\circ(D^*) \cup D^\circ(R^*)\}$ will carve a strip along $\overset{\frown}{C'G}$ much wider than $\overline{D'D^*}$. This again makes the geometric possibility of remaining quadruple of Σ_2 necessarily a small deformation of that of a 5□-one. This completes the proof. □

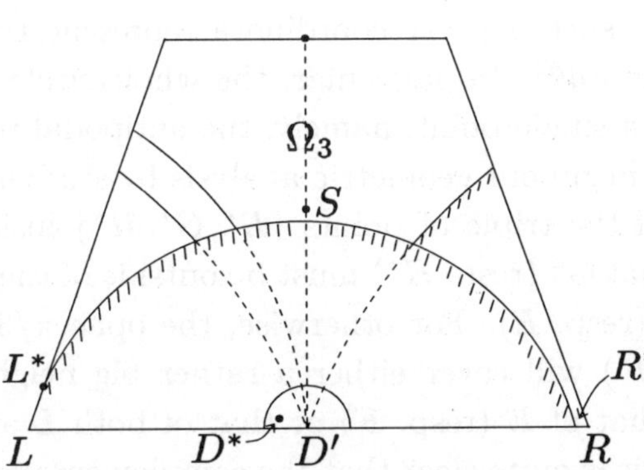

Figure 7.16

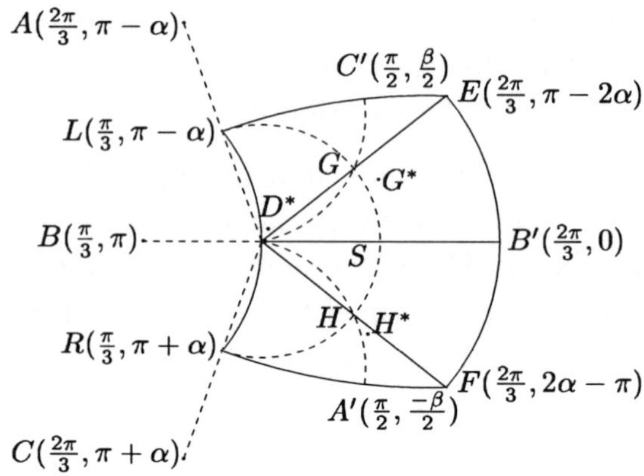

Figure 7.17

Finally, we shall proceed to obtain a much better upper bound on $\overline{D'D^*}$ by the following much more refined geometric analysis. The main result is that $\overline{D'D^*} < 0.018$.

In the polar coordinate representation of the geometry of Ω as indicated in Figure 7.17, the region Ω is bounded by the three (spherical) $\pi/3$-circular arcs centered at $A(\frac{2\pi}{3}, \pi-\alpha)$, $B(\frac{\pi}{3}, \pi)$, $C(\frac{2\pi}{3}, \pi+\alpha)$, respectively and the $\frac{2\pi}{3}$-circular arc centered at D', while the three dotted arcs inside Ω represent the (sphercial) $\frac{\pi}{3}$-circular arcs centered at L, D' and R, respectively. Notice that the actual spherical distance in such a polar coordinate representation is given by $ds^2 = dr^2 + \sin^2 r\, d\theta^2$. In particular, the whole circle of $r = \pi$ actually shrinks to a single point, namely, the antipodal point of D'.

Now, let us begin our geometric analysis by studying the possible arrangement of the triple of points $\{D^*, G^*, H^*\}$ inside Ω. First of all, we claim that G^* (resp. H^*) must be outside of the *open* $\pi/3$-disk centered at L (resp. R). For otherwise, the open $\pi/3$-disk centered at G^* (resp. H^*) will cover either a rather big neighborhood of L (resp. R), or that of E (resp. F) or that of both L and E (resp. R and F). Thus, it is quite clear that the complementary region of such an open $\pi/3$-disk in Ω can no longer accommodate the remaining six points. Secondly, we claim that D^* must be very close to D'. Set δ to be the distance between D^* and D'. We shall prove an upper bound estimate of $\delta < 0.018$ in the following.

Suppose on the contrary that there exists such a septuple with $\delta \geq 0.018$. We shall divide our derivation of the contradiction into two cases according to the angular coordinate of D^* is less than $(\pi - 2\alpha)$ or at least $(\pi - 2\alpha)$. Set (δ, θ) to be the polar coordinate of D^*.

Case 1: $0 \leq \theta < (\pi - 2\alpha)$, $\delta \geq 0.018$. Then,

$$\cos\rho(D^*, E) = \frac{\sqrt{3}}{2}\sin\delta\cos(\pi - 2\alpha - \theta) - \frac{1}{2}\cos\delta\,, \quad \delta \geq 0.018$$

$$\Rightarrow \rho(D^*, E) \leq \frac{2\pi}{3} - 0.01403636\,. \tag{16}$$

Note that the base angles of an isosceles spherical triangle with side-lengths $\{\frac{\pi}{3}, \frac{\pi}{3}, \frac{2\pi}{3} - 2\varepsilon\}$ is equal to

$$\cos^{-1}\left(\frac{1}{\sqrt{3}}\tan\left(\frac{\pi}{3} - \varepsilon\right)\right)\,. \tag{17}$$

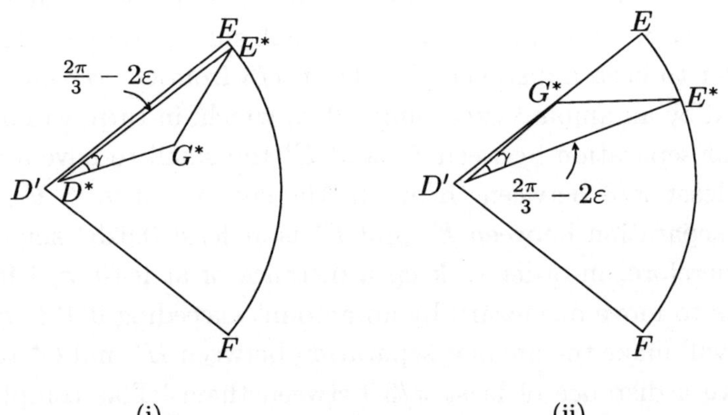

Figure 7.18

Thus, in the case $2\varepsilon \geq 0.0143636$, the base angles are at the least equal to $\gamma_0 = 0.1792$. Set the polar coordinates of G^* (resp. E^*) to be (r_1, θ_1) (resp. (r_2, θ_2)). At this juncture, we shall further subdivide this case into two subcases according to $\theta_2 > \theta_1$ or otherwise. The relative positions among $\{D^*, G^*, E^*\}$ for the two subcases are as indicated in Figure 7.18(i) and 7.18(ii).

Note that the above lower bound of 2ε will also become larger as δ becoming larger than 0.018. Therefore, in estimating the lower bound of $\{\theta_1, \theta_2\}$, it suffices to consider the most critical case of $\delta = 0.018$. Moreover, in estimating the lower bound of γ, as indicated in Figure 7.18(i) and (ii), it suffices to consider the most critical case that both $\rho(D^*, G^*)$ and $\rho(G^*, E^*)$ are equal to $\pi/3$. Thus, γ_0 is given in term of ε by formula (17) for such a critical case, while γ is bounded below by that of (17) in general. Using the method of area estimate, it is not difficult to see that

$$|\theta_2 - \theta_1| \geq \gamma_0 - 0.018 = 0.1612\,. \tag{18}$$

Thus, $\theta_1 \leq \pi - 2\alpha - 0.1612$ in the subcase of $\theta_2 > \theta_1$, while in the subcase of $\theta_1 > \theta_2$, $\theta_2 \leq \pi - 2\alpha - 0.1612 + 0.0074 = \pi - 2\alpha - 0.1538$. However, in both subcases, it is impossible to have the $\frac{\pi}{3}$-separation for $\{G^*, E^*, H^*, F^*\}$. In the subcase of $\theta_2 > \theta_1$, the angular separation between G^* and H^* is at least 0.0328 smaller than α. Therefore, in order to keep a distance of at least $\pi/3$ from G^*, H^* has to move upward by an amount exceeding 0.025, which, in turn, will make the angular separation between F^* and E^* too small to have a distance of at least $\pi/3$ between them. In the subcase of $\theta_1 > \theta_2$, the angular separation between E^* and F^* is at least 0.0254 smaller than α. Therefore, in order to keep a distance of at least $\pi/3$ from E^*, F^* has to move downward by an amount exceeding 0.018, which, in turn, will make the angular separation between H^* and G^* too small to have a distance of least $\pi/3$ between them. This completes the derivation of contradiction for Case 1.

Case 2: $\theta \geq \pi - 2\alpha$, $\delta \geq 0.018$. Since D^* may be assumed to be quite close to D', i.e. δ is quite small, θ can be, at most, slightly larger than $\pi/2$. Thus, θ can only varying within a range of 0.9, namely, between $(\pi - 2\alpha)$ and $\frac{\pi}{2} + 0.01$. In fact, for the range of $\pi - 2\alpha \leq \theta \leq 2\pi - 4\alpha$, Eq. (16) will again imply that $\rho(D^*, E) \leq \frac{2\pi}{3} - 0.01403636$. Thus, the same proof as in that of Case 1 will show such a subconfiguration is *impossible*. In the remaining case of $2\pi - 4\alpha \leq \theta < \frac{\pi}{2} + 0.01$, one has $\cos(\pi - \alpha - \theta) \geq 0.851851\cdots$. Thus,

$$\cos\rho(D^*, L) = \frac{\sqrt{3}}{2}\sin\delta\cos(\pi - \alpha - \theta) - \frac{1}{2}\cos\delta\,, \quad \delta \geq 0.018$$

$$\Rightarrow \rho(D^*, L) \leq \frac{2\pi}{3} - 0.0153585\cdots. \tag{19}$$

Therefore, again from (17), the θ-coordinate of L^* must be at least 0.186 smaller than that of L, namely, $\theta(L^*) \leq \pi - \alpha - 0.186$ which will, in turn, make the θ-coordinate of G^* decreasing by at least 0.18, namely, $\theta(G^*) = \theta_1 \leq \pi - 2\alpha - 0.18$. Hence, the same reason as that of Case 1 will show that such a subconfiguration is again *impossible.* This completes the proof that the upper bound estimate that $r(D^*) = \rho(D', D^*) < 0.018$. Moreover, the above proof not only proves that the radial parameter of D^* can only vary in a very small range, it *also* proves that the *radial parameters of the other six vertices* of such a subconfiguration *can only vary in a very small range.* Futhermore, the above rather precise estimates on the radial parameters of all the seven vertices of such a type I subconfiguration clearly implies that the *angular separations* between the following pair, namely $\{L^*, G^*\}, \{G^*, H^*\}, \{H^*, R^*\}, \{E^*, F^*\}$ can only be minutely smaller than $\alpha = \cos^{-1}(\frac{1}{3})$. All in all, this not only proves that such a type I subconfiguration is always a small deformation of those 5□-ones, but also provides a rather precise quantitative control of all such possible deformations. This completes the proof of Lemma 7.1.2. □

Remarks (i) In the slightly more general case of a 4-fork vertex with $(\alpha, \alpha, \beta, \beta)$-angular arrangement but with some of the four forks being slightly longer than $\pi/3$, the complementary region becomes a proper subset of the one as indicated in Figure 7.11. Therefore, it follows directly from Lemma 7.1.2 that $\mathcal{S}(\Sigma)$ must be also congruent to a small deformation of the $5\square$-ones.

(ii) Moreover, it follows from the proof of Lemma 7.1.2 that a 4-fork vertex with $(\alpha, \alpha, \beta, \beta)$-angular arrangement is possible to occur in a type I configuration *only* when its four forks are at most minutely longer than $\pi/3$.

Now, let us proceed to study the possibilities of 4-fork vertices with other angular arrangements that can be included in a type I configuration. Geometrically speaking, one studies what the types of 4-fork arrangements are whose complementary regions can accommodate the remaining seven vertices of a type I configuration (i.e. a $\frac{\pi}{3}$-separated septuple). Therefore, our first step is, naturally, the investigation of the geometric structure of the complementary region associated to a 4-fork arrangement. The geometry of a given 4-fork arrangement is, of course, completely determined by the lengths of its four forks and their *angular arrangement.* Since the special cases of $(\alpha, \beta, \alpha, \beta)$- and $(\alpha, \alpha, \beta, \beta)$-angular arrangements have already been thoroughly treated in Lemmas 7.1.1 and 7.1.2, we shall always assume that the angular arrangements are different from the above two in the following discussion of this subsection. Let us first consider the case that *all* the four forks are of the minimal length of $\pi/3$. For such a special case, the congruence class of a given 4-fork arrangement is completely determined by the four angles between the forks. It is, again, convenient to place the given 4-fork vertex at the north pole and represent the complementary region by stereographic projection.

Example 7.1.1 Suppose that the north pole N is a 4-fork vertex of a type I configuration with all the four forks of the minimal length

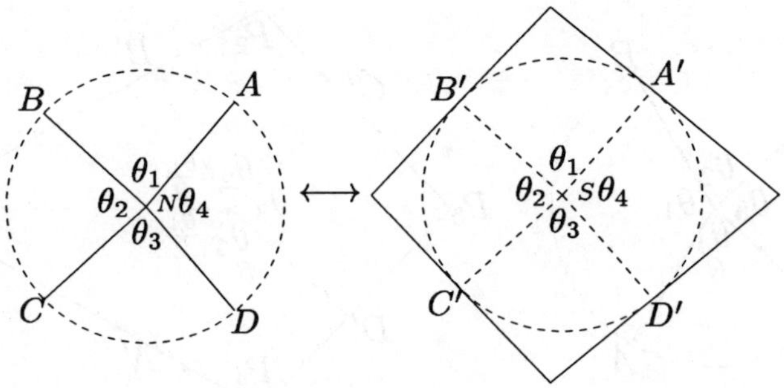

Figure 7.19

of $\pi/3$ and all the four angles less than or equal to $\beta = \cos^{-1}(-\frac{1}{3})$. Then, the remaining seven vertices of such a type I configuration all belong to the associated complementary region whose stereographic representation is the circumscribing "quadrilateral" as indicated in Figure 7.19.

The left-hand one is the south pole stereographic representation of the given 4-fork vertex with $\{\theta_i; 1 \le i \le 4\}$ as the angular arrangement, while the right-hand one is the north-pole stereographic representation of the associated complementary region, Ω. The four sides of the "quadrilateral" actually represent the (spherical) $\pi/3$-circular arcs centered at A, B, C and D, respectively, while the two dotted circles represent the (spherical) $\pi/3$-circles centered at N and S respectively.

Example 7.1.2 Let the north pole N be a 4-fork vertex of a type I configuration with all the four forks of the minimal length of $\pi/3$. Suppose that one of the four angles actually exceeds $\beta = \cos^{-1}(-\frac{1}{3})$, say $\theta_1 = \angle ANB > \beta$. Then, the remaining seven vertices of such a type I configuration all belong to the complementary set of the union of the five open $\pi/3$-disks centered at the five vertices *plus the interior of the circumcircle of* $\overset{\frown}{\triangle ANB}$. The stereographic representation of such a complementary region, Ω, is a truncated circumscribing quadrilateral as indicated in Figure 7.20, while the truncated part

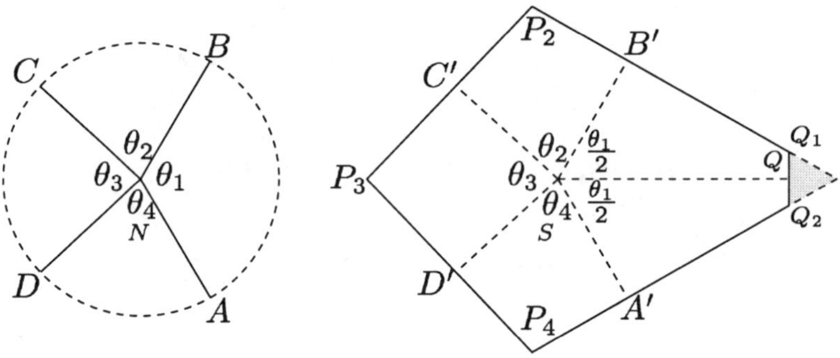

Figure 7.20

represents those points of the quadrilateral that lie inside the interior of the circumscribing circle of $\widehat{\triangle ANB}$, thus should be excluded by the "*convexity property*" of spherical configuration.

[Note that there can be at most one angle exceeding β, because each of the four angles must be at least equal to $\cos^{-1}(\frac{1}{3})$.]

Again, the left-hand one is the south-pole stereographic representation of the given 4-fork configuration at N, while the right-hand one is the north-pole stereographic representation of its associated complementary region, Ω, which is a circumscribing quadrilateral with a "triangular" tips truncated, i.e. the part lying inside the interior of the circumcircle of $\widehat{\triangle ANB}$. The circumradius of the $\pi/3$-isosceles spherical triangle $\widehat{\triangle ANB}$ with θ_1 as its top angle is given by

$$\tan R = \frac{1}{\sqrt{3}\cos\frac{\theta_1}{2}}\,. \tag{20}$$

The geometry of the complementary region, as indicated in Figure 7.20, can be computed as follows.

(i) Set the two diagonal lengths (resp. two pair of angles) of an equilateral $\pi/3$-quadrilateral to be d and d' (resp. 2γ and $2\gamma'$). Then, the pair of diagonals cut it into four right-angle triangles with $\{\frac{\pi}{3}, \frac{d}{2}, \frac{d'}{2}\}$

(resp. $\{\frac{\pi}{2}, \gamma, \gamma'\}$) as the side-lengths (resp. the angles). Then, one has the following functional relationships among them, namely

$$\cos\frac{d}{2}\cdot\cos\frac{d'}{2}=\frac{1}{2}\Leftrightarrow(1+\cos d)(1+\cos d')=1\,,$$
$$\tan\frac{d'}{2}=\sqrt{3}\cos\gamma\,,\quad \tan\frac{d}{2}=\sqrt{3}\cos\gamma'\,, \tag{21}$$
$$\sin\frac{d}{2}=\frac{\sqrt{3}}{2}\sin\gamma\,,\quad \sin\frac{d'}{2}=\frac{\sqrt{3}}{2}\sin\gamma'\,.$$

In particular, the two diagonals (resp. two pairs of angle) become equal when and only when the quadrilateral is cocircular, namely, $d=d'=\frac{\pi}{2}, \gamma=\gamma'=\frac{\beta}{2}$. Moreover, direct numerical computations of (21) will give the following list of corresponding values of $\{\gamma-\frac{\beta}{2}, \frac{\beta}{2}-\gamma', d-\frac{\pi}{2}, \frac{\pi}{2}-d'\}$:

$\gamma-\frac{\beta}{2}$	0.01	0.02	0.03	0.04	0.05
$\frac{\beta}{2}-\gamma'$	0.01007	0.20208	0.03065	0.04116	0.05183
$d-\frac{\pi}{2}$	0.01409	0.02808	0.04196	0.05574	0.06940
$\frac{\pi}{2}-d'$	0.01429	0.02889	0.04380	0.05903	0.07458

$\gamma-\frac{\beta}{2}$	0.06	0.07	0.08	0.09	0.10
$\frac{\beta}{2}-\gamma'$	0.06265	0.07364	0.08478	0.09610	0.10758
$d-\frac{\pi}{2}$	0.08295	0.09637	0.10968	0.122863	0.13592
$\frac{\pi}{2}-d'$	0.09046	0.10667	0.12323	0.14013	0.15738

(ii) Using the spherical polar coordinate system with the south pole S as the origin and $\overrightarrow{SQ}$ as the reference of direction, the positions of

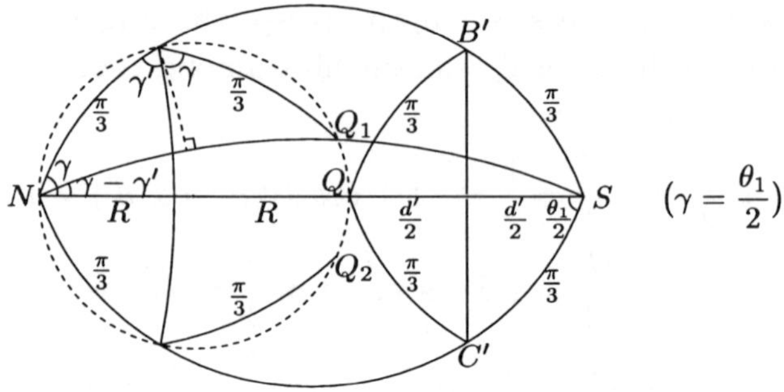

Figure 7.21

$Q, Q_1, Q_2, P_i, i = 2, 3, 4$, are given in terms of central angles $\{\theta_i, 1 \leq i \leq 4\}$ as follows (cf. Figure 7.21):

$$\begin{aligned}
&\rho(S, Q) = 2\tan^{-1}\left(\sqrt{3}\cos\frac{\theta_1}{2}\right), \quad \left[\text{by (20)},\ R + \frac{d'}{2} = \frac{\pi}{2}\right], \\
&\rho(S, Q_1) = \rho(S, Q_2) = \pi - 2\sin^{-1}\left(\frac{\sqrt{3}}{2}\sin\frac{\theta_1}{2}\right), \\
&\rho(S, P_i) = \tan^{-1}\sqrt{3}\sec\frac{\theta_i}{2}, \\
&\theta(Q_1) = -\theta(Q_2) = \frac{\theta_1}{2} - \sin^{-1}\left(\frac{2}{\sqrt{3}}\sin\tan^{-1}\left(\sqrt{3}\cos\frac{\theta_1}{2}\right)\right).
\end{aligned} \tag{22}$$

Next let us proceed to study the geometric possibilities of accommodating the remaining seven vertices of a type I configuration inside the above types of "complementary" regions (cf. Figure 7.19 and Figure 7.20). Suggested by the results of Lemmas 7.1.1 and 7.1.2 of this section, it is rather natural to divide such a study into two cases according to whether the pair of larger central angles are *adjacent* to each other or are separated by the pair of smaller central angles in the angular arrangement at the 4-fork vertex N. Let us first consider the latter, assuming that $\theta_1 \geq \theta_3$ is the larger pair among the four central angles.

Lemma 7.1.3 *Let N be a 4-fork vertex of a type I configuration $\mathcal{S}(\Sigma)$. If the pair of larger angles at N are separated by the pair of smaller angles at N, then the angular arrangement at N is exactly $(\beta, \alpha, \beta, \alpha)$ and $\mathcal{S}(\Sigma)$ is either the f.c.c. or the h.c.p. configuration.*

Proof Lemma 7.1.3 is a further strengthening of Lemma 7.1.1. Thus, it suffices to prove the *non-existence* of a septuple of $\pi/3$-separated points inside of the corresponding complementary region for any such angular arrangement other than $(\beta, \alpha, \beta, \alpha)$. We may assume without loss of generality for such a proof of non-existence that all the four forks at N are actually of the minimal length of $\pi/3$ and the pairs of larger (resp. smaller) angles satisfy $\theta_1 \geq \theta_3$ (resp. $\theta_2 \geq \theta_4$). We shall divide the proof of non-existence into the following cases according to the angular distribution as follows.

Case 1: $\theta_1 = \beta$, $\theta_4 = \alpha$. In this case, $\theta_2 > \alpha$ and $\theta_3 < \beta$, lest the distribution becomes $(\beta, \alpha, \beta, \alpha)$, which has already been studied in Lemma 7.1.1. Set $\theta_2 = \alpha + \varepsilon, \theta_3 = \beta - \varepsilon$. Let us first consider the subcase that ε is small. Then, the associated complementary region of such an angular arrangement can be considered as an ε-deformation of that of the $(\beta, \alpha, \beta, \alpha)$ case studied in Lemma 7.1.1. As indicated in Figure 7.22, its stereographic representation is the circumscribing quadrilateral which moves the tangent line at D' to that of D_ε. Thus, Ω_ε adds a small wedge $D'_\varepsilon F F_\varepsilon$ while subtracting a larger wedge $D'_\varepsilon W W_\varepsilon$ to Ω.

Suppose on the contrary that there exists a septuple, Σ_1, of $\frac{\pi}{3}$-separated points inside such a small deformation of Ω. Then, it follows from the quantitative analysis of the proof of Lemma 7.1.1 that Σ_1 must include both E and W_ε and moreover, Σ_1 must be a small deformation of one of the three possibilities of $\frac{\pi}{3}$-separated septuples inside of Ω, cf. Remark (i) following the proof of Lemma 7.1.1. Therefore, Σ_1 must also include a point in the near vicinity of either F_ε or D', as indicated by P_1 (resp. P_2). Suppose that Σ_1 contains a point P_2 in the near vicinity of D', then we may assume that Σ_1

Ω and Ω_ε :

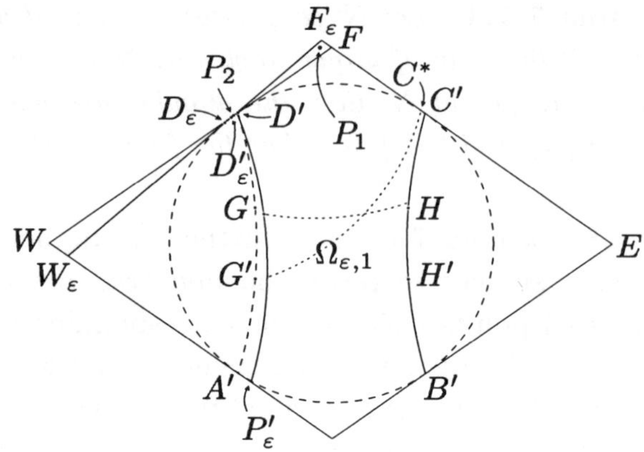

Figure 7.22

also contains a point, C^*, in the near vicinity of C', without loss of generality for the proof of *non-existence* of such a Σ_1. However, it follows easily from the magnified buckling effect that such a reduced complementary region $\Omega_\varepsilon \setminus \{D^\circ_{\frac{\pi}{3}}(P_2) \cup D^\circ_{\frac{\pi}{3}}(C^*)\}$ can no longer accommodate the remaining triple of Σ_1. Therefore, such a Σ_1 does *not* exist and hence, Σ_1 must contain a point P_1 in the near vicinity of F_ε. Again, it follows from the magnified buckling effect that the reduced complementary region (i.e. $\Omega_{\varepsilon,1} \setminus D^\circ_{\frac{\pi}{3}}(P_1)$) can no longer accommodate the remaining quadruple of Σ_1. Notice that the distances between E and points along the west boundary of $\Omega_{\varepsilon,1} \setminus D^\circ_{\frac{\pi}{3}}(P_1)$ are at most equal to $\frac{2\pi}{3} - \frac{1}{2}\varepsilon$.

Next let us consider the remaining subcase that ε is not so small, say at least 0.12. Suppose on the contrary that there exists a septuple, Σ_1, of $\frac{\pi}{3}$-separated points inside Ω_{ε_0} and assume that ε_0 is the smallest one among such ε's. Then, Σ_1 must contain a point, P, along the boundary circular arc $\overset{\frown}{D'_\varepsilon F_\varepsilon}$. If P is in the close vicinity of either F_ε or D', then the same proof as in the above case of small ε also proves the non-existence of such a Σ_1. Therefore, it suffices to consider the remaining situation that P is situated in the middle

Ω, Ω_ε and Ω'_ε :

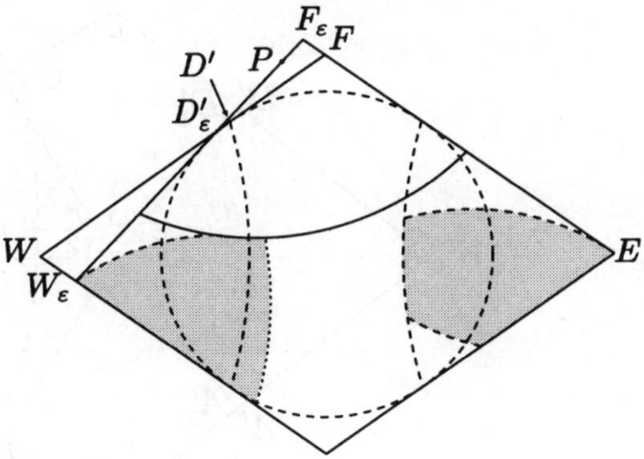

Figure 7.23

portion of $\widehat{D'_\varepsilon F_\varepsilon}$. Set $\Omega'_\varepsilon = \Omega_\varepsilon \backslash D^\circ_{\frac{\pi}{3}}(P)$. The replacement sets $R(E, \Omega'_\varepsilon)$ and $R(W_\varepsilon, \Omega'_\varepsilon)$ are as indicated in Figure 7.23 by the shaded regions. Therefore, it is not difficult to see that one may assume, without loss of generality for the proof of non-existence, that both E and W_ε belongs to such a Σ_1. Set

$$\Omega''_\varepsilon = \Omega'_\varepsilon \setminus \{D^\circ_{\frac{\pi}{3}}(E) \cup D^\circ_{\frac{\pi}{3}}(W_\varepsilon)\},$$

then it is quite obvious that Ω''_ε cannot accommodate the remaining quadruple of points of Σ_1. This proves the non-existence of such a Σ_1 in Ω_ε for $\varepsilon \geq 0.12$ and hence also completes the proof of Lemma 7.1.3 for Case 1.

Case 2: $\theta_1 > \beta$, $\theta_2 = \theta_4 = \alpha$. Set $\theta_1 = \beta + 2\varepsilon$, $\theta_3 = \beta - 2\varepsilon$. Then, the associated complementary region of this case is as indicated in Figure 7.24.

Suppose on the contrary that there exists a septuple Σ_1 of $\pi/3$-separated points inside of Ω_ε. Let us first consider the case that Σ_1 contains W_ε and a point E^* on the arc $\widehat{Q_1 Q_2}$, e.g. $E^* = Q$ or Q_1. It follows from (21) and (22) that the distances between Q and the tangent points C' and B' (cf. Figure 7.21), (i.e. that of the $\frac{\pi}{3}$-circles centered at S and that of A and D) are exactly equal to

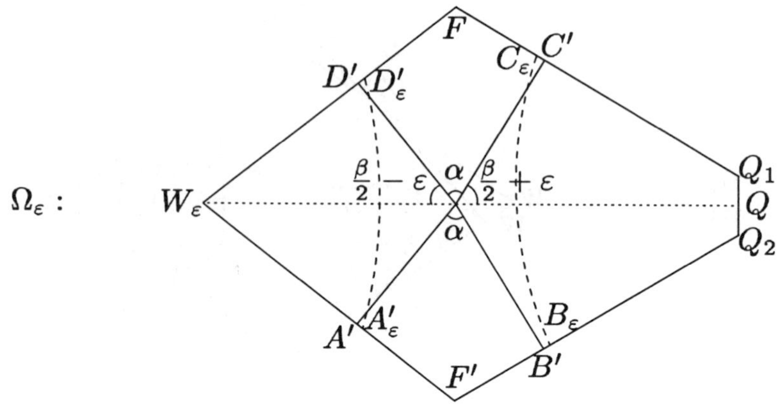

Figure 7.24

$\pi/3$, while the angles $\angle QSQ_1$ and $\angle QSQ_2$ are only slightly larger than 2ε (i.e. 2ε add a minute term of ε^2-order). Moreover, set D'_ε and A'_ε to be respectively the points on $\overset{\frown}{W_\varepsilon F}$ and $\overset{\frown}{W_\varepsilon F'}$ with $\pi/3$-distance to W_ε. Then, again from (21), $\angle D'SD'_\varepsilon = \angle A'SA'_\varepsilon$ is only slightly smaller than 2ε (i.e. 2ε subtract a minute term of ε^2-order). Therefore, the reduced complementary regions $\Omega\backslash\{D^\circ(W_\varepsilon) \cup D^\circ(Q)\}$ (resp. $\Omega\backslash\{D^\circ(W_\varepsilon) \cup D^\circ(Q_1)\}$) are the ones bounded between the $\pi/3$-circular arcs of $\overset{\frown}{D'_\varepsilon A'_\varepsilon}$ and $\overset{\frown}{C'B'}$ (resp. $\overset{\frown}{C_\varepsilon B_\varepsilon}$) while the maximal width of the second one, namely, $\overline{A'_\varepsilon B_\varepsilon}$ is only minutely larger than $\pi/3$ (i.e. by a tiny amount of ε^2-order). Notice that the $\pi/3$-circular arc $\overset{\frown}{C_\varepsilon B_\varepsilon}$ is "essentially" a rotation of $\overset{\frown}{C'B'}$ by an angle equal to $\angle QSQ_1$, while the boundary arc for other choices of $E^* \in \overset{\frown}{QQ_1}$ are that of smaller rotation angles. From the above geometric data on the possible reduced complementary regions, it is quite straightforward to show the *non-existence* of a quintuple of $\pi/3$-separated points in such a reduced complementary region, which is essentially a proper subset of that of Lemma 7.1.1, i.e. the case of $\varepsilon = 0$.

Moreover, the non-existence of septuples of $\pi/3$-separated points inside Ω which do not include W_ε or points on $\overset{\frown}{Q_1 Q_2}$ can again be

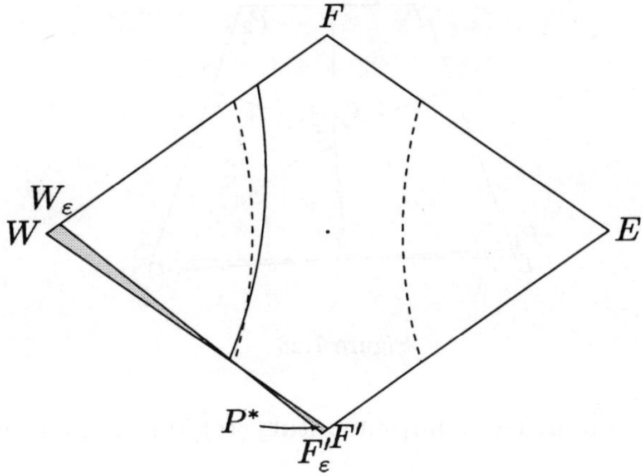

Figure 7.25

proved by the *technique of replacement*, just the same way as in the proof of *non-existence* of a septuple does not include both E and W in Lemma 7.1.1.

Case 3: $\theta_1 = \beta$, $\theta_4 > \alpha$. One may consider such a case as a deformation of Case 1, i.e. with $\theta_4 = \alpha$. Therefore, it suffices to show the non-existence of a septuple of $\pi/3$-separated points which includes a point P^* inside the added tiny wedge (cf. Figure 7.25). The proof is essentially the same as that of Case 1, but considerably simpler because the upper portion of the reduced complementary region becomes narrower than that of the Case 1-type.

Case 4: $\theta_1 > \beta$, $\theta_2 > \alpha$. Again, one may consider such a case as a deformation of another one of Case 2-type, i.e. with $\theta_2 = \alpha$. Thus, the proof of this case is actually the same as that of Case 2. This completes the proof of Lemma 7.1.3. □

Next let us study the remaining case that the pair of larger angles are adjacent to each other, say $\{\theta_1, \theta_4\}$. Let us first consider the special case that both of the smaller two are actually equal to the minimal of $\alpha = \cos^{-1}\frac{1}{3}$. Thus, $\theta_1 + \theta_4 = 2\beta$. Set $\theta_1 = \beta + 2\varepsilon$ and

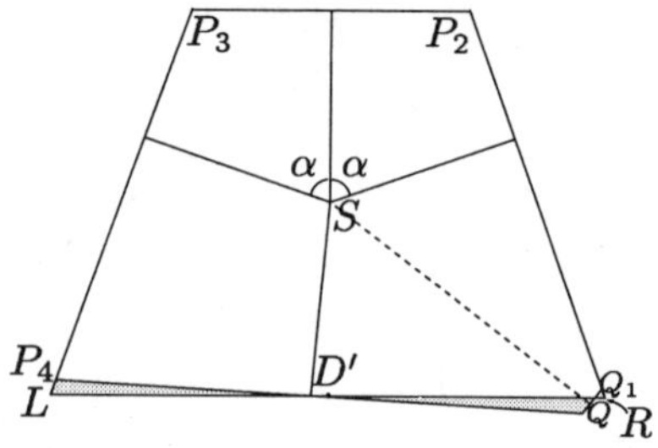

Figure 7.26

$\theta_4 = \beta - 2\varepsilon$. Then, the complementary region Ω_ε is as indicated in Figure 7.26.

Again, one may regard the above complementary region Ω_ε as a deformation of Ω_0, which is exactly the one that has already been thoroughly analyzed in Lemma 7.1.2. The difference sets between them are as indicated in Figure 7.26 by two shaded wedges and one tiny corner tip; $\Omega_\varepsilon \backslash \Omega_0$ consists of the right-hand side wedge while $\Omega_0 \backslash \Omega_\varepsilon$ consists of the left-hand side wedge and the tiny corner tip. In particular, a septuple of $\pi/3$-separated points inside of Ω_ε which does not contain point in $\Omega_\varepsilon \backslash \Omega_0$ is, of course, also inside Ω_0, and thus must be a smaller deformation of a 5□-one, by Lemma 7.1.2. Therefore, it suffices to analyze the remaining possibility that Σ_1 actually includes some points of $\Omega_\varepsilon \backslash \Omega_0$. Let Σ_1 be such a septuple. It is obvious that $\Omega_\varepsilon \backslash \Omega_0$ can have at most two points of such a Σ_1. In the case that the thin wedge actually contains a pair of points of Σ_1, one of them must be very close to D' and the other must be very close to Q. From that, it is easy to show that Σ_1 must be a small deformation of a 5□-one. Thus, it suffices to study the remaining case that $\Omega_\varepsilon \backslash \Omega_0$ contains exactly one point of Σ_1.

If such a point belongs to $D^\circ_\delta(D')$, $\delta = 0.15$, then the same proof as that of Lemma 7.1.2 will show that its distance to D' is at most 0.018 and Σ_1 must again be a small deformation of a 5□-one. In the remaining case that such a point is situated outside of $D^\circ_\delta(D')$, it is

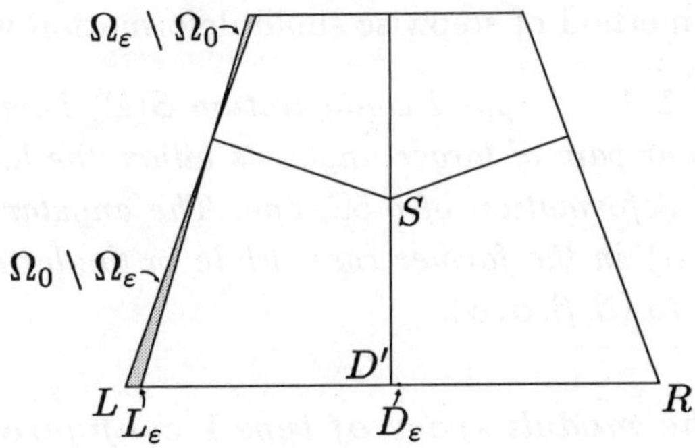

Figure 7.27

not difficult to see that such a point of Σ_1 inside $\Omega_\varepsilon \backslash \Omega_0$ must be in the vicinity of $\widehat{QQ_2}$. Again, by the same kind of proof as that of Lemma 7.1.2, one first shows that Σ_1 still contains a point D^* inside $D_\delta^\circ(D')$ and then proceeds to show that $\overline{D'D^*} < 0.018$ and Σ_1 must be a small deformation of a 5□-one. Moreover, such a proof will also show that ε itself must be smaller than 0.1.

Next let us study the case that $\theta_1 = \beta$, $\theta_2 = \alpha$ while $\theta_3 = \alpha + 2\varepsilon$, $\theta_4 = \beta - 2\varepsilon$. In such a case, the complementary region Ω_ε is as indicated in Figure 7.27.

We shall again regard Ω_ε as a deformation of Ω_0. The difference sets, $\Omega_\varepsilon \backslash \Omega_0$ and $\Omega_0 \backslash \Omega_\varepsilon$, are shown by the shaded wedges in Figure 7.27. Thus, it suffices to consider those septuples which contain a point of $\Omega_\varepsilon \backslash \Omega_0$. Set L_ε to be the left corner point of Ω_ε and D_ε to be the middle point of the $\pi/3$-circular arc $\widehat{L_\varepsilon R}$. Let Σ_1 be such a septuple, i.e. $\pi/3$-separated and containing one point of $\Omega_\varepsilon \backslash \Omega_0$. Then, the same proof as that of Lemma 7.1.2 will again first show that such a Σ_1 must contain a point D^* inside of $D_\delta^\circ(D_\varepsilon), \delta = 0.15$, and then proceed to show that $\overline{D^* D_\varepsilon} < 0.018$ and Σ_1 must be a small deformation of a 5□-one. Moreover, such a proof will also show that ε, itself, must be smaller than 0.1, otherwise Ω_ε cannot contain such a septuple Σ_1.

Such a method of stepwise small deformation will show that

Lemma 7.1.4 *A type I configuration $\mathcal{S}(\Sigma)$ having a 4-fork vertex with adjacent pair of larger angles is either the h.c.p. configuration or a small deformation of a 5□-one. The angular distribution must be $(\beta, \beta, \alpha, \alpha)$ in the former case, while in the latter case, it must be quite close to $(\beta, \beta, \alpha, \alpha)$.*

7.1.3 *The moduli space of type* I *configuration*

The totality of all congruence classes of type I configurations can be naturally organized into a space we shall call the *moduli space of type* I *configurations.* Let us first determine its dimension. Recall that the edge-lengths of the unique symmetric one are uniformly equal to $\tan^{-1} 2 = \pi/3 + 0.05995\cdots$. Thus, each of its twelve vertices has two degrees of freedom to vary while they remain at least $\pi/3$-separated. Therefore, there are altogether 24 degrees of freedom for local deformations of such a symmetric configuration within the set of type I configurations. This proves that the dimension of the moduli space is, in fact, equal to 21. In this subsection, we shall study certain aspects of its geometry that will play an important role in the proof of Main Theorem I on locally averaged density.

7.1.3.1 *Deformations of the f.c.c. and the h.c.p. configurations*

It contains two special points representing the only two congruence classes of type I configurations with the maximal possible number of six quadrilaterals, namely, the f.c.c. configuration and the h.c.p. configuration. Their corresponding single-layer local packings are the *only* ones that attain the *optimal locally averaged density of* $\pi/\sqrt{18}$ (i.e. in their optimal double-layer extensions). Thus, it is natural to expect that those local packings *very close* to one of them will be able to attain locally averaged densities very close to the optimal one of $\pi/\sqrt{18}$. Therefore, in the proof of Main Theorem I, it is necessary to produce rather a accurate estimate on the upper bound of those locally averaged densities associated to their double-layer extensions, in order to ensure that they are, indeed, always smaller than $\pi/\sqrt{18}$. Naturally, such an accurate estimate must be based upon some rather precise control on the geometry of local deformations of these two special local packings, especially those deformations that keep all the twelve neighbors touching the central sphere. The following lemma provides such a geometric understanding of the local deformations of these two important special points in the moduli space.

Lemma 7.1.5 *The h.c.p. configuration is an isolated point in the moduli space, while the f.c.c. configuration is a cusp point whose tangent cone consists of a single direction represented by the following one-parameter local deformations, namely*

$$\Sigma(\varepsilon) = \left\{ \begin{array}{l} \left(\pm\cos\left(\frac{\pi}{4}+\varepsilon\right),\ \pm\sin\left(\frac{\pi}{4}+\varepsilon\right),0\right) \\ \left(\pm\sin\left(\frac{\pi}{4}+\varepsilon\right),\ 0,\ \pm\cos\left(\frac{\pi}{4}+\varepsilon\right)\right) \\ \left(0,\pm\cos\left(\frac{\pi}{4}+\varepsilon\right),\ \pm\sin\left(\frac{\pi}{4}+\varepsilon\right)\right) \end{array} \right\}. \tag{23}$$

Proof Let us first consider the simple geometry of the local deformations among $\pi/3$-equilateral quadrilaterals in the vicinity of the cocircular one. Such a quadrilateral is uniquely determined by one of its angles. Thus, the above local deformation can conveniently be parameterized by setting one of its angles to be $\beta + 2\varepsilon$. The pair of diagonals subdivide such a quadrilateral into four congruent right-angle triangles with $\frac{\beta}{2}+\varepsilon$ and γ' as the other two angles. Thus,

$$\sin\frac{d}{2} = \frac{\sqrt{3}}{2}\sin\left(\frac{\beta}{2}+\varepsilon\right), \quad \cos\gamma' = \frac{1}{\sqrt{3}}\tan\frac{d}{2}. \tag{24}$$

Hence, its area is given by

$$A(\varepsilon) = 4\left(\frac{\beta}{2} + \varepsilon + \gamma' - \frac{\pi}{2}\right) = (4\beta - 2\pi) + \varphi(\varepsilon)\,,$$

$$\varphi(\varepsilon) = 4\cos^{-1}\left\{\frac{1}{\sqrt{3}}\tan\sin^{-1}\left[\frac{\sqrt{3}}{2}\sin\left(\frac{\beta}{2}+\varepsilon\right)\right]\right\} - 2\beta + 4\varepsilon \quad (25)$$

$$\doteq -(2.8\cdots)\varepsilon^2\,.$$

Therefore, under small ε-deformations (measured by the change of angles) the area of each quadrilateral can only decrease by an amount less than $3\varepsilon^2$, thus a total decrement of at most $18\varepsilon^2$. On the other hand, the total area increment of the eight triangles will be roughly 0.4 times the total increment of the twenty-four edge-lengths. Hence, the total increment of the twenty-four edge-lengths has an upper bound of $45\varepsilon^2$. In fact, the above upper bound estimate can be considerably improved if one *also* accounts for the *area-increasing effects* of those increments of edge-lengths for the *six quadrilaterals.* Anyhow, the above upper bound estimate on the total increment of the twenty four edge-lengths, of course, implies that each of the twenty four edge-lengths can have, at most, a second order increment. Therefore, *up to the first order of infinitesimal*, the geometric effect of such an ε-deformation on each triangle (resp. quadrilateral) is an ε-order rotation (resp. an ε-order deformation of the $\pi/3$-equilateral quadrilaterals). However, such individual deformations of the eight triangles and the six quadrilaterals cannot be fitted together to obtain an ε-deformation of the h.c.p. configuration, while in the case of the f.c.c. configuration, they can be fitted together in the way explicitly described by (23). Geometrically, the deformation is as indicated in Figure 7.28. □

7.1.3.2 *Type* I *configurations with quadrilaterals or* 4*-fork vertices*

Algebraically, the twelve vertices of a type I configuration can be represented by twelve unit vectors $\{\mathbf{a}_j; 1 \le j \le 12\}$ with their

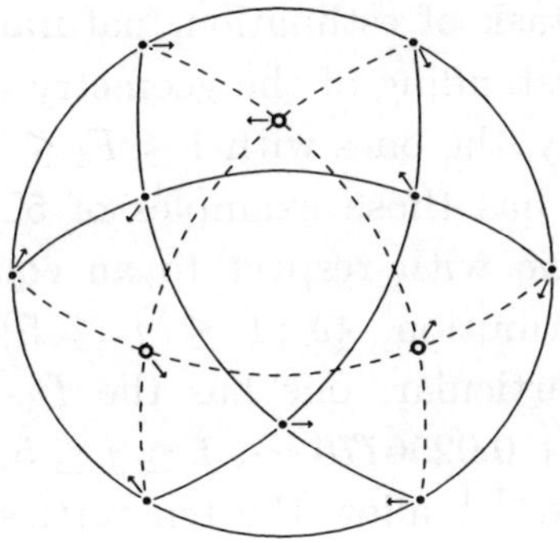

Figure 7.28

inner products $\{\mathbf{a}_j \cdot \mathbf{a}_k; 1 \leq j < k \leq 12\}$ satisfying a collection of equalities and inequalities, such as $\mathbf{a}_j \cdot \mathbf{a}_k \leq \frac{1}{2}$, $1 \leq j < k \leq 12$, etc. From the viewpoint of geometric invariant theory, the moduli space of type I configurations can be represented as a 21-dimensional real semi-algebraic set in 66-dimensional affine space with the above 66 inner products as the coordinates. However, such an algebraic representation is very complicated and hence not very useful. For one thing, the "convexity conditions" of spherical configurations already involve a collection of inequalities that are rather difficult to handle. Anyhow, the generic points are those configurations where *none of such inequalities becomes an equality.* In particular, such *generic configurations* all have *no* quadrilaterals and hence, twenty triangular faces and thirty edges. Thus, on the one hand, those type I configurations containing some quadrilaterals always belong to the *non-generic* part of the moduli space; and on the other hand, those type I local packings with "big-holes" (which correspond to quadrilaterals) often have higher locally averaged densities than otherwise. Therefore, for the purpose of proving an upper bound estimate smaller than $\pi/\sqrt{18}$ for those type I local packings with "big-holes" but other than f.c.c. and h.c.p., it again requires higher accuracy than that of the case of type I local packings without big-hole. Such a more demanding task of estimation, naturally, needs a considerable degree of understanding of the geometry of this kind of type I configuration, namely, the ones with $1 \le F_4 \le 5$.

First of all, one has those examples of 5□-ones which are reflectionally symmetric with respect to an equator, and are given by the angular distribution, $\{\delta_j; 1 \le j \le 5\}$, at both poles (cf. Figure 7.10). In particular, one has the D_5-symmetric one with $\delta_j = \frac{2\pi}{5} = \cos^{-1}(\frac{1}{3}) + 0.0256776\cdots$, $1 \le j \le 5$. The small amounts of δ_j in excess of $\cos^{-1}\frac{1}{3}$ allow the ten vertices of the five quadrilaterals to have some small room to vary. It is not difficult to arrange such small variations so as to obtain small deformations with $F_4 = 1$, 2, 3, 4 or 5, such as the ones indicated schematically in Figure 7.29. It follows from Euler's formula that the number of edges is given by $E = 30 - F_4$. Therefore, for type I configurations with $F_4 \ge 1$, the number of forks at some vertices must be less than five. On the other hand, it is easy to show that 3-fork vertices are

$F_4 = 4$:

$F_4 = 3$:

$F_4 = 2$:

$F_4 = 1$:

$F_4 = 0$:

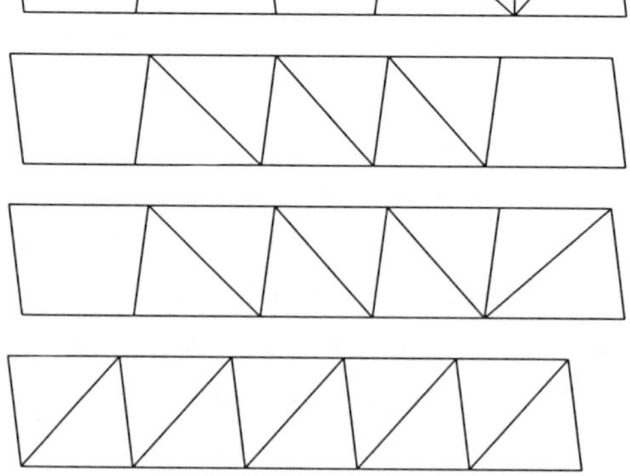

Figure 7.29

geometrically impossible for type I configurations. Thus, a type I configuration with at least one quadrilateral must have some 4-fork vertices. Therefore, by Lemmas 7.1.3 and 7.1.4, all those type I configurations with $1 \leq F_4 \leq 5$ are necessarily small deformations of those 5□-ones (cf. Figure 7.10).

Corollary 7.1.1 *If $\mathcal{S}(\Sigma)$ is a type I configuration with $1 \leq F_4 \leq 5$, then it is a small deformation of a 5□-one as indicated in Figure 7.10.*

The following are some of the relevant geometric properties of those small deformations of the 5□-ones.

(i) Each of the 5□-ones, as indicated in Figure 7.10, consists of ten triangles and five quadrilaterals; the latter are arranged like a belt around the equator while the former are grouped into two clusters of five triangles around the two poles, each of the triangles having its area very close to the lower bound of $\triangle_{\pi/3} = 2\tan^{-1}\frac{\sqrt{2}}{5}$. It is not difficult to see that a small deformation of such 5□-ones will still have two such clusters of five triangles around the two poles, each of the triangles having its area still quite close to $\triangle_{\pi/3}$. However, the effect of such a small deformation will, in general, buckle some of its five quadrilaterals into pairs of triangles whose shapes and sizes are quite close to the one with $(\pi/3,\ \pi/3,\ \pi/2)$ as the side-lengths.

(ii) Each of the 5□-ones has twenty-five edges; fifteen of them are of the minimal length of $\pi/3$ while ten of them can be slightly longer than $\pi/3$, their lengths consisting of five pairs given in terms of the angular distributions $\{\delta_j,\ 1 \leq j \leq 5\}$, namely, $2\sin^{-1}(\frac{\sqrt{3}}{2}\sin\frac{\delta_j}{2})$, $1 \leq j \leq 5$, where each δ_j is at least equal to $\alpha = \cos^{-1}\frac{1}{3}$ and their total sum is, of course, equal to 2π. Therefore, corresponding to each angular excess, i.e. $\varepsilon_j = \delta_j - \alpha$, one has an edge-excess given by

$$2\sin^{-1}\left(\frac{\sqrt{3}}{2}\sin\frac{1}{2}(\alpha+\varepsilon_j)\right) - \frac{\pi}{3}, \quad 1 \leq j \leq 5, \qquad (26)$$

where $\Sigma\varepsilon_j = 2\pi - 5\alpha = 0.128388\cdots$. Correspondingly, in the geometry of the type I local packing associated to such a 5□-configuration, one has ten *gully-type gaps* whose widths are given by $4 \cdot \frac{\sqrt{3}}{2} \sin \frac{\delta_j}{2} - 2$, $1 \le j \le 5$. Thus, the total amount of the above ten gap-widths is equal to

$$4\sqrt{3}\sum_{j=1}^{5} \sin \frac{\delta_j}{2} - 20\,, \quad \Sigma\delta_j = 2\pi\,, \quad \delta_j \ge \alpha\,. \tag{27}$$

Geometrically, the sum $\Sigma \sin \frac{\delta_j}{2}$ can be interpreted as half of the total length of an inscribing pentagon of the unit circle. It follows from the concavity of $y = \sin x$ for the range $\frac{\alpha}{2} \le x \le \pi - 2\alpha$,

$$4 \sin \frac{\alpha}{2} + \sin 2\alpha \le \sum_{j=1}^{5} \sin \frac{1}{2}\delta_j \le 5 \sin \frac{\pi}{5}\,. \tag{28}$$

Therefore, the total amount of the ten gap-widths is bounded between the following two numbers, namely

$$\left(\frac{16\sqrt{2}}{3\sqrt{3}} - 4\right) = 0.35464843\cdots$$

and

$$20\left(\sqrt{3}\sin\frac{\pi}{5} - 1\right) = 0.361478418\cdots\,. \tag{29}$$

In the case of small deformations of 5□-ones, the above twenty-five edges remain quite close to $\pi/3$ while there may add at most five edges of lengths close to $\pi/2$, namely those diagonals of the buckled quadrilaterals. Correspondingly, those type I local packings associated to such small deformations of the 5□-ones always have *gully-type* gaps corresponding to those small edge-excesses of the twenty-five short ones. It is not difficult to see that the total amount of such gap-widths is again bounded below by 0.35464843 for such a small deformation of the 5□-ones. In fact, it suffices to account the total "edge-excess" of the two Euclidean "space-pentagons" (i.e. not necessarily planary ones), it is easy to see that their sum total is already at least equal to the above lower bound.

7.1.4 *Geometry of type* I *icosahedrons*

The moduli space of the type I configuration is a 21-dimensional real semi-algebraic set. Thus, there is a natural 21-dimensional volume that provides the canonical measurement of the sizes of its various subsets. The results of Subsec. 7.1.3 show that the "great majority" of type I configurations are of the *combinatorial* type of the regular icosahedron, namely, the type of having 20 triangles, 30 edges while the stars of its 12 vertices are all of $5\triangle$-type. We shall call such type I configurations simply *type* I *icosahedrons* and their corresponding local packings *icosahedron type* I *local packings.* In this subsection, we shall study those geometric aspects of type I icosahedrons that will be useful in the upper bound estimation of the locally averaged densities of those double-layer extensions of icosahedron type I local packings.

7.1.4.1 *Total area excess and its distribution*

A type I icosahedron has 20 triangles while the area of each of them is bounded below by that of the $\pi/3$-equilateral triangle, namely, $\triangle_{\pi/3} = 2\tan^{-1}\frac{\sqrt{2}}{5}$. Anyhow, the difference between the total spherical area and $20\triangle_{\pi/3}$, i.e. $4\pi - 20\triangle_{\pi/3} = 1.540658646$ can be considered as the overall measurement of the *non-tightness* of type I local packings in general. Geometrically, the above total amount of "area-excess" will be manifested as a considerable amount of *edge-excess*, meaning the amount of edge-length in excess of $\pi/3$. In fact, it is quite simple to estimate the lower bound of the sum of the three edge-excess of a triangle in terms of its area excess. Let $\widehat{\triangle ABC}$ be a spherical triangle with $\{a, b, c\}$ as its side-lengths and $\triangle_{\pi/3} + \varepsilon$ as its area. Then,

$$a + b + c - \pi \geq 3\cos^{-1}\left(\frac{\cos(\alpha + \frac{\varepsilon}{3})}{1 - \cos(\alpha + \frac{\varepsilon}{3})}\right) - \pi\,. \tag{30}$$

For example, in the case that the 20 triangles are of equal area of $\pi/5$, the total amount of edge-excess is at least equal to that of regular spherical icosahedron, namely, $30 \cdot (\tan^{-1} 2 - \pi/3) = 1.798535$. However, for the purpose of achieving a good enough collective volume estimation for the thirteen local cells of such an icosahedron type I local packing, it is more important to know the *distribution of edge-excess* rather than the total amount of edge-excess. Anyhow, let us begin with a study on the *distribution of area excess* (whose total amount is always equal to $1.540658646\cdots$).

Definition To each vertex A_i of a given type I icosahedron $\mathcal{S}(\Sigma)$, the total area of the five triangles of the star of A_i (i.e. the ones with A_i as a vertex) is defined to be the *spread of* $\mathcal{S}(\Sigma)$ *at* A_i and shall be denoted by $\sigma(A_i)$.

Remarks (i) The amount of area excess contained in the star of A_i can be regarded as a natural measurement of the *non-tightness* around the touching neighbor S_i in the corresponding type I local packing.

(ii) In the total sum of the twelve *spreads* of a given type I icosahedron, i.e. $\Sigma\sigma(A_i)$, the area of each triangle has been counted three times. Thus,

$$\sum_{i=1}^{12} \sigma(A_i) = 12\pi \tag{31}$$

and hence, the *mean value* of the spreads of any type I icosahedron is always equal to π.

(iii) Using the technique of area-decreasing deformations, it is not difficult to show that the spreads of type I icosahedrons are bounded below by the area of the $5\triangle$-star with uniform $\frac{\pi}{3}$-radial edge-lengths and with four of its central angles equal to α. It consists of four $\pi/3$-equilateral triangles and a $\pi/3$-isosceles triangle with $2\pi - 4\alpha$ as its top angle. Straightforward computation will show that its total area is given by

$$4\triangle_{\pi/3} + 0.591342948 = 2.796485342\,. \tag{32}$$

The above lower bound estimate shows that, on the one hand, a spread can become 0.3451 smaller than the mean value of π, and on the other hand, each spread contains at least 0.040057 area excess.

7.1.4.2 *Geometry of* $5\triangle$*-star configurations*

The local geometry of a type I icosahedron in the vicinity of a given vertex is that of a $5\triangle$-star configuration whose ten edge-lengths are at least $\pi/3$. The total area of such a $5\triangle$-star is exactly the spread while the *localized area excess* is the difference between the spread and $5\triangle_{\pi/3}$. Such a $5\triangle$-star configuration has ten edges (i.e. five radial edges and five boundary edges) and five central angles. Thus, it is natural to study the geometric correlations between the area, edge-lengths and angular distribution of such a geometric structure, such as the *correlation between* the *area excess* and the *edge-excess*, etc. Moreover, for the purpose of estimating the volumes of local cells, it is advantageous to have separate accountings of the edge-excess, namely, to account the radial edge-excess and the boundary edge-excess separately.

Example 7.1.3 Suppose all the five radial edge-lengths of a $5\triangle$-star are equal to the lower bound of $\pi/3$. Then, by the Lexell theorem, its area becomes the maximal one if the five central angles are equal while its area becomes the minimal one if the five central angles consists of four α and one $(2\pi - 4\alpha)$. Therefore, the areas of this family of $5\triangle$-stars are bounded between that of (32) and the following:

$$10\cos^{-1}\left\{\frac{1}{\sqrt{3}}\tan\sin^{-1}\left(\frac{\sqrt{3}}{2}\sin\frac{\pi}{5}\right)\right\} - 3\pi = 2.7987\cdots\,. \tag{33}$$

The above result shows, in particular, that any 5$\triangle$-star with total area exceeding the above number must contain certain amount of *radial edge-excess.* Thus, it is natural to estimate the *lower bound* of the *total radial edge-excess* in terms of the total area of the 5$\triangle$-star *in excess of* 2.7987. For example, what is the least amount of total *radial* edge-excess for a 5$\triangle$-star with its total area equal to the mean value of π, which is roughly 0.34289 in excess of 2.7987?

Example 7.1.4 Suppose that the radial edge-excesses are uniformly the same. Then, the total area of such a 5$\triangle$-star is maximal when the central angles are all equal to $\frac{2\pi}{5}$. Therefore, the maximal area can be given as the function of the uniform radial edge-excess as follows, namely

$$f(\varepsilon) = 10\cos^{-1}\left\{\frac{1}{\tan(\frac{\pi}{3}+\varepsilon)}\tan\sin^{-1}\left(\sin\frac{\pi}{5}\sin\left(\frac{\pi}{3}+\varepsilon\right)\right)\right\} - 3\pi\,,$$

$$f(0) = 2.798700252\cdots. \tag{34}$$

The following is the computer plot of the graph of $f(\varepsilon) - f(0)$:

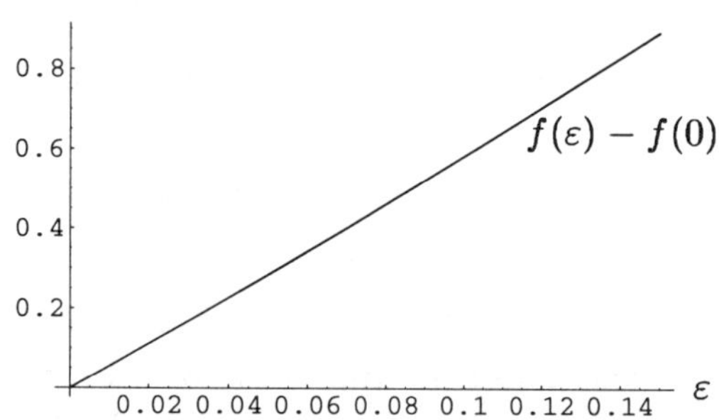

In the general case of *non-uniform* edge-excesses, it is natural to study the following geometric problem.

Problem *For a given set of radial edge-lengths of* 5$\triangle$*-star configurations, how does one determine the set of central angles which makes the area maximal?*

Let us first study the simpler corresponding problem of Euclidean geometry. Let $\{\ell_i,\ 1 \le i \le 5\}$ be the radial lengths in cyclic order of a $5\triangle$-star Euclidean configuration and $\{\theta_i;\ 1 \le i \le 5\}$ be the central angles in cyclic order. Then, its total area is given by

$$A(\theta_i) = \frac{1}{2}\sum_{i=1}^{5} \ell_i \ell_{i+1} \sin\theta_i\,, \quad \sum_{i=1}^{5} \theta_i = 2\pi\,. \tag{35}$$

Thus, it is analytically a problem of optimization with the constraint of $\Sigma\theta_i = 2\pi$. Using the method of the Lagrange multiplier, one has the following set of equations:

$$\frac{1}{2}\ell_i\ell_{i+1}\cos\theta_i = \lambda\,, \quad 1 \le i \le 5\,, \quad \sum_{i=1}^{5}\theta_i = 2\pi\,. \tag{36}$$

The geometric meaning of the first set of five equations (namely, $\ell_1\ell_2\cos\theta_1 = \ell_2\ell_3\cos\theta_2 = \cdots = \ell_5\ell_1\cos\theta_5$) is that the pairs of diagonals of $\square OA_iA_{i+1}A_{i+2}$ are all mutually perpendicular, as indicated in Figure 7.30.

Now let us proceed to study the spherical case of $5\triangle$-star configurations with $\{\ell_i = \pi/3 + \varepsilon_i,\ 1 \le i \le 5\}$ as the set of radial edge-lengths in cyclic order. Set

$$\begin{aligned}
&c_i = \cos\ell_i\,, \quad s_i = \sin\ell_i = \sqrt{1 - c_i^2}\,,\\
&x_i = 1 + \cos b_i = 1 + c_ic_{i+1} + s_is_{i+1}\cos\theta_i\,, \quad u_i = c_i + c_{i+1} + x_i\,,\\
&D_i = \{-(c_i + c_{i+1})^2 + 2(c_ic_{i+1} + 1)x_i - x_i^2\}^{1/2}\,, \quad \triangle_i = 2\tan^{-1}\frac{D_i}{u_i}\,,
\end{aligned} \tag{37}$$

where $\{b_i\}$ (resp. $\{\theta_i\}$, $\{\triangle_i\}$) are the boundary edge-lengths (resp. central angles, areas of triangles) in cyclic order. Thus, the analytical problem is the maximization of

$$A = \sum_{i=1}^{5}\triangle_i = 2\sum_{i=1}^{5}\tan^{-1}\frac{D_i}{u_i} \tag{38}$$

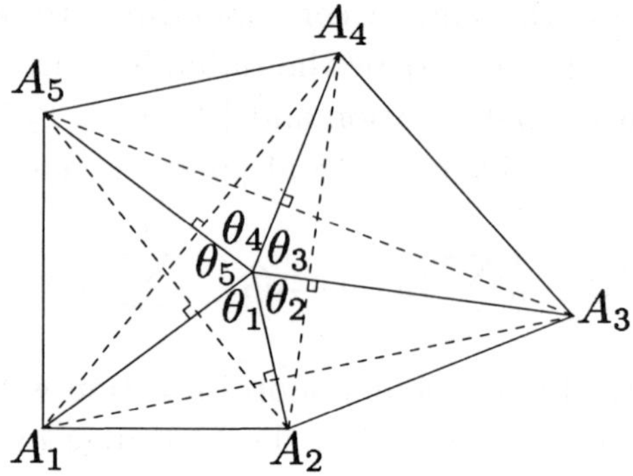

Figure 7.30

with the constraint of $\sum_{i=1}^{5} \theta_i = 2\pi$. Let us again analyze the above analytical problem by the method of the Lagrange multiplier. Notice that $\{c_i\}, \{s_i\}$ are given constants and x_i, u_i, D_i are all functions of θ_i only. Therefore, differentiation of (37) will give

$$\begin{aligned}\frac{\partial A}{\partial \theta_i} &= \frac{d\triangle_i}{d\theta_i} = \frac{d\triangle_i}{dx_i} \cdot \frac{dx_i}{d\theta_i} \\ &= \frac{c_i + c_{i+1} - x_i}{x_i D_i}(-s_i s_{i+1} \sin \theta_i) = \frac{x_i - c_i - c_{i+1}}{x_i} .\end{aligned}$$

Hence, the equations of the Lagrange multiplier are simply as follows, namely

$$(1 - \lambda)x_i = c_i + c_{i+1} , \quad 1 \leq i \leq 5 , \quad \sum_{i=1}^{5} \theta_i = 2\pi . \tag{39}$$

Analytically, one can use the ith equation of (39) to solve for $\cos \theta_i$ in terms of $\{c_i, c_{i+1}\}$ and the multiplier λ and then use the condition, i.e. the last equation of (39) to determine the value of λ. Geometrically, one has the following set of equalities, namely

$$\frac{c_1 + c_2 - x_1}{x_1} = \frac{c_2 + c_3 - x_2}{x_2} = \frac{c_3 + c_4 - x_3}{x_3}, \text{ etc} . \tag{40}$$

Thus, one would like to find out what is the *geometric meaning* of the above set of equalities in the cosines and sines of lengths and angles. Notice that each of the above equalities is exactly the analytical condition of the optimization of $\triangle_i + \triangle_{i+1}$ under the constraint of $\theta_i + \theta_{i+1} =$ constant. Applying Lexell's theorem to such a geometric problem, such an *optimal spherical geometric configuration* is characterized by the following geometric condition.

The circle passing through A_{i+1} and the *antipodal points* of A_i and A_{i+2}, i.e. the *Lexell circle*, and the ℓ_{i+1}-circle centered at the central vertex O are touching at A_{i+1}. Conversely, among the family of Lexell circles passing through A'_i and A'_{i+2}, there is a unique one which is tangent to the ℓ_{i+1}-circle centered at O. That touching point is exactly the optimal position for A_{i+1}.

Remark The above geometric interpretation of the analytical condition of (40) not only provides a conceptual proof of the *uniqueness* of solution for each given constant of $\theta_i + \theta_{i+1}$, it also shows the *monotonic-decreasing* properties for deformations away from the optimal position of A_{i+1}. In the situation of 5$\triangle$-stars of type I configurations, there are boundary (i.e. inequality) conditions which limit the deformations of the central angles $\{\theta_i\}$ for 5$\triangle$-stars with a given set of radial edge-lengths $\{\ell_i = \pi/3 + \varepsilon_i\}$. Therefore, the above monotonicity enables us to solve the optimization problem with such boundary conditions even when the optimal one given by (39) happens to be outside of the region specified by the boundary conditions (i.e. inequalities).

With the above set of analytical equations (i.e. (39)) together with the geometric understanding both of the *uniqueness* of optimal solution and of the *monotonicity* of angular deformations of 5$\triangle$-star configurations, it is then quite straightforward to use a computer to study the numerical values of the maximal areas for 5$\triangle$-star type I subconfigurations with given sets of *small radial edge-excess* $\{\varepsilon_i,\ 1 \leq i \leq 5\}$ and then compare their values among the ones with the *same small amount* of *total radial edge-excess*.

7.1.4.3 5△-*star configurations with a lopsided distribution of edge-excess*

The average value of the twelve *spreads* of a type I icosahedron is always equal to π. Therefore, a majority of them always exceed 2.7987 by a considerable amount and hence a majority of the twelve 5△-stars always have considerable amount of edge-excess, *both* in the *radial part* and in the *boundary part.* However, in the study of the local extensions corresponding to such a 5△-star configuration and the *lower bound volume estimates* of the resulting local cells (cf. next section), it is not only the *total amount* of radial (resp. boundary) edge-excess, but also the distribution of them that will play an important role. For example, if the distribution of radial edge-excess of such a 5△-star configuration is *not* very lopsided, then, as one shall see in the next section, it is quite simple to provide a lower bound of the volume of the resulting local cell that already exceeds $4\sqrt{2}$ by some comfortable margin. Thus, for local extensions corresponding to 5△-stars of sizes close to π, the crucial volume estimation task mainly lies in the rather special case whose distributions of radial edge-excess are *very lopsided*, namely, largely concentrated into a single radial edge.

Let us first study the geometric structures of those 5△-stars with four $\pi/3$-radial-edges, namely, with all their radial edge-excess concentrated into a single edge. Set A to be the central vertex and $\{A_i, 1 \leq i \leq 5\}$ to be the five boundary vertices of a given 5△-star. Let $\overline{AA_2} = \ell$ be the only radial edge longer than $\pi/3$ and $\{\theta_i, 1 \leq i \leq 5\}$ be the set of central angles in cyclic order. Such a 5△-star is uniquely determined by the length ℓ and the above set of central angles. We shall first study its area as a function of $\{\ell, \theta_i; 1 \leq i \leq 5\}$ (or rather, of $\{\ell, \theta_i; 1 \leq i \leq 4\}$ by setting $\theta_5 = 2\pi - (\theta_1 + \theta_2 + \theta_3 + \theta_4)$), and denote it simply by $A(\ell, \theta_i)$.

As indicated in Figure 7.31, A_i, $i \neq 2$, are points on the $\pi/3$-circle centered at A while the central angles satisfy the following inequalities, namely, $\theta_3, \theta_4, \theta_5 \geq \alpha = \cos^{-1}\frac{1}{3}$, $\theta_1, \theta_2 \geq \cos^{-1}(\frac{1}{\sqrt{3}}\tan\frac{\ell}{2})$.

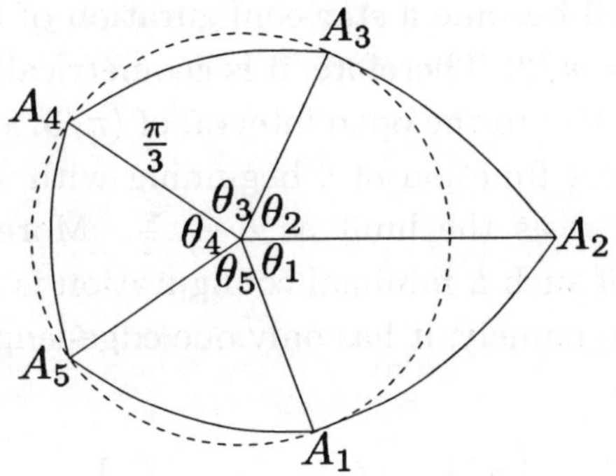

Figure 7.31

First of all, for a given pair of $\{\theta_1, \theta_2\}$ satisfying $\theta_1, \theta_2 \geq \cos^{-1}(\frac{1}{\sqrt{3}} \tan\frac{\ell}{2}), \theta_1 + \theta_2 < 2\pi - 3\alpha$, the above area function $A(\ell, \theta_i)$ is maximal (resp. miminal) when $\{\theta_3, \theta_4, \theta_5\}$ are equal (resp. two of them are equal to α), namely

$$A(\ell, \theta_1, \theta_2, \hat{\theta}, \alpha, \alpha) \leq A(\ell, \theta_i) \leq A(\ell, \theta_1, \theta_2, \bar{\theta}, \bar{\theta}, \bar{\theta}), \qquad (41)$$

where $\bar{\theta} = \frac{1}{3}(2\pi - \theta_1 - \theta_2)$, $\hat{\theta} = 2\pi - 2\alpha - \theta_1 - \theta_2$. Furthermore, it follows from the monotonic decreasing property of central angle deformations away from the optimal one that both $A(\ell, \theta_1, \theta_2, \bar{\theta}, \bar{\theta}, \bar{\theta})$ and $A(\ell, \theta_1, \theta_2, \hat{\theta}, \alpha, \alpha)$ are maximal (resp. minimal) when $\theta_1 = \theta_2$ (resp. one of them equal to $\cos^{-1}(\frac{1}{\sqrt{3}} \tan\frac{\ell}{2})$). Therefore, for a given ℓ, $A(\ell, \theta_i)$ is minimal when

$$\begin{aligned} &\theta_1 = \theta_2 = \cos^{-1}\left(\frac{1}{\sqrt{3}} \tan\frac{\ell}{2}\right), \\ &\theta_3 = \theta_4 = \alpha,\ \theta_5 = 2\pi - 2\alpha - 2\cos^{-1}\left(\frac{1}{\sqrt{3}} \tan\frac{\ell}{2}\right) \end{aligned} \qquad (42)$$

or any interchange of the above $\{\theta_3, \theta_4, \theta_5\}$, because $A(\ell, \theta_i)$ is symmetric with respect to the last three variables. The above minimal value of $A(\ell, \theta_i)$ for each given ℓ is, of course, just a function of ℓ, which will be denoted by $\breve{A}(\ell)$. Notice that the above "minimal configuration" will become a star configuration of the type of one $\square$ and three $\triangle$ if $\ell = \pi/2$. Therefore, it is geometrically natural to limit the range of ℓ in $\breve{A}(\ell)$ to the open interval of $(\pi/3, \pi/2)$. It is a monotonically increasing function of ℓ beginning with $\breve{A}(\frac{\pi}{3}) = 2.7964\cdots$ and approaching π as the limit as $\ell \to \frac{\pi}{2}$. Moreover, the boundary edge-excess of such a minimal configuration is also concentrated into a single edge, namely, it has only one edge-length exceeding $\pi/3$ which is given by

$$\ell^* = 2\sin^{-1}\left\{\frac{\sqrt{3}}{2}\sin\left(\alpha + \cos^{-1}\left(\frac{1}{\sqrt{3}}\tan\frac{\ell}{2}\right)\right)\right\}. \tag{43}$$

The following are the computer plots of the graphs of $\breve{A}(\ell)$ and $\ell^*(\ell)$ for $\frac{\pi}{3} \le \ell \le \frac{\pi}{2}$:

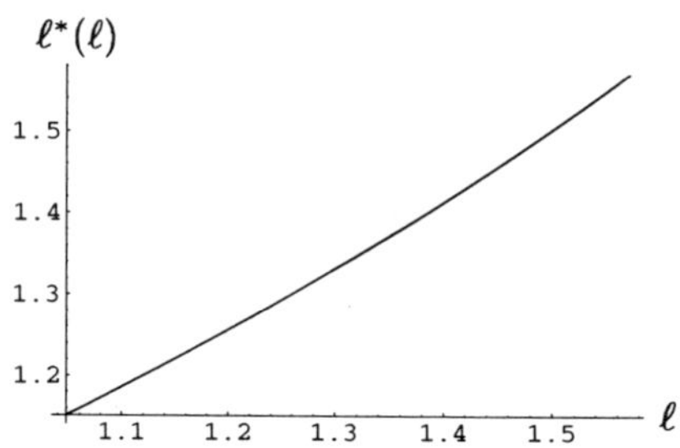

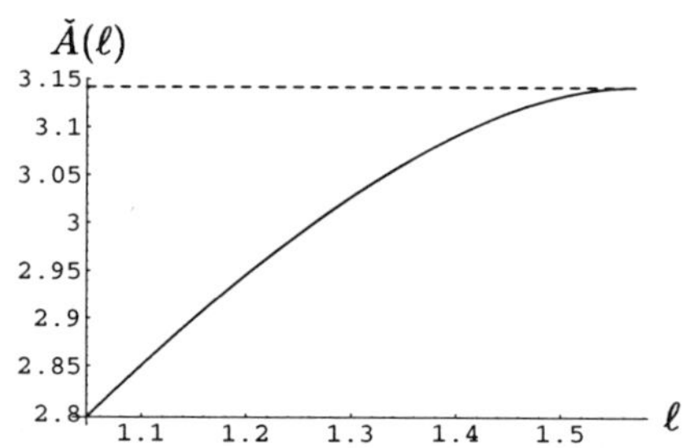

Remarks (i) For each given length ℓ between $\pi/3$ and $\pi/2$, there are exactly two geometrically distinct such $5\triangle$-stars with minimal area of $\breve{A}(\ell)$, as indicated in Figure 7.32.

(ii) If the area of a type I $5\triangle$-star is at most equal to $\breve{A}(\ell)$, $\pi/3 < \ell < \pi/2$, then its *longest radial edge* is *at most equal to* ℓ.

(iii) Let ℓ be the maximal radial edge length of a given type I $5\triangle$-star. If its area exceeds $\breve{A}(\ell)$ by a considerable amount, then either the distribution of its boundary edge-excess is not very lopsided or its total radial edge-excess is considerably larger than $(\ell - \pi/3)$, or both.

(iv) In the limiting case of $\ell = \pi/2$, ℓ^* is also equal to $\pi/2$ and the "minimal" configurations of Figure 7.32 become 4-fork star configurations as indicated in Figure 7.33. Therefore, the two kinds of configurations indicated in Figure 7.32 with ℓ close to $\pi/2$, or small deformations of them are respectively small deformations of the two configurations of Figure 7.33. Thus, the angles and edge-lengths of such small deformations are, of course, respectively close to that of Figure 7.33.

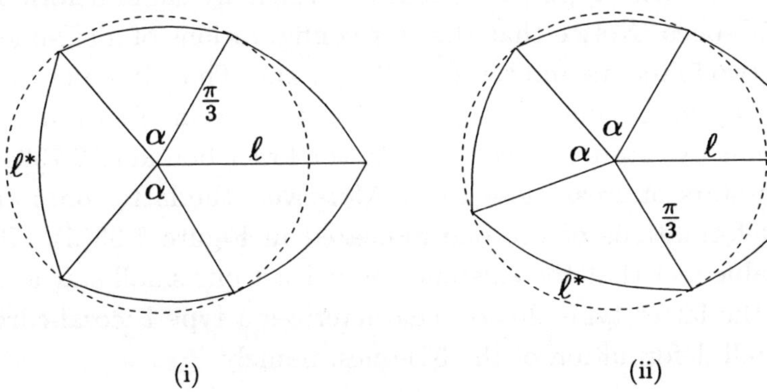

Figure 7.32

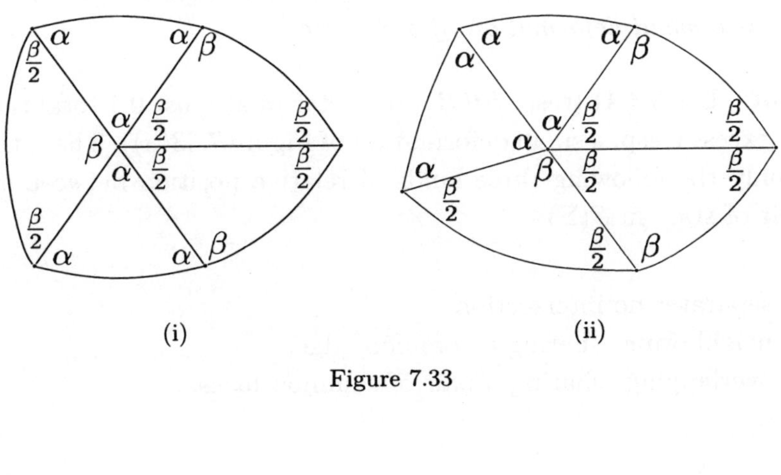

Figure 7.33

7.1.4.4 *A characterization of type* I *icosahedrons which are small deformations of the* 5□*-ones*

In Subsecs. 7.1.2 and 7.1.3, we prove that all those type I configurations with at least one 4-fork vertex or at least one quadrilateral are small deformations of those 5□-ones, except the two 6□-ones, i.e., the f.c.c. and the h.c.p. configurations. Of course, there are also small deformations of the 5□-ones which are of the combinatorial type of regular icosahedron. Thus, it is natural (and useful for the purpose of proving Main Theorem I) to study the geometric characterization of those type I icosahedrons which are small deformations of the 5□-ones. Notice that the star configurations of a 5□-one consists of two 5△-stars and ten (2□, 2△)-stars. Thus, it is not difficult to see that an icosahedron type small deformation of such a 5□-one always has two stars of areas close to the lower bound of $2.796485\cdots$ and ten stars of areas close to π. Moreover, the latter ones are all small deformations of the one indicated in Figure 7.33(ii). It is a remarkable fact that the existence of at least one small star and one star of the latter type already characterizes a type I icosahedron to be a small deformation of the 5□-ones, namely

Lemma 7.1.6 *Let $\mathcal{S}(\Sigma)$ be a type I icosahedron. If it contains a star of at most* 0.1 *total radial edge-excess and another star which is a small deformation of the one indicated in Figure* 7.33(*ii*), *then $\mathcal{S}(\Sigma)$ is a small deformation of a* 5□*-one.*

Proof Let $St(A)$ (resp. $St(B)$) be a star of at most 0.1 total radial edge excess (resp. a small deformation of Figure 7.33(ii)). Then, there are only the following three kinds of relative position between such a pair of stars in $\mathcal{S}(\Sigma)$:

(i) separate: no intersection,
(ii) neighboring: sharing a common edge,
(iii) overlapping: sharing a pair of common faces.

Note that it is impossible to have the case of sharing only a single vertex. For otherwise, this single common vertex will be necessarily of 6-fork type.

Case 1: $St(A) \cap St(B) = \phi$. This case is actually impossible. The complementary region of $St(B)$ is a small deformation of the one indicated in Figure 7.34, which is a subregion of the one indicated in Figure 7.11.

On the other hand, it follows from the small upper bound of the total radial edge-excess of $St(A)$, $St(A)$ is necessarily a small deformation of a 5Δ-star with five $\pi/3$ radial edges. It is not difficult to show that the size and shape of the complementary region of Figure 7.34 is far from able to accommodate the six vertices of such a small deformation of a 5Δ-star with $\pi/3$ radial edges.

Case 2: Neighboring Note that the local geometries at the five boundary edges of $St(A)$ are more or less the same while the local geometries at the five boundary edges of $St(B)$ are quite different. Therefore, we shall subdivide our discussion on the geometric possibilities of matching the above two star configurations along a common edge according to which of the five boundary edges of $St(B)$ is matched with that of $St(A)$. Thus, we shall first label the boundary edges of $St(B)$ as indicated in Figure 7.35 and subdivide our discussion accordingly. Since ℓ^* is of length close to $\pi/2$, it is much

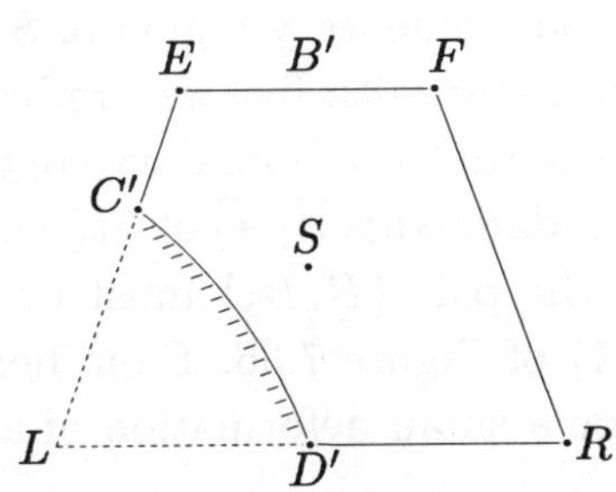

Figure 7.34

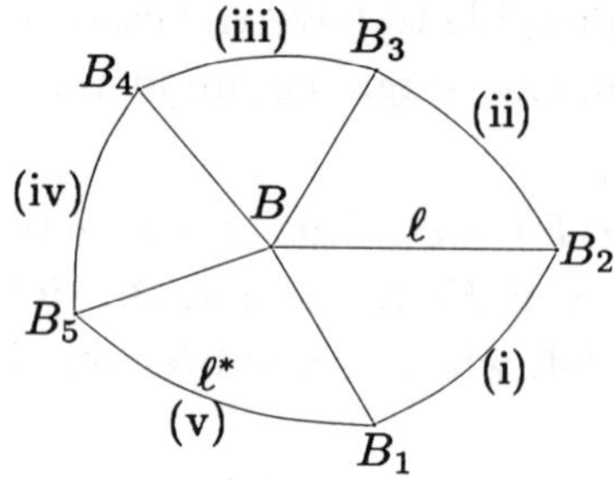

Figure 7.35

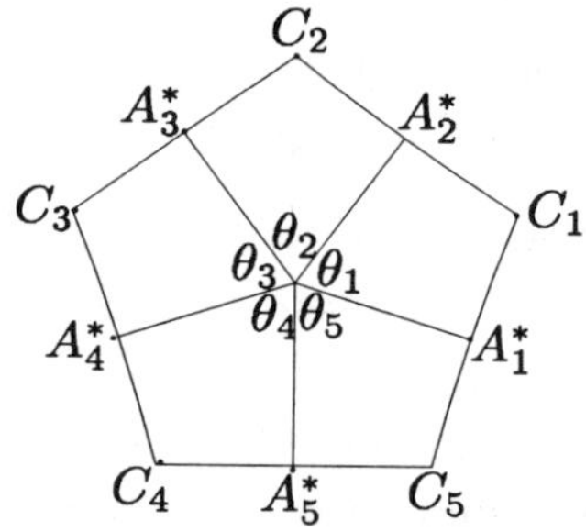

Figure 7.36

too long to be matched with any boundary edge of $St(A)$. Thus, it suffices to discuss the remaining four subcases.

Subcase (i): The complementary region of $St(A)$ is a small deformation of that of a $5\triangle$-star with five $\pi/3$ radial edges, as indicated in Figure 7.36. Suppose that the boundary edge (i) of Figure 7.35 is matched with the boundary edge $\overline{A_1A_2}$ of $St(A)$. Then, the geometry of $St(B)$ shows that the pair $\{B, B_3\}$ must be a small deformation of the pair of $\{A_1^*, A_2^*\}$ of Figure 7.36. From here, it is quite easy to see that $\mathcal{S}(\Sigma)$ must be a small deformation of a $5\square$-one.

Subcase (ii): Suppose that $\overline{B_2B_3}$ is matched with $\overline{A_2A_3}$. Then, the five radial edges of $St(A_3)$ are all close to $\pi/3$ in length while

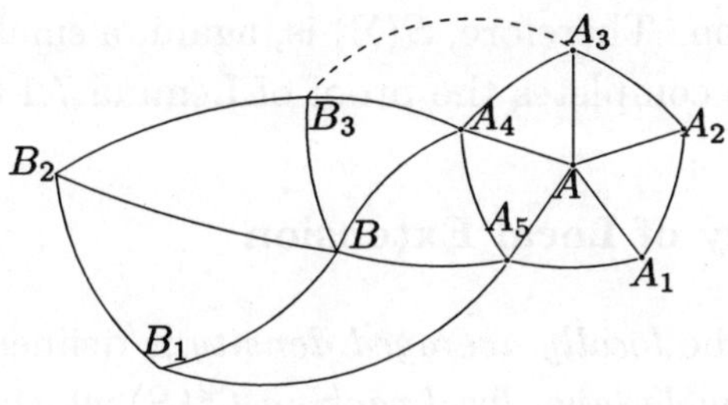

Figure 7.37

four of the central angles at A_3 are respectively small deformations of $\alpha, \alpha, \alpha, \beta$. Thus, the remaining fifth angle at A_3 is close to $\pi - 2\alpha = \alpha - 0.55128\cdots$, which is far too small for such a 5$\triangle$-star. Hence, the occurrence of Subcase (ii) is *impossible*.

Subcase (iii): Suppose that $\overline{B_3B_4}$ is matched with $\overline{A_3A_4}$. Then, $St(A_3)$ is again of the geometric type of that of the Subcase (ii). Hence, the occurrence of this subcase is again *impossible*.

Subcase (iv): Suppose that $\overline{B_4B_5}$ is matched with $\overline{A_4A_5}$. As indicated in Figure 7.37, $St(A_4)$ is again a 5$\triangle$-star with its five radial edges of lengths close to $\pi/3$. Moreover, $\angle A_4BB_1$ and $\angle A_4B_3B_2$ are both close to π. Therefore, the same geometric structure as that of Subcase (i) again occurs here for $St(A_4)$ and $\{B_2, B_1\}$ as that of $St(A)$ and $\{B, B_3\}$ in Subcase (i). Hence, on the one hand, such a geometric structure implies that $\mathcal{S}(\Sigma)$ should be a small deformation of a 5$\square$-one; but on the other hand, $\mathcal{S}(\Sigma)$ has a pair of overlapping small 5$\triangle$-stars, namely, $St(A)$ and $St(A_4)$, which clearly contradict to each other. This proves that the occurrence of Subcase (iv) is, again, *impossible*.

Case 3: Overlapping Note that $St(B)$ has only two triangles of small sizes and hence, there is only one way for $St(A)$ and $St(B)$ to share two triangles, namely, that pair of small sizes (i.e. area-wise). Moreover, such a matching will have the geometric structure of $St(A)$ and $\{B_1, B_2\}$ be the same as that of $St(A)$ and $\{A_1^*, A_2^*\}$ up to a small deformation. Therefore, $\mathcal{S}(\Sigma)$ is, again, a small deformation of a 5□-ones. This completes the proof of Lemma 7.1.6. □

7.2 Geometry of Local Extension

Geometrically, the *locally averaged density* is defined in terms of the structure of a *double-layer local packing* $\mathcal{L}^2(S)$ which can be regarded as an extension of the single-layer local packing $\mathcal{L}(S)$ it contains. Let S_j be a close neighbor of the central sphere S and $\mathcal{L}(S_j)$ be the (single-layer) local packing of S_j contained in $\mathcal{L}^2(S)$. Then, the restriction of the extension (i.e. $\mathcal{L}(S) \subset \mathcal{L}^2(S)$) to $\mathcal{L}(S_j)$ constitutes a *local extension*, namely, from the partial surrounding $\mathcal{L}(S_j) \cap \mathcal{L}(S)$ to a full-fledged local packing $\mathcal{L}(S_j)$. It is exactly the *localization* of the extension, $\mathcal{L}(S) \subset \mathcal{L}^2(S)$, to the vicinity of S_j. In this section, we shall study the geometry of individual local extensions as well as the *interaction* between *neighboring* local extensions. The main purpose of such a geometric study is to facilitate the collective volume estimation of the cluster of local cells of the spheres of $\hat{\mathcal{L}}(S)$ that is needed in the proof of Main Theorem I. Therefore, we shall focus our attention mainly on those *volume influential* aspects of local extenstions, such as the *buckling effects*, and the *volume estimates* of some more critical kinds of local extensions.

7.2.1 *Basic geometry of local extensions*

Let S' be a close neighbor of S. Then, the partial local surrounding of S' consists of S and the collection of common neighbors of S and S' which is exactly the same kind of geometric structure that one encounters in Subsec. 4.3.1. Therefore, we shall use the same system of notations such as

(i) denoting the set of common neighbors by $\{S_j, 1 \leq j \leq k\}$,
(ii) setting A, A' to be the intersection point of $\overline{OO'}$ and S, S' respectively and A_j (resp. A'_j) to be the intersection point of $\overline{OO_j}$ (resp. $\overline{O'O_j}$) and S (resp. S').

Then, the geometry of the partial surrounding of S' can be concisely represented by $St(A), \{h_j\}$ and $St(A'), \{h'_j\}$ which are functionally related by those equations of Subsec. 4.3.1.

For the purpose of providing some efficient collective volume estimates of the cluster of local cells, good enough for the proof of Main Theorem I, those local extensions with the volumes of their resulting local cells considerably larger than $4\sqrt{2}$ will play very little role, except in making the task of collective volume estimates much easier than otherwise. Notice that those local extensions which *add* a considerable amount of total *buckling height* will always make the resulting local cells considerably larger than $4\sqrt{2}$. Therefore, it suffices to focus our attention on studying the geometry of those local extensions that add at most a small amount of total buckling height; the *additional neighbors* of the local extension are all touching or almost touching neighbors of S'.

7.2.1.1 *Complementary regions and type* I *local extensions*

Definition Let $St(A')$ and $\{h'_j\}$ be the geometric data of (i). Its *associated complementary region* is defined to be the set of points on S' whose spherical distance to A' (resp. A'_j) is at least equal to $\cos^{-1}\frac{1+h}{2}$ (resp. $\cos^{-1}\frac{1+h'_j}{2}$).

Definition A local extension is called a *type I local extension* if all *additional* neighbors of the local extension are *touching neighbors* of S'.

Remarks (i) The set of touching points of a given *type I local extension* is a $\frac{\pi}{3}$-separated subset of its associated complementary region. Conversely, such a subset can also be realized as the set of touching points of a type I local extension.

(ii) Those local extensions with at most a small amount of *additional buckling height* corresponding to those point sets satisfying the following modification, namely, within a very close vicinity of the above associated complementary region, may have some pairwise distance slightly shorter than $\pi/3$. Moreover, it is quite simple to estimate the lower bound of the amount of *additional buckling height* in terms of the actual *edge-shortages* and the distances between those *outside* points and the complementary region.

(iii) In the special but also important crucial case that S and S' touch each other and moreover, where all of $\{S_j, 1 \le j \le k\}$ are *common touching neighbors* of S and S', it is easy to see that $St(A)$ and $St(A')$ are isometric to each other and the associated complementary region of $St(A')$ on S' is also isometric to the complementary region of $St(A)$ on S that has already been discussed in Sec. 7.1.

Example 7.2.1 Let $\mathcal{L}(S)$ be a local packing of 5□-type and S' be the neighbor touching S at the north (or south) pole. Then, the associated complementary region on S' is isometric to that of $St(A)$ on S as indicated in Figure 7.36. Thus, it is not difficult to see that it can accommodate a large variety of type I local extensions ranging from those 5□-ones and their small deformations to those type I icosahedrons with a rather different distribution of gaps. Moreover, the volumes of the resulting local cells of such various local extensions actually cover a considerable range in the vicinity of $4\sqrt{2}$. Therefore, one needs to study the correlation between some simple geometric invariants of the local extension and the volume of its resulting local cell (cf. Subsec. 7.2.2) in order to obtain a better control of the collective volume estimate than otherwise.

The following are some explicit computations on the volumes of the resulting local cells of some specific local extensions:

1. Suppose that the local extension adds six touching neighbors of S' with five of them touching S' at the vertices $\{C_i, 1 \le 1 \le 5\}$ of the complementary region as indicated in Figure 7.36. Then, the spherical configuration $\mathcal{S}(\Sigma')$ is a type I icosahedron with fifteen of its twenty triangles given as follows, namely

(i) five pairs of $\pi/3$-isosceles triangles with $\theta_i, 1 \le i \le 5$ as their respective top angles,

(ii) five $\pi/3$-isosceles triangles with $\psi_i, 1 \leq i \leq 5$, as their respective top angles while ψ_i is given by

$$\psi_i = 2\left(\pi - \sin^{-1}\frac{2}{\sqrt{3}}\sin\tan^{-1}\sqrt{3}\cos\frac{\theta_i}{2}\right.$$
$$\left. - \sin^{-1}\frac{2}{\sqrt{3}}\sin\tan^{-1}\sqrt{3}\cos\frac{\theta_{i+1}}{2}\right). \tag{1}$$

In the special case that all five θ_i are equal to $\frac{2\pi}{5}$, the above fifteen triangles comprise ten triangles with side-lengths $\{\pi/3, \pi/3, \cos^{-1} 0.481762745\}$ and five triangles with side-lengths $\{\pi/3, \pi/3, \cos^{-1} 0.382059631\}$, while their areas are respectively equal to $\Delta_1 = 0.55974005$, $\Delta_2 = 0.60109144$. Hence, the average area of the remaining five triangles of $\mathcal{S}(\Sigma')$ is equal to $\overline{\Delta} = \frac{1}{5}(2\pi - 10\Delta_1 - 5\Delta_2) = 0.792702582$. Thus, straightforward computation gives

$$\begin{aligned} f(\sigma_1) &= 0.240193696\,, \\ f(\sigma_2) &= 0.263126682\,, \\ f(\overline{\sigma}_{\overline{\Delta}}) &= 0.374682715\,. \end{aligned} \tag{2}$$

Therefore, the volume of the resulting local cell of such a local extension, no matter how one places the sixth sphere, is always bounded below by

$$\begin{aligned} 10f(\sigma_1) + 5f(\sigma_2) + 5f(\overline{\sigma}_{\overline{\Delta}}) = 5.590984 &= L_{12} + 0.040693 \\ &= 4\sqrt{2} - 0.06587025\,. \end{aligned} \tag{3}$$

However, the above lower bound is *only* a lower bound for those local extensions with five additional spheres *already* fixed at the touching positions of $\{C_i, 1 \leq i \leq 5\}$.

In fact, it is not difficult to construct other kinds of local extensions with considerably smaller local cells by moving the touching points of the five additional neighbors *inwards*. The following is such an explicit family of examples:

2. Choose the (spherical) polar coordinate system on S' with A' as the origin and $\overline{A'A'_1}$ as the reference direction. Then, the coordinates of A'_i and C_i are given by

$$A'_i : \left(\pi/3, (i-1)\frac{2\pi}{5}\right), \quad C_i : \left(d, (2i-1)\frac{\pi}{5}\right), \quad 1 \leq i \leq 5, \qquad (4)$$

where $d = 2\tan^{-1}\sqrt{3}\cos\frac{\pi}{5} = 1.90194354$. Set Σ'_ε to be the extension of $\{A', A'_i, ; 1 \leq i \leq 5\}$ by adding

$$C' : (\pi, *), \quad C'_i : \left(d+\varepsilon, (2i-1)\frac{\pi}{5}\right), \quad 1 \leq i \leq 5. \qquad (5)$$

Then, $\mathcal{S}(\Sigma'_\varepsilon)$ consists of four sets of congruent quintuples of triangular faces whose cosines of side-lengths are respectively as follows:

$$\sigma_1 : \left(\frac{1}{2}, \frac{1}{2}, \cos a = 0.481762745\right),$$

$$\sigma_2 : \left(\cos\frac{a}{2}\cos\left(\frac{d}{2}+\varepsilon\right), \cos\frac{a}{2}\cos\left(\frac{d}{2}+\varepsilon\right), \cos a\right),$$

$$\sigma_3 : \left(\cos\frac{a}{2}\cos\left(\frac{d}{2}+\varepsilon\right), \cos\frac{a}{2}\cos\left(\frac{d}{2}+\varepsilon\right), \cos c\right),$$

$$\sigma_4 : (-\cos(d+\varepsilon), -\cos(d+\varepsilon), \cos c), \qquad (6)$$

where $\cos c = 1 - 2\sin^2\frac{\pi}{5}\sin^2(d+\varepsilon)$. Hence, it is straightforward to compute the volume of the resulting local cell of such a local extension, namely

$$\operatorname{vol} T(\Sigma'_\varepsilon) = 5\{f(\sigma_1) + f(\sigma_2) + f(\sigma_3) + f(\sigma_4)\} \qquad (7)$$

and the following is the computer plot of $\operatorname{vol} T(\Sigma'_\varepsilon)$ for $0 \leq \varepsilon \leq (\frac{2\pi}{3} - d) = 0.19245$:

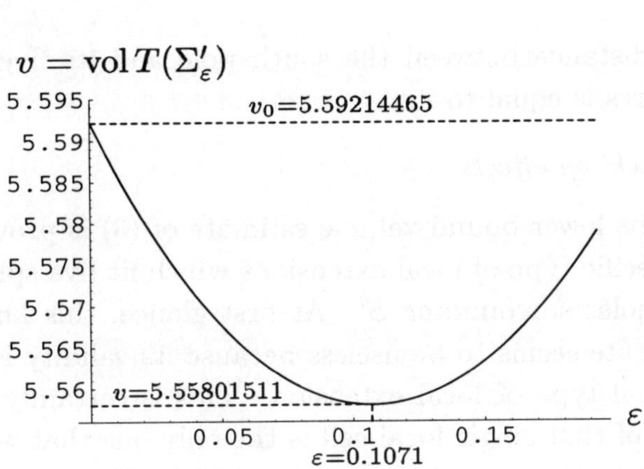

Example 7.2.2 Suppose that S' is a touching neighbor of S, i.e. $A = A'$, and it is surrounded by another five touching neighbors of S, $\{S_j, 1 \le j \le 5\}$, with at most a small amount of total gap-width between S' and S_j, namely, $St(A)$ is a small 5Δ-star. Let r_j (resp. θ_j) be the radial edge-lengths (resp. central angle) of $St(A)$. Then, the corresponding radial edge-lengths of $St(A')$ are given by

$$r'_j = \frac{1}{2}(\pi - r_j), \quad 1 \le j \le 5 \tag{8}$$

and the associated complementary region of $St(A')$ on S' can be represented by stereographic projection as indicated in Figure 7.38,

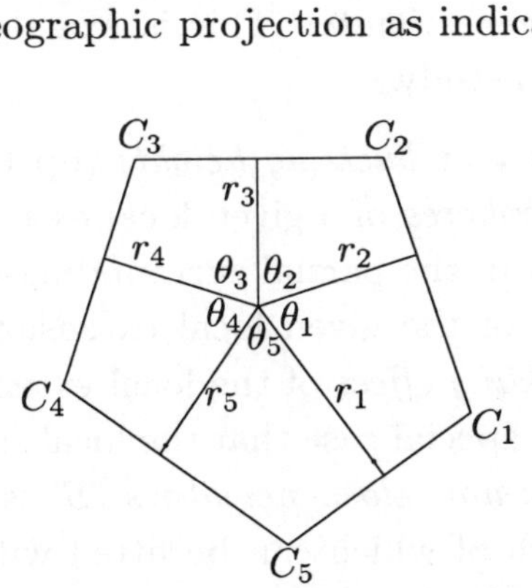

Figure 7.38

where the distance between the south pole and its five r_j'-circular boundary arcs is equal to r_j.

7.2.1.2 *Buckling effects*

Note that the lower bound volume estimate of (3) is proved only for the very specific type of local extensions which fit five spheres to the triangular holes surrounding S'. At first glance, this kind of lower bound estimate seems to be useless because its validity is limited to a very special type of local extension. This is certainly the case if the volume of that single local cell is the only one that we are going to account for. However, in the proof of Main Theorem I, what one needs is the collective accounting of the total volume of a cluster of local cells. Therefore, it is natural to study the "*effects*" of those local extensions with *smaller* local cells (e.g. smaller than the estimate of (3)) on the volume of the neighboring local cells (or rather, on the volume estimates of them). For example, certain local extensions of a given partial local surrounding of S', such as the second one in Example 7.2.1, may make the resulting local cells smaller than some others, but such tighter local extensions often introduce a considerable amount of total *buckling height* for the local extensions of its neighboring ones, thus inevitably producing an overall increment in the *collective volume estimate.* This is exactly the *buckling effect* that we are going to study.

Definition The set of *buckling heights* (up to 0.3 in magnitude) between the *added spheres* of a given local extension at S' and those *close neighbors* of S in the partial surrounding of S' is defined to be the *buckling effects* of the given local extension. The total sum is called the *total buckling effect* of the local extension.

In the important special case that the local surrounding of $\{S, S'\}$ consists of five *common close neighbors*, S' is surrounded by five triangular holes each of which can be fitted with a touching sphere and without any conflicting of space, namely, with the five vertices of the associated complementary region $\{C_i, 1 \leq i \leq 5\}$ as their touching points. In the next subsection, we shall provide a lower bound volume estimate for the resulting local cells of those local extensions with *at most a small amount of total buckling effect.*

Next let us study another kind of buckling effect. Recall that a quadrilateral $\widehat{\square ABCD}$ is called a *slightly buckled quadrilateral* if its four vertices are *close to cocircular* but *not* cocircular (from the Euclidean viewpoint, they are close to coplanar but not coplanar). Therefore, it is actually the union of two adjacent triangles of a spherical configuration matching along one of its diagonals, say $\overline{AC}$. In the case that $\{A, B, C, D\}$ are actually the touching points of four neighbors to the central sphere, say $\{S_A, S_B, S_C, S_D\}$, the outside sphere S^* (resp $\hat{S}$) which touches S_A, S_C and S_B (resp. S_D) does *not* touch S_D (resp. S_B).

Definition The smaller of the buckling heights between $\{S^*, S_D\}$ and $\{\hat{S}, S_B\}$ is defined to be the buckling effect of $\widehat{\square ABCD}$.

For the purpose of proving Main Theorem I, it suffices to study the buckling effects of those *slightly buckled* quadrilaterals whose four side-lengths are all quite close to $\pi/3$.

Lemma 7.2.1 *Let $\widehat{\square ABCD}$ be a buckled quadrilateral with side-lengths all equal to $\pi/3$ and the pair of diagonal lengths equal to ℓ and λ. Then, the buckling effect of $\widehat{\square ABCD}$ is equal to*

$$(3 - 4\cos\lambda + 2\cos 2\lambda)^{1/2} - 1 = -2\cos\lambda = \frac{2\cos\ell}{1+\cos\ell}\,. \tag{9}$$

Proof Set $\overline{AC} = \ell$ and $\overline{BD} = \lambda$. Then, the $\pi/3$-equilateral assumption implies that

$$(1+\cos\lambda)(1+\cos\ell) = 1 \quad \text{and} \quad \cos\frac{\lambda}{2}\cos\frac{\ell}{2} = \frac{1}{2}\,. \tag{10}$$

Set r to be the circumradius of $\widehat{\triangle ABC}$. Then,

$$\begin{aligned} &\cos\left(\frac{\lambda}{2} - r\right)\cos\frac{\ell}{2} = \cos r\,, \\ &\tan r = \frac{1}{2\sin\frac{\lambda}{2}\cos\frac{\ell}{2}} = \cot\frac{\lambda}{2} \Rightarrow r = \frac{\pi}{2} - \frac{\lambda}{2}\,. \end{aligned} \tag{11}$$

Observe that the center distance between S^* and S_D is equal to twice the buckling effect plus two. One has

$$\begin{aligned}(1+h)^2 &= 1 + 4\cos^2 r - 4\cos r\cos(\lambda - r)\\ &= 1 + 4(1-\cos\lambda)\cos^2 r - 2\sin\lambda\sin 2r\\ &= 3 - 4\cos\lambda + 2\cos 2\lambda\,,\end{aligned} \tag{12}$$

namely

$$h = (3 - 4\cos\lambda + 2\cos 2\lambda)^{1/2} - 1 = -2\cos\lambda = \frac{2\cos\ell}{1+\cos\ell}\,. \tag{13}$$

□

The following are some specific values of the buckling effects of the $\pi/3$-equilateral quadrilaterals with given diagonal lengths:

ℓ	1.56	1.55	1.54	1.53	1.52
h	0.02136	0.04074	0.05974	0.07837	0.09664

Remark If some of the side-lengths are slightly longer than $\pi/3$, then the buckling effect of such a buckled quadrilateral with ℓ as its diagonal length will also become slightly larger. Thus, the above values as well as the value given by (9) can also be regarded as a *lower bound estimate* of the buckling effect of such quadrilaterals in terms of its diagonal length.

Example 7.2.3 Set Σ_ε to be the set of twelve points on $S^2(1)$ whose Cartesian coordinates are given as

$$\Sigma_\varepsilon = \left\{\begin{array}{c}\left(\pm\cos\left(\frac{\pi}{4}+\varepsilon\right), \pm\sin\left(\frac{\pi}{4}+\varepsilon\right), 0\right)\\ \left(\pm\sin\left(\frac{\pi}{4}+\varepsilon\right), 0, \pm\cos\left(\frac{\pi}{4}+\varepsilon\right)\right)\\ \left(0, \pm\cos\left(\frac{\pi}{4}+\varepsilon\right), \pm\sin\left(\frac{\pi}{4}+\varepsilon\right)\right)\end{array}\right\}. \tag{14}$$

Notice that $\mathcal{S}(\Sigma_0)$ is exactly the spherical configuration of the f.c.c. local packing. Thus, the type I local packing with Σ_ε as the set of touching points is an ε-deformation of the f.c.c. local packing. By

Lemma 7.1.5, all the other ε-deformations of the f.c.c. local packing are, in fact, ε^2-perturbations of the above explicitly given ε-deformation. Anyhow, all ε-deformations of the f.c.c. configuration have six buckled quadrilaterals whose two diagonals are approximately equal to $\frac{\pi}{2} \pm 2\varepsilon$ (i.e. modulo ε^2). Therefore, the buckling effect of each of the six buckled quadrilaterals is approximately (i.e. up to the first order of ε) given by

$$h \equiv \frac{2\sin 2\varepsilon}{1+\sin 2\varepsilon} \equiv 4\varepsilon \pmod{\varepsilon^2}. \tag{15}$$

Thus, $\mathcal{S}(\Sigma_\varepsilon)$ contains a total amount of 24ε in the buckling effects of its six buckled quadrilaterals.

7.2.2 *Local extensions and volume estimations*

7.2.2.1 *Collective volume estimate and localized estimates*

Let $\hat{\mathcal{L}}(S_0) = \{S_j; 0 \le j \le m\}$ be the core packing of S_0 contained in a given double-layer local packing $\mathcal{L}^2(S_0)$. The *locally averaged density* $\overline{\rho}(S_0, \mathcal{L}^2)$ is defined to be the following specific weighted average, namely

$$\overline{\rho}(S_0, \mathcal{L}^2) = \sum_{j=0}^{m} w_j \rho(S_j, \mathcal{L}^2) \Big/ \sum_{j=0}^{m} w_j\,, \tag{16}$$

where $w_j = \lambda_j \operatorname{vol} C(S_j, \mathcal{L}^2)$, $\lambda_j = 13/\#\hat{\mathcal{L}}(S_j)$. Therefore, the proof of the *optimal upper bound estimate of* $\pi/\sqrt{18}$ for the locally averaged density amounts to the following *collective volume estimate*, namely

$$\begin{aligned}
&\overline{\rho}(S_0, \mathcal{L}^2) \le \pi/\sqrt{18} \\
&\Leftrightarrow \frac{\pi}{\sqrt{18}} \sum_{j=0}^{m} \lambda_j(\operatorname{vol} C(S_j, \mathcal{L}^2) - 4\sqrt{2}) \ge 0 \\
&\Leftrightarrow \sum_{j=0}^{m} \lambda_j(\operatorname{vol} C(S_j, \mathcal{L}^2) - 4\sqrt{2}) \ge 0\,.
\end{aligned} \tag{17}$$

Geometrically, it is natural to regard $\mathcal{L}^2(S_0)$ as an extension of $\hat{\mathcal{L}}(S_0)$ (i.e. a single-layer coating of $\hat{\mathcal{L}}(S_0)$), while the *local extension* of $\hat{\mathcal{L}}(S_0) \cap \mathcal{L}(S_j)$ to $\mathcal{L}(S_j)$ can be regarded as the localization of the extension, $\hat{\mathcal{L}}(S_0) \subset \mathcal{L}^2(S_0)$, to the vicinity of S_j. Such a geometric localization naturally leads to the following question on the extent of *localizability* of the collective volume estimate of (17): Roughly speaking, to what extent, can the estimate of (17) be *reducible* to that of *individual localized estimates*? In retrospect, one might say that the task of proving the collective estimate of (17) is, in fact, *to a large extent localizable.* For example, in one of the most critical cases that $\hat{\mathcal{L}}(S_0)$ is the f.c.c. core packing, the proof of the estimate of (17) can be completely localized to the proof that the volume of the resulting local cell of each individual local extension is at least equal to $4\sqrt{2}$ and it is *equal* to $4\sqrt{2}$ when and only when the individual local extension is either the f.c.c. or the h.c.p. local packings (cf. Lemma 7.2.2). In hindsight, it is, indeed, a wonderful situation that the proof of the estimate of (17) can largely be localized to that of the individual local extensions which, of course, greatly simplifies the complexity of such a task. This remarkable *geometric reality* also provides futher convincing evidence of the "naturality" of this approach to the proof of Kepler's Conjecture and the least action principle on crystal formation.

In the proof of the Dodecahedron Conjecture that achieves the *optimal upper bound estimate of local density*, one encounters the geometric complexity of *single-layer local packings*, while in the proof of the *optimal upper bound estimate of* $\pi/\sqrt{18}$ for the *locally average density*, one encounters the geometric complexity of *double-layer local packings.* Of course, the geometric complexity of the latter is much, much more complicated than that of the former. Naturally, one expects that the task of *optimal* collective estimate of the volumes of a cluster of local cells should be much, much more difficult than the *optimal* estimate of the volume of a single local cell. However, thanks to the reality of localizability and the geometric understanding of type I configurations, such complexities and difficulties are still quite manageable.

Remarks (i) For each given type of partial local surrounding, the degree of freedom of its local extensions is much, much smaller than that of single-layer local packing *without* constraint.

(ii) The volume estimation of the local cells of all possible local extensions of a *given type of partial local structure* is, in fact, much simpler than the volume estimates for the proof of the Dodecahedron Conjecture. However, to begin with, one needs some kind of control on the geometry of the given partial local structure.

(iii) In the collective estimate of (17), the contribution of rectilinear slabs will often play a major role. Moreover, it is natural to distinguish the following two kinds of rectilinear slabs and to account for their contributions towards the collective volume estimate separately, namely

(a) those rectilinear slabs that are already contained in the given core packing $\hat{\mathcal{L}}(S)$,
(b) those rectilinear slabs that are "created" by the extension of $\hat{\mathcal{L}}(S_0)$ to $\mathcal{L}^2(S)$.

We shall call the former *build-in slabs* and the latter *additional slabs.* Geometrically, those *build-in slabs* correspond to the *buckling heights* and the *gully-widths* of the given core packing $\hat{\mathcal{L}}(S)$, while those *additional slabs* correspond to the two kinds of *buckling effects* discussed in Subsec. 7.2.1.

7.2.2.2 *Local extensions with local cells of volumes close to* $4\sqrt{2}$

Let us first study an important special case that the given partial local surrounding consists of five touching neighbors with the $(\alpha, \beta, \alpha, \beta)$-arrangement of Lemma 7.1.1.

Lemma 7.2.2 *Let $\mathcal{L}(S')$ be a local extension of the partial local packing consisting of five touching neighbors with the $(\alpha, \beta, \alpha, \beta)$-arrangement. Then, the volume of the local cell $C(\mathcal{L})$ is at least equal to $4\sqrt{2}$ and it is equal to $4\sqrt{2}$ when and only when $\mathcal{L}(S')$ is the f.c.c. or the h.c.p. local packing.*

Proof The associated complementary region of such a partial local packing is the same as the one indicated in Figure 7.3. If $\mathcal{L}(S')$ is, in fact, a type I local packing, then it follows from Lemma 7.1.1 that $\mathcal{L}(S')$ is either the f.c.c. or the h.c.p. local packing, thus having $C(\mathcal{L})$ of volume equal to $4\sqrt{2}$.

In the remaining case of non-type I local packings, it suffices to consider those local extensions with exactly twelve close neighbors and with at most 0.03 total buckling heights. For otherwise, it is straightforward to apply the volume estimation techniques of Chap. 4 to show that $\operatorname{vol} C(\mathcal{L})$ exceeds $4\sqrt{2}$. (Note that $\mathcal{S}'(\Sigma')$ always contains quite a few triangles with edge-lengths close to $\{\pi/3, \pi/3, \pi/2\}$.)

Therefore, the additional seven close neighbors of the local extension $\mathcal{L}(S')$ are all almost touching S'. Hence, its associated spherical configuration $\mathcal{S}(\Sigma')$ contains seven points within the very, very near vicinity of the complementary region as indicated in Figure 7.3 and with separation almost equal to $\pi/3$. It follows from the quantitative analysis that we carried out in the proof of Lemma 7.1.1 that such a configuration, $\mathcal{S}(\Sigma')$, must be a tiny deformation of either the f.c.c. or the h.c.p. configuration. In fact, it follows from the combination of such quantitative analysis and the *magnified buckling phenomenon* (cf. (16) of Sec. 7.1) that the order of magnitude of the deformation of $\mathcal{S}(\Sigma')$ (i.e. from either the f.c.c. or the h.c.p. configuration) is roughly the same as that of the maximal buckling heights. On the other hand, we can estimate the possible *decrement* of $\operatorname{vol} T(\Sigma)$ for such a small deformation of f.c.c. or h.c.p. configurations by the formulas of gradient vector field (i.e. (35), (35′) and (36) of Sec. 4.2). It follows from such a straightforward computation and direct application that the possible decrement (i.e. $4\sqrt{2} - \operatorname{vol} T(\Sigma'))$ is bounded above by a *second order* smallness in terms of the small magnitude of the deformation. Therefore, the volume estimate of $\sum L(h_j)$ is always of a larger order of magnitude than the possible decrement of the volume of the core part. Hence, the volume of the local cell of a non-type I local extension always exceeds $4\sqrt{2}$, in fact, by an amount proportional to $\sum h_j$. This completes the proof of Lemma 7.2.2. □

Lemma 7.2.3 *Let $\mathcal{L}(S')$ be a local extension of the partial local packing consisting of five touching neighbors with $(\alpha, \alpha, \beta, \beta)$-arrangement. Then, the volume of the local cell $C(\mathcal{L})$ is less than or equal to $4\sqrt{2}$ only when $\mathcal{L}(S')$ is either the h.c.p. local packing or a small deformation of a 5□-one.*

Proof The associated complementary region of such a partial local packing is the same as the one indicated in Figure 7.11. In the case that the local extension is, in fact, a type I local packing, the above assertion follows directly from Lemma 7.1.2. In the remaining case of non-type I extension, it again suffices to consider those local extensions with exactly twelve close neighbors and with very small amount of total buckling height, namely, all the seven additional close neighbors of the local extension are almost touching. Therefore, it again follows from the combination of the magnified buckling phenomenon and the quantitative analysis in the proof of Lemma 7.1.2 that its associated configuration $\mathcal{S}(\Sigma')$ must be either a small deformation of the h.c.p. configuration or that of the 5□-type I configuration. Moreover, $\mathcal{L}(S')$ must be itself the h.c.p. core packing in the former subcase, otherwise the volume of $C(\mathcal{L})$ would exceed $4\sqrt{2}$. This completes the proof. □

Remarks (i) For the purpose of proving the collective volume estimate of (17), it suffices to study those local extensions whose resulting local cells are of volumes close to $4\sqrt{2}$, namely, at most exceeding $4\sqrt{2}$ by a small amount. Such local extensions always have at most twelve close neighbors and can have at most a small amount of *additional buckling height.*

(ii) Although the *buckling effects* of a given local extension $\mathcal{L}(S')$ do not produce rectilinear slabs belonging to $C(S')$ itself, they *do* produce slabs of thicknesses equal to the buckling effects belonging to its neighboring local cells. Therefore, the proof of the collective estimate of (17) can largely be reduced to the study of those double-layer local extensions which contain at most a small amount of total buckling effect. Anyhow, the existence of considerable amount of total buckling effect in a given $\mathcal{L}^2(S)$ will certainly make the collective estimate much simpler than otherwise.

(iii) For a *given* type of partial local packing, the degree of freedom of its local extensions is already much more limited than that of single layer local packings *without* constraint, while the *condition of at most a small amount of total buckling effect* will further reduce their degree of freedom much more. This basic geometric fact makes the volume estimation of local cells of such local extensions much simpler than otherwise. Thus, the volume estimation of the local cell of each *individual local extension* is always much, much simpler than that of the proof of the Dodecahedron Conjecture (cf. Chap. 6). This is another basic reason that makes the collective volume estimate in the proof of Main Theorem I, at all, manageable.

7.2.2.3 *Local extensions with at most a small amount of total buckling effect and volume estimation of their local cells*

Let us begin with the study of a few specific examples of importance.

Example 7.2.1′ Let S' be a touching neighbor of S surrounded by a quintuple of common touching neighbors with symmetric angular arrangement. Then, both the stars of A on S (resp. S') consist of five $\frac{\pi}{3}$-isosceles triangles of top angles equal to $\frac{2\pi}{5}$. Suppose that $\mathcal{L}(S')$

is a type I local extension around such an S' *without buckling effect.* Then, the volume of its resulting local cell is minimal when and only when $\mathcal{S}(\Sigma)$ consists of ten $\frac{\pi}{3}$-isosceles triangles with their top angles equal to $\frac{2\pi}{5}$, five $\frac{\pi}{3}$-isosceles triangles with their base-lengths equal to $\cos^{-1} 0.38205963$ and five a-isosceles triangles with their top angles equal to $\frac{2\pi}{5}$ with $a = 1.23964911$. Thus, the optimal lower bound of the volume of such a local cell is given as follows, namely

$$\operatorname{vol} C(\mathcal{L}(S')) \geq 10f(\sigma_1) + 5f(\sigma_2) + 5f(\sigma_3) = 4\sqrt{2} - 0.064709599\,, \tag{18}$$

where the edge-lengths of σ_1 (resp. σ_2, σ_3) are equal to $\{\frac{\pi}{3}, \frac{\pi}{3}, \cos^{-1} 0.481762745\}$ (resp. $\{\frac{\pi}{3}, \frac{\pi}{3}, \cos^{-1} 0.38205963\}$, $\{a, a, \cos^{-1} 0.38205963\}$).

Proof Set Ω to be the *complementary region* on S' and Ω' be its stereographic representation with A as the north-pole. Then, Ω' is a regular circumscribing pentagon of the $\frac{1}{\sqrt{3}}$-circle centered at the origin. Set Σ to be the set of twelve touching points of S' and its touching neighbors of such an extension $\mathcal{L}(S')$. Let the additional six points of Σ_0 be exactly the five vertices of Ω and the antipodal point of A in S'. Then, $\mathcal{S}(\Sigma_0)$ consists of ten σ_1's, five σ_2's and five σ_3's and hence the volume of $C(\mathcal{L}(S'))$ is given by

$$\begin{aligned}\operatorname{vol} C(\mathcal{L}(S')) &= \operatorname{vol} T(\Sigma_0) = 10f(\sigma_1) + 5f(\sigma_2) + 5f(\sigma_3) \\ &= 4\sqrt{2} - 0.064709599\,. \end{aligned} \tag{19}$$

Due to the fact that the total buckling effect *only* accounts for those buckling heights of at most 0.3 between the added spheres and those neighboring close neighbors, the assumption of *no* buckling effect means that those additional points of Σ along the boundary of Ω are either situated at the vertices or situated at the middle portions of those boundary arcs. Therefore, Σ either contains the five vertices of Ω or contains five points situated at the five middle portions. However, it is easy to show that the volume of $C(\mathcal{L}(S'))$ for the latter is always considerably larger than $4\sqrt{2} - 0.04$. Hence, it suffices to consider the former that Σ includes the five vertices of Ω. For such a tangent polyhedron $T(\Sigma)$, we claim that its volume is minimal when the twelveth point of Σ is located at the antipodal point of A. Let P be the position of the twelveth point of Σ which makes $\mathrm{vol}\, T(\Sigma)$ minimal. It follows from the geometric meaning of the differential condition that P is the *center of gravity* of the twelveth face of $T(\Sigma)$. Thus, it follows from the 5-fold symmetry of the arrangement of the other eleven points of Σ, P is situated at the antipodal point of A. □

Example 7.2.1″ Let S' be a touching neighbor of S surrounded by a quintuple of common touching neighbors with the most lopsided angular arrangement, i.e. $\{\alpha, \alpha, \alpha, \alpha, 2\pi - 4\alpha\}$. Suppose that $\mathcal{L}(S')$ is a type I local extension *without buckling effect.* Then, the volume of its resulting local cell is at least equal to $(4\sqrt{2} - 0.063841199)$.

Proof Again, it suffices to consider the geometric situation that Σ includes the *five vertices* of the *complementary region* Ω. Therefore, $\mathcal{S}(\Sigma)$ already has the following fifteen triangular faces among its twenty faces, namely

(i) eight $\frac{\pi}{3}$-equilateral triangles,
(ii) five $\frac{\pi}{3}$-isosceles triangles with $(2\pi - 4\alpha)$ as their top angles,
(iii) a pair of $\frac{\pi}{3}$-isosceles triangles with 1.44767851 as their top angles.

Hence, it is straightforward to use Lemmas 4.2.2 and 4.2.3 to obtain the following lower bound estimate, namely

$$\mathrm{vol}\, T(\Sigma) > 8f(\sigma_0) + 5f(\sigma_1) + 2f(\sigma_2) + 5f(\overline{\sigma}_{\bar{\triangle}}) = 4\sqrt{2} - 0.063841199\,, \tag{20}$$

where the cosines of their side-lengths are respectively given as follows, and $\bar{\Delta}$ is the averaged area of the remaining quintuple:

$$\begin{aligned}
&\sigma_0 : \left\{\frac{1}{2}, \frac{1}{2}, \frac{1}{2}\right\}, \\
&\sigma_1 : \left\{\frac{1}{2}, \frac{1}{2}, 0.407407407\right\}, \\
&\sigma_2 : \left\{\frac{1}{2}, \frac{1}{2}, 0.342105261\right\}, \\
&\bar{\Delta} = \frac{1}{5}\{4\pi - 8\Delta(\sigma_0) - 5\Delta(\sigma_1) - 2\Delta(\sigma_2)\}.
\end{aligned} \qquad (20')$$

□

Example 7.2.1‴ Let S' be a touching neighbor of S surrounded by a quintuple of common touching neighbors and $\mathcal{L}(S)$ be a type I local extension around such an S' *without buckling effect.* For the purpose of lower bound volume estimation of the local cell of such a local extension, it again suffices to consider the case that the additional six points of Σ consists of the five vertices of Ω and a point in the vicinity of the antipodal point of A. Thus, such a Σ (resp. $\mathcal{S}(\Sigma)$) is always quite a small deformation of Σ_0 (resp. $\mathcal{S}(\Sigma_0)$) of Example 7.2.1′. Therefore, we may use the gradient vector field $\nabla \mathrm{vol}\, T(\Sigma)|_{\Sigma_0}$ (or rather, the negative of it) to obtain a highly accurate estimate of the possible *decrement* of $\mathrm{vol}\, T(\Sigma)$ as compared to $\mathrm{vol}\, T(\Sigma_0)$ (cf. (36) of Sec. 4.2).

It follows from the symmetry property of Σ_0 that the *negative gradient vector field*, $-\nabla \mathrm{vol}\, T(\Sigma)|_{\Sigma_0}$, is as indicated in Figure 7.39.

Note that the deformations at the A_i's due to the change of angular arrangements are always in the direction of $\frac{\pi}{3}$-latitude circle, while the negative gradient vectors at the A_i's are in the southward longitudal direction, thus having no contribution whatsoever. Therefore, the only contributions are coming from the *longitudal components* of those deformations at the B_i's. Moreover, the above longitudal

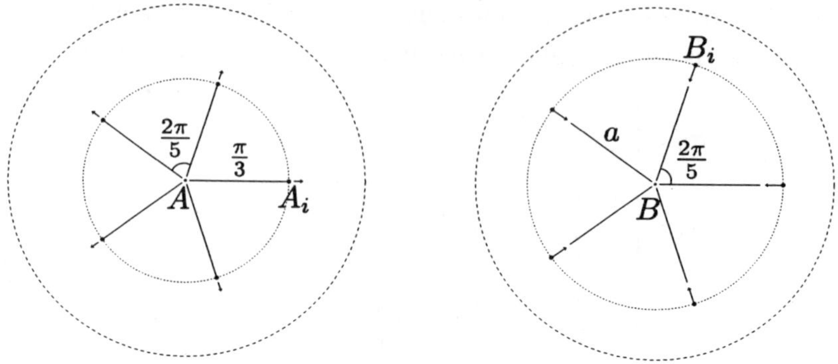

Figure 7.39

components can be computed in terms of the angular deformations as follows. Set $\theta_i = \frac{2\pi}{5} - 2\varepsilon_i$, $(\sum_{i=1}^{5} \varepsilon_i = 0)$. Then, the southward longitudal component of the deformation at B_i is given by

$$2\tan^{-1}\left(\sqrt{3}\cos\left(\frac{\pi}{5} - \varepsilon\right)\right) - 2\tan^{-1}\left(\sqrt{3}\cos\frac{\pi}{5}\right). \qquad (21)$$

Therefore, it follows from the fact that $\sum \varepsilon_i = 0$, the possible decrement of vol $T(\Sigma)$ is bounded by $\sum \varepsilon_i^2$ times a rather small constant, because the linear terms automatically cancel each other.

Example 7.2.4 Let S' be a touching neighbor of S surrounded by a quintuple of common touching neighbors and $\mathcal{L}(S')$ be a type I local extension around such an S' *with* a small amount of *total buckling effect*, say at most 0.1. Then, the volume of $C(\mathcal{L}(S'))$ is bounded below by

$$4\sqrt{2} - 0.065 - (0.1) \cdot \text{total buckling effect}. \qquad (22)$$

Proof Again, it suffices to consider those local extensions whose Σ include a quintuple of points in the vicinities of the five vertices of the associated complementary region Ω. For otherwise, the assumption on the total buckling effect, again, implies that Σ includes a quintuple of points in the vicinities of the middle portions of the five boundary

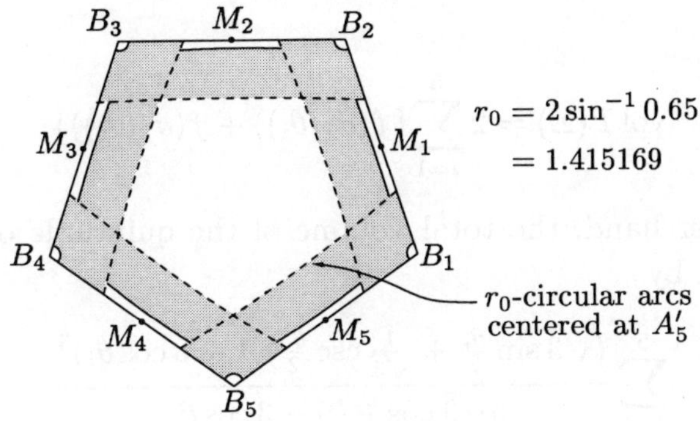

Figure 7.40

arcs as indicated in Figure 7.40, which makes the volume of $T(\Sigma)$ considerably larger than the bound of (22).

Therefore Σ is again a small deformation of Σ_0 and the possible *decrement* of $\operatorname{vol} T(\Sigma)$ as compared to Σ_0 can, again, be estimated by the gradient formula with high accuracy. Hence, the maximal amount of decrement will be produced by the southward deformations at the B_i's. On the one hand, the lengths of the gradient vectors at the B_i's are equal to 0.129228, and on the other hand, a southward ε_i-deformation at B_i will produce a pair of buckling effects slightly larger than $(0.7)\varepsilon_i$. Hence, the total amount of decrement is at most (0.1)-times the total buckling effect. □

Example 7.2.5 Let $\mathcal{S}(\Sigma)$ be a $5\square$-type I-configuration with $\{\theta_i, 1 \le i \le 5\}$ as the angular distribution at the two poles, $\alpha \le \theta_i \le 2\pi - 4\alpha$. Set $\sigma_1(\theta)$ (resp. $\sigma_2(\theta)$) to be the triangle whose triple of cosines of side-lengths are given as

$$\begin{aligned} \sigma_1(\theta) &: \left\{\frac{1}{2}, \frac{1}{2}, \frac{3}{4}\cos\theta + \frac{1}{4}\right\}, \\ \sigma_2(\theta) &: \left\{\frac{1}{2}, \frac{3}{4}\cos\theta - \frac{1}{4}, \frac{3}{4}\cos\theta + \frac{1}{4}\right\}. \end{aligned} \tag{23}$$

Then,

$$\operatorname{vol} T(\Sigma) = 2\sum_{i=1}^{5}\{f(\sigma_1(\theta_i)) + f(\sigma_2(\theta_i))\}\,. \tag{24}$$

On the other hand, the total volume of the quintuple of truncated tips is given by

$$\sum_{i=1}^{5}\frac{(\sqrt{3}\sin\frac{\theta_i}{2} + \frac{1}{\sqrt{3}}\csc\frac{\theta_i}{2})(1-3\cos\theta_i)^3}{6\sqrt{3}\cos\frac{\theta_i}{2}(5-3\cos\theta_i)}\,. \tag{25}$$

Set

$$\begin{aligned} F(\theta) &= 2f(\sigma_1(\theta)) + 2f(\sigma_2(\theta)) \\ &\quad - \frac{(\sqrt{3}\sin\frac{\theta}{2} + \frac{1}{\sqrt{3}}\frac{\theta}{2})(1-3\cos\theta)^3}{6\sqrt{3}\cos\frac{\theta}{2}(5-3\cos\theta)}\,. \end{aligned} \tag{26}$$

Then,

$$\operatorname{vol}\widehat{T}(\Sigma) = \sum_{i=1}^{5} F(\theta_i) \tag{27}$$

while the following computer plots of $F(\theta)$ and $F''(\theta)$, $\cos^{-1}\frac{1}{3} \le \theta \le 2\pi - 4\cos^{-1}\frac{1}{3}$, show that $F(\theta)$ is a convex function for θ up until $2\pi - 4\cos^{-1}\frac{1}{3} - 0.0053$.

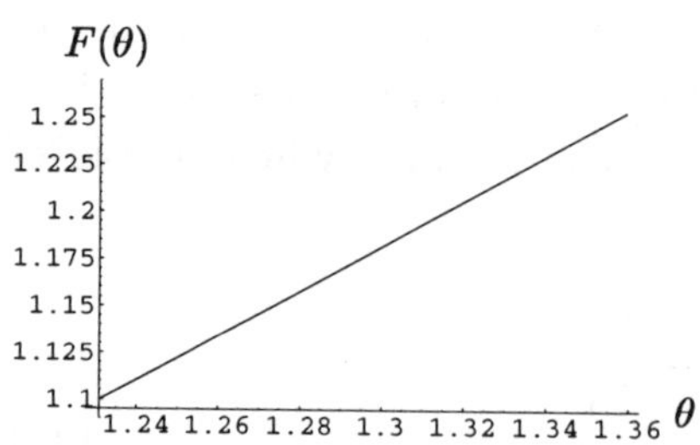

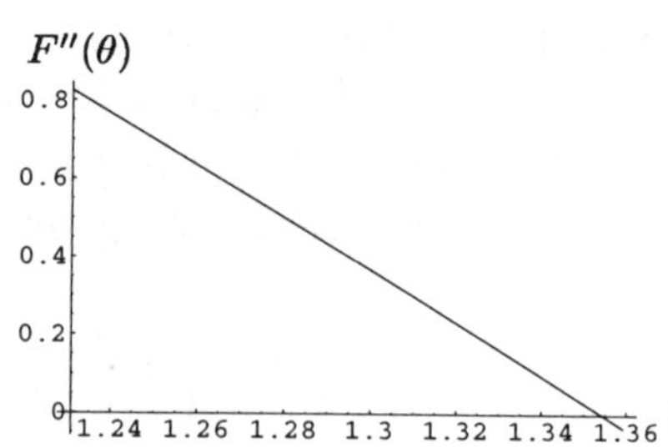

Therefore,

$$
\begin{aligned}
(4\sqrt{2} - 0.00609) = 5F\left(\frac{2\pi}{5}\right) &\le \operatorname{vol}\widehat{T}(\Sigma) \\
&\le 4F(\alpha) + F(2\pi - 4\alpha) \\
&= (4\sqrt{2} + 0.00289)\,. \qquad (28)
\end{aligned}
$$

Example 7.2.5′ Let $\mathcal{S}(\Sigma)$ be a small deformation of a 5□-one, namely, a 5 ⧄-type I-configuration with a small amount of total buckling effect, say with $\sum_{i=1}^{5} h(⧄_i)$ at most equal to 0.1. Then,

$$\operatorname{vol}\widehat{T}(\Sigma) \ge 4\sqrt{2} - 0.00909\,. \qquad (29)$$

Proof Such an $\mathcal{S}(\Sigma)$ can also be regarded as a small deformation of the 5□-configuration with $\theta_i = \frac{2\pi}{5}$, $1 \le i \le 5$, say denoted by $\mathcal{S}(\Sigma^*)$. It follows from the D_5-symmetry of $\mathcal{S}(\Sigma^*)$ that gradient vector field $\nabla T(\Sigma)|_{\Sigma^*}$ also has a D_5-symmetry. Therefore, its vectors at the poles are zeros, while its vectors at the other ten points of Σ^* are in the longitudal direction. On the other hand, such a small deformation only has tiny longitudal components at the above ten points of Σ^*. Hence, it is not difficult to see that $\operatorname{vol}\widehat{T}(\Sigma)$ is at least equal to $\operatorname{vol}\widehat{T}(\Sigma^*) - 0.003$, namely, $4\sqrt{2} - 0.00909$. □

Example 7.2.6 Let $\hat{\mathcal{L}}(S)$ be a type I core packing of 5□-type and S' be one of its touching neighbors with a $(\triangle, \triangle, \square, \square)$-type local surrounding. Set $C'(S')$ to be the *reduced* local cell of S' which *discounts* the volumes of those slabs, if any, between S' and its neighboring touching neighbors of S. If the *total buckling effect* of the local extension at S' is at most equal to 0.1, then the volume of $C'(S')$ is at least equal to $4\sqrt{2} - 0.01$.

Proof Set A to be the touching point of S and S', and $St(A)$ (resp. $St'(A)$) to be the star configuration of A in S (resp. S'). Let us first consider the special (and actually also the most critical) case that $St(A)$ consists of a pair of $\frac{\pi}{3}$-equilateral triangles and a pair of $\frac{\pi}{3}$-squares. In such a situation, there are no slabs between S'

and its neighboring touching neighbors of S and hence $C'(S') = C(S)$. By Lemma 7.1.2 and its proof, a (non h.c.p.) type I local extension at S' *without buckling effect* is a small deformation of the 5□-one with $(\alpha, \alpha, \alpha, \alpha, 2\pi - 4\alpha)$-angular arrangement. As indicated in Figure 7.41, it only moves G (resp. H) by a small amount towards the center.

Therefore, the volume of such a $C'(S')$ is very, very close to $4F(\alpha) + F(2\pi - 4\alpha)$.

In general, $St(A)$ consists of a pair of $\frac{\pi}{3}$-isosceles triangles and a pair of "rectangles" with $\frac{\pi}{3}$ as their "width," while the cosines of the two base-lengths (resp. two lengths) of the pair of triangles (resp. rectangles) are given by

$$\frac{3}{4}\cos\theta_i + \frac{1}{4} \text{ and } \frac{3}{4}\cos\theta_{i+1} + \frac{1}{4}, \quad \alpha \le \theta_i, \theta_{i+1} \le 2\pi - 4\alpha\,. \tag{30}$$

In such a case, $St'(A)$ becomes smaller and hence the complementary region Ω' (i.e. of $St'(A)$ in S') becomes larger than that of Figure 7.41 (cf. Figure 7.42). It is not difficult to see that the volume of the *reduced* local cell $C'(S')$ corresponding to a local extension *without* buckling effect also becomes larger than that of the special case.

Again, it follows from the same kind of estimation via the gradient vector field that the volume of the reduced local cells with respect to local extensions with at most 0.1 total buckling effect can only be a tiny bit smaller than that of local extensions *without* buckling effect. □

Ω :

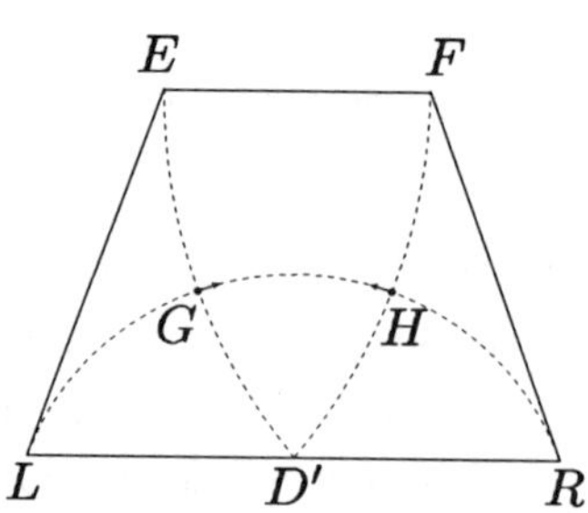

Figure 7.41

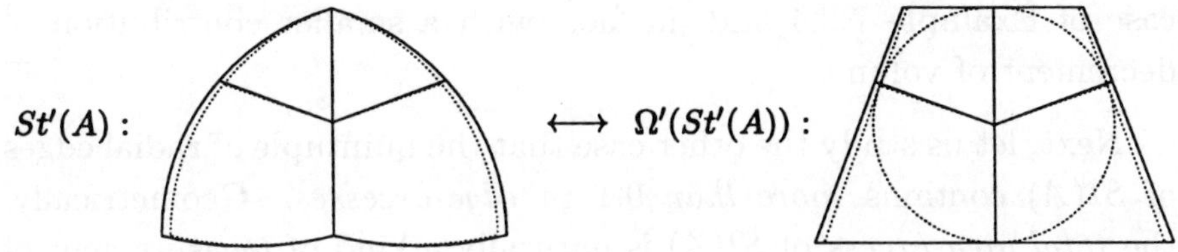

Figure 7.42

Remark It is not difficult to improve the above lower bound estimate of $\operatorname{vol} C'(S')$ to $4\sqrt{2} - 0.0075$ by the following slightly better way of geometric accounting, namely, one may regard such a $\mathcal{S}(\Sigma)$ as a small deformation of $\mathcal{S}(\Sigma')$ which is the 5□-configuration with $\{\alpha, \alpha, \frac{2}{3}\beta, \frac{2}{3}\beta, \frac{2}{3}\beta\}$ as its angular arrangement. Such a replacement of *initial configuration* will make the upper bound on the magnitude of deformation considerably smaller, with the initial value of $\operatorname{vol} \widehat{T}(\Sigma') = 2F(\alpha) + 3F(\frac{2}{3}\beta) = 4\sqrt{2} - 0.00534$. However, the smaller lower bound estimate of $4\sqrt{2} - 0.01$ already suffices for the purpose of proving the collective volume estimate of (17) (cf. Chap. 8).

Example 7.2.7 Let S' be a touching or almost touching neighbor of S which is surrounded by five touching or almost touching neighbors of S in a given $\hat{\mathcal{L}}(S)$. If the associated 5△-star contains *at most* 0.1 *radial edge-excesses* and the local extension at S' contains at most 0.1 buckling effects, then the volume of the *reduced local cell* $C'(S')$, which discounts the volume of those slabs, if any, between S' and its quintuple of surrounding neighbors, is bounded below by

$$4\sqrt{2} - 0.065 - (0.1) \cdot \text{total buckling effect}\,. \tag{31}$$

Proof In the special case of zero amount of total radial edge-excess, it is just Example 7.2.4. Otherwise, the small amount of total radial edge-excess only makes the corresponding configuration $\mathcal{S}(\Sigma)$ have smaller triangles for the cluster of fifteen faces around A. Therefore, it can also be regarded as a small deformation of that of the special case of Example 7.2.4 and, in fact, with a smaller contribution of decrement of volume. □

Next, let us study the other case that the quintuple of radial edges of $St(A)$ contains *more than* 0.1 *in edge-excesses.* Geometrically, the *total area-excess* of $St(A)$ is naturally a kind of measurement of the *local non-tightness* around S', while the total amount of *radial edge-excess* can also be regarded as another kind of measurement of the local non-tightness around S'. Roughly speaking, both kinds of local non-tightness will have their contributions toward the volume estimate of $C(S')$. However, the contribution of the latter kind of local non-tightness is much more direct and substantial than that of the former. Therefore, we shall use the total amount of radial edge-excesses as the kind of measurement of the local non-tightness around S'.

Definition A 5△-star is called a *small* (resp. *non-small*) *star* if its *total radial edge-excess* is at most (resp. more than) 0.1.

Remarks (i) To each individual radial edge-excess, ε_j, of small or medium size (say, up to 0.36), there corresponds a slab of thickness slightly smaller than the edge-excess. For example, in the case that both S' and S_j are touching neighbors of S, the thickness of the slab between them is given by

$$h(\varepsilon_j) = 2\sin\left(\frac{1}{2}\overline{AA_j}\right) - 1 = 2\sin\left(\frac{\pi}{6} + \frac{\varepsilon_j}{2}\right) - 1 \equiv \frac{\sqrt{3}}{2}\varepsilon_j\,. \quad (32)$$

Thus, such an individual radial edge-excess, ε_j, will contribute an amount of $L(h(\varepsilon_j))$ to the *volume estimate* of $C(S')$.

(ii) It follows from the concavity of $L(h)$ that the total contribution of such a volume estimate (i.e. $\sum_{j=1}^{5} L(h(\varepsilon_j))$) for a *given amount* of *total radial edge-excess* (i.e. $\sum_{j=1}^{5} \varepsilon_j$) will be smaller for a more lopsided distribution. For example, in the case that the total radial edge-excess is at most equal to 0.36, the estimate $\sum_{j=1}^{5} L(\varepsilon_j)$ will be minimal when all the radial edge-excesses are concentrated into a single one, namely, when S' touches four of its surrounding neighbors of S.

Let $St(A)$ be a non-small 5Δ-star in a type-I configuration with exactly one non-zero radial edge-excess. Then, such a star configuration necessarily contains a certain amount of boundary edge-excess which can be estimated in terms of the given radial edge-excess. Anyhow, the total amount of the *volume estimate* of those slabs corresponding to its *boundary edge-excesses* becomes minimal in the geometric situation that all of its boundary edge-excesses are, again, concentrated into a single one. Therefore, for a *given amount of total radial edge-excess*, the contribution towards the collective volume estimate of (17) will be minimal in the case that $St(A)$ is of the type as indicated in Figure 7.32(i) or (ii).

Example 7.2.8 Let S' be a touching neighbor in a given type I core packing $\hat{\mathcal{L}}(S)$ with its associated star configuration, $St(A)$, as indicated in Figure 7.32(i) or (ii) and with ℓ lying between $\frac{\pi}{3} + 0.1$ and 1.4. The complementary region Ω' of $St(A)$ in S' is as indicated in Figure 7.43(i) or (ii).

Note that the above complementary region, Ω', is geometrically the same as the complementary region, Ω, of a quadruple of $\frac{\pi}{3}$-forks with the angular arrangement of $\{\alpha, \beta - \varepsilon, \alpha, \beta + \varepsilon\}$

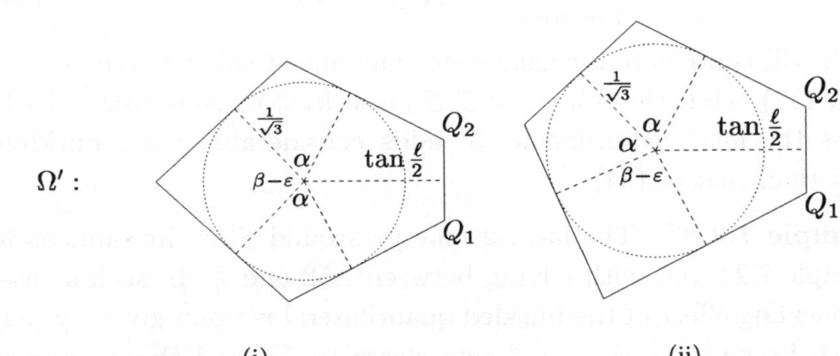

Figure 7.43

(resp. $\{\alpha, \alpha, \beta - \varepsilon, \beta + \varepsilon\}$) that we discussed in the proofs of Lemmas 7.1.3 and 7.1.4. Therefore, the above Ω' is far from able to accommodate a septuple of $\frac{\pi}{3}$-separated points. Hence, a local extension at S' can add at most six touching neighbors of S', while any local extension at S' with seven additional neighbors of S' always containing a considerable amount of total buckling height which, in fact, makes the resulting local cell of S' larger than that of six additional neighbors *without* buckling heights. Moreover, the separation between the pair of vertices as indicated by $\{Q_1, Q_2\}$ in Figure 7.43 is always considerable smaller than $\frac{\pi}{3}$. Hence, a local extension with six additional touching neighbors of S' is always accompanied by a certain amount of total buckling effect, while the volume of the resulting local cell, $C(S')$, of a local extension with only a moderate amount of total buckling effect is at least equal to

$$L(h(\varepsilon)) + (4\sqrt{2} - 0.065) - (0.1) \cdot \text{total buckling effect}\,, \qquad (33)$$

where $h(\varepsilon) = 2\sin\frac{\ell}{2} - 1$.

Example 7.2.8′ The local geometry around S' is the same as in Example 7.2.8 but with ℓ lying between 1.4 and 1.52. In such a case, the buckling effect of the $\frac{\pi}{3}$-quadrilateral with ℓ as one of its diagonal length is given by

$$\frac{2\cos\ell}{1+\cos\ell}, \quad 1.4 \le \ell \le 1.52\,, \qquad (34)$$

which will contribute a considerable amount of volume estimate towards (17), while the volume of $C(S')$ is at least equal to $(4\sqrt{2}-0.01)$ unless the local extension at S' adds considerably more buckling effect than that of (34).

Example 7.2.8″ The local geometry around S' is the same as in Example 7.2.8 but with ℓ lying between 1.52 and $\frac{\pi}{2}$. In such a case, the buckling effect of the buckled quadrilateral is again given by (34) (which becomes smaller as ℓ gets closer to $\frac{\pi}{2}$), and the volume of $C(S')$ is at least equal to $(4\sqrt{2} - 0.01)$.

Chapter 8

The Proof of Main Theorem I

In this chapter, we shall present the proof of Main Theorem I on sphere packings, namely

Main Theorem I (cf. [Hsi-2]) *Let $\bar{\rho}(S_0, \mathcal{L}^2)$ be the locally averaged density of S_0 with respect to a given double-layer local packing $\mathcal{L}^2(S_0)$. Then,*

$$\bar{\rho}(S_0, \mathcal{L}^2) \leq \pi/\sqrt{18} \tag{1}$$

and the equality holds when and only when $\mathcal{L}^2(S_0)$ is isometric to a double-layer local packing (of a sphere) in a hexagonal close packing.

As has already been shown in Chap. 3, both the *Kepler Conjecture* and the following *least action principle of crystal formation of dense-packing type* follow readily from the above main theorem.

The Least action principle of crystal formation of dense-packing type Let $\mathcal{C}$ be a given (finite) cluster of spheres and $\bar{\rho}(\mathcal{C})$ be its *global density of the second kind* (cf. Subsec. 3.3.2). Then,

$$\bar{\rho}(\mathcal{C}) \leq \pi/\sqrt{18} \tag{2}$$

and the equality holds when and only when $\mathcal{C}$ has a double-layer coating $\mathcal{C}^{**}$ which is a hexagonal cluster.

Thus, the crystal structures of the hexagonal close packing type can be regarded as the mathematical consequence of the optimization of the global density of the second kind.

As has already been pointed out in Subsec. 7.2.2 (cf. (17)), the proof of Main Theorem I amounts to the proof of the estimate

$$\sum \lambda_j(\operatorname{vol} C(S_j, \mathcal{L}^2) - 4\sqrt{2}) \geq 0 \qquad (\star)$$

for all possible double-layer local packings and the equality holds when and only when $\mathcal{L}^2(S_0)$ is a subcluster of a hexagonal close packing. In the most critical case that $\#\mathcal{L}(S_j), 0 \leq j \leq 12$, are uniformly equal to 13, all λ_j are equal to 1 and the needed collective volume estimate is that the *total volume of the thirteen local cells is at least equal to* $52\sqrt{2}$ and it is *equal to* $52\sqrt{2}$ *when and only when* $\mathcal{L}^2(S_0)$ is a *subcluster of a hexagonal close packing.*

The task of such an optimal collective volume estimate is a formidable one which will heavily rely on both the *volume estimation techniques* of Chap. 4 and the *geometric understanding* of Chaps. 5 and 7 on *single-layer local packings* and *local extensions.* Actually, the results of Chaps. 4, 5 and 7 were developed mainly for the purpose of proving the collective volume estimate of $(\star)$.

Geometrically, the *locally averaged density* $\bar{\rho}(S_0, \mathcal{L}^2)$ is defined in terms of the structure of a *double-layer local packing* $\mathcal{L}^2(S_0)$ which is much, much more complex than the geometric structure of a single-layer local packing that one encounters in the proof of the Dodecahedron Conjecture (cf. Chap. 6). However, the second layer spheres only play the role of making the local cells $C(S_j, \mathcal{L}^2)$ defined and hence they are only of secondary importance compared to the geometry of $\mathcal{L}(S_0)$, which is clearly of primary importance for the proof of $(\star)$. Therefore, it is natural to regard a double-layer local packing $\mathcal{L}^2(S_0)$ as an *extension* of the single layer local packing

$\mathcal{L}(S_0)$ and to organize the proof of $(\star)$ according to the geometry of $\mathcal{L}(S_0)$, or the core packing $\hat{\mathcal{L}}(S_0)$ it contains. The major geometric feature of a core packing with *at most twelve* close neighbors is its *non-tightness*, while that of a core packing with *at least thirteen* close neighbors is its "*overcrowdedness.*" They are, of course, fundamentally different and it is natural to divide the proof into two major cases accordingly. In fact, the subcase of core packings with twelve touching neighbors, called *type* I *core packings* in Chap. 7, turns out to be the most critical case for the proof of $(\star)$. In particular, it includes the only two *optimal* core packings (i.e. f.c.c. and h.c.p.) which can actually be extended to double-layer packings of maximal locally averaged density of $\pi/\sqrt{18}$. The geometric results of Sec. 7.1 on type I configurations will play a major role in the proof of this most crucial case.

8.1 The Proof of Main Theorem I for Core Packings with at Least Thirteen Close Neighbors

In this section, we shall present the proof of Main Theorem I for double-layer extensions of core packings with at least thirteen close neighbors, namely, thirteen, fourteen or fifteen close neighbors. The proof for such an overcrowded case is much simpler than that of the untight case, mainly due to the rather strong lower bound estimates on the total buckling heights of such overcrowded core packings (cf. Lemmas 5.3.1, 5.3.2 and 5.3.3).

Case 1: Core packings with fifteen close neighbors Let $\hat{\mathcal{L}}(S_0)$ be a core packing with fifteen close neighbors. Then, by Lemma 5.3.3, the total buckling height Σh_j of such a core packing is at least 1. Corresponding to each close neighbor S_j with non-zero buckling height $h_j (\leq 0.09)$, there are a *pair of rectilinear* slabs of thickness h_j whose volumes are at least equal to $L(h_j)$. Therefore, the total volume of such slabs lying between the central sphere S_0 and the close neighbors in $\hat{\mathcal{L}}(S_0)$ is at least equal to

$$\sum_{j=1}^{15} 2L(h_j) \geq 2(11L(0.09) + L(0.01))\,, \tag{1}$$

which is at least 1.92. On the other hand, the volume of the *core part* of $C(S_0)$ is bounded below by

$$L_{15} = 4\sqrt{2} - 0.4381\,. \tag{2}$$

Set $C'(S_j)$ to be the reduced local cell of $S_j, 1 \leq j \leq 15$, which subtracts the h_j-slab lying between S_j and S_0 from the local cell $C(S_j)$. Assuming that the buckling effect of the local extension $\mathcal{L}(S_j)$ is at most 0.1, one has

$$\operatorname{vol} C'(S_j) \geq 4\sqrt{2} - 0.075\,. \tag{3}$$

Therefore,

$$\sum_{j=0}^{15} \lambda_j(\operatorname{vol} C(S_j, \mathcal{L}^2) - 4\sqrt{2}) \geq \frac{13}{16}\left(\sum_{j=1}^{15} L(h_j) - 0.4381\right) + \sum_{j=1}^{15} L(h_j) - 15(0.075) > 0.259\,. \tag{4}$$

Case 2: Core packings with fourteen close neighbors Let $\hat{\mathcal{L}}(S_0)$ be a core packing with fourteen close neighbors. Then, by Lemma 5.3.2, the total buckling height Σh_j of such a core packing is at least 0.56. Therefore, on the one hand, the volume of the central local cell $C(S_0)$ is bounded below by $L_{14} + \Sigma L(h_j)$, namely

$$\operatorname{vol} C(S_0) > L_{14} + \sum_{j=1}^{14} L(h_j) > 4\sqrt{2} + 0.19 \tag{5}$$

and on the other hand, one has

$$\sum_{j=1}^{14} \operatorname{vol} C(S_j) > \sum_{j=1}^{14} L(h_j) + \sum_{j=1}^{14} \operatorname{vol} C'(S_j)\,, \tag{6}$$

where $\{C'(S_j), 1 \leq j \leq 14\}$ are the reduced local cells as in Case 1. Moreover, among the seven vertices of a 6Δ-star subconfiguration of $\mathcal{S}'(\Sigma)$, at most one of their star-configuration can be of the geometric type of a small star, namely, 5Δ-star with minimal or almost minimal radial edges. Therefore, $\mathcal{S}'(\Sigma)$ contains at most a triple of such small stars if the total buckling height is not much larger than 0.56. Hence, at most three *reduced* local cells can be smaller than $4\sqrt{2} - 0.02$, while the volume of the smallest triple of *reduced* local cells are at least equal to $4\sqrt{2} - 0.075$. Anyhow, the following collective volume estimate

$$\sum_{j=0}^{14} \lambda_j (\operatorname{vol} C(S_j) - 4\sqrt{2}) > 0.2615$$

shows that the collective volume estimate of $(*)$ holds with a comfortable margin.

Case 3: Core packings with thirteen close neighbors Let $\hat{\mathcal{L}}(S_0)$ be a core packing with thirteen close neighbors. Then, by Lemma 5.3.1, the total buckling height Σh_j of such a core packing is at least 0.325. Therefore,

$$\begin{aligned} \operatorname{vol} C(S_0) > L_{13} + \sum_{j=1}^{13} L(h_j) &> L_{13} + 3L(0.09) + L(0.055) \\ &= 4\sqrt{2} + 0.07533\,. \end{aligned} \tag{7}$$

On the other hand, one has

$$\sum_{j=1}^{13} \operatorname{vol} C(S_j) > \sum_{j=1}^{13} L(h_j) + \sum_{j=1}^{13} \operatorname{vol} C'(S_j)\,, \tag{8}$$

where $\{C'(S_j), 1 \leq j \leq 13\}$ are the reduced local cells as in Cases 1 and 2.

It is not difficult to show that $\mathcal{S}'(\Sigma)$ can have at most one *small star* unless the total buckling height is considerably larger than the lower bound of 0.325. Therefore, the volumes of $\{C'(S_j), 1 \leq j \leq 13\}$ are at least $4\sqrt{2} - 0.02$ except possible only one of them, which is still larger than $4\sqrt{2} - 0.075$ provided the *buckling effect of its local extension is at most* 0.1. Hence,

$$\sum_{j=0}^{13} \lambda_j (\operatorname{vol} C(S_j) - 4\sqrt{2}) > \frac{13}{14}(0.07533) + \sum_{j=1}^{13} L(h_j)$$
$$- 12 \cdot (0.02) - 0.075 > 0.06878 \,. \qquad (9)$$

This completes the proof of Main Theorem I for core packings of at least thirteen close neighbors. □

8.2 The Proof of Main Theorem I for Core Packings with at Most Twelve Close Neighbors

In this section, we shall present the proof of $(\star)$ for those double-layer extensions of core packings with *at most twelve* close neighbors. The basic common geometric feature of this type of core packing is the *non-tightness* of its coverage. For the purpose of achieving the estimate of $(\star)$ for this type of $\mathcal{L}^2(S_0)$, the relevent geometric features are, (i) the distribution of non-tightness into various types of gaps, and (ii) the volume contributions of such gaps. In case the given core packing $\hat{\mathcal{L}}(S_0)$ contains some *non-touching* close neighbors, namely, with *non-zero total buckling heights* $\sum_{j=1}^{m} h_j$, the cluster of local cells $\{C(S_j, \mathcal{L}^2), 0 \leq j \leq m\}$ always contains a collection of rectlinear slabs of total thickness of $2\sum_{j=1}^{m} h_j$, sandwiching between the central sphere and its close neighbors. Moreover, the gap-width between a neighboring pair of close neighbors $\{S_i, S_j\}$ is roughly corresponding to $\sqrt{3}$-times of the amount of $(\ell(\widehat{A_i A_j}) - \ell(h_i, h_j))$. Therefore, the increment of total buckling heights not only increases the amount of gaps between the central sphere and its close neighbors, but also increases the amount of gaps between close neighbors. Thus, the case of core packings with zero total buckling heights is always more critical than that of non-zero total buckling heights, while core packings with twelve close neighbors always has a smaller amount of non-tightness than core packings with less than twelve close neighbors. Hence, the special case of type I core packings is the most critical case and the major portion of this section is devoted to the proof of this case.

8.2.1 *Case* I: *Type* I *core packing* (*i.e. with twelve touching neighbors*)

The geometry of a type I core packing is completely determined by the geometry of its twelve touching points $\Sigma = \{A_j; 1 \leq j \leq 12\}$ which can be concisely represented and fully revealed by the geometry of its associated *spherical configuration* $\mathcal{S}(\Sigma)$, a type I configuration. The totality of congruence classes of type I configurations constitutes a moduli space of 21 dimensions. Let $\mathcal{S}(\Sigma)$ be a given type I configuration and $\mathcal{S}'(\Sigma)$ be one of its triangulation by adding a diagonal to each of its quadrilateral faces, if any. Then, $\mathcal{S}'(\Sigma)$ consists of twenty triangular faces $\{\sigma_i; 1 \leq i \leq 20\}$ with areas of at least $\triangle_{\pi/3} = 2\tan^{-1}\frac{\sqrt{2}}{5}$. The *area-excess* of σ_i is defined to be $(A(\sigma_i) - \triangle_{\pi/3})$ while the *edge-excess* of $\widehat{A_iA_j}$ is defined to be $(\ell(\widehat{A_iA_j}) - \pi/3)$. The total sum of area-excesses is given by

$$\sum_{i=1}^{20}(A(\sigma_i) - \triangle_{\pi/3}) = 4\pi - 20\triangle_{\pi/3} = 1.54065\cdots\,, \qquad (1)$$

which can be regarded as a rough measurement of the *total amount of non-tightness*, while the individual area-excesses $\{A(\sigma_i) - \triangle_{\pi/3}\}$ can be considered as a *distribution of non-tightness*. Moreover, the amount of edge-excess contained in the three edges of σ_i can be estimated in terms of its area-excess by the isoperimetric inequality of spherical triangles, while the edge-excesses are directly correlated to the widths of gaps and hence more closely related to the volume contributions of individual non-tightness. For example, both the area-excesses and the edge-excesses are uniformly distributed in the case of spherical regular icosahedron, namely

$$\begin{aligned} &20 \text{ area-excesses of: } A(\sigma_i) - \triangle_{\pi/3} = \pi/5 - \triangle_{\pi/3} = \frac{1}{20}(1.54065\cdots), \\ &30 \text{ edge-excesses of: } \ell_k - \pi/3 = \tan^{-1} 2 - \pi/3 = 0.05995\cdots. \end{aligned} \quad (2)$$

On the other hand, the area-excesses of both the f.c.c. configuration and the h.c.p. configuration are concentrated into the six squares, while their edge-excesses are also concentrated into the six diagonals added to the six "squares", namely

$$\begin{aligned} &6 \text{ area-excesses of: } \square_{\pi/3} - 2\triangle_{\pi/3} = \frac{1}{6}(1.54065\cdots), \\ &6 \text{ edge-excesses of: } \pi/2 - \pi/3 = \pi/6. \end{aligned} \quad (3)$$

Roughly speaking, the former is a type I core packing with *uniform distribution of non-tightness*, while the latter are type I core packings with the *most lopsided distribution of non-tightness.* Anyhow, let us briefly analyze the manifestation of the above two opposite types of distributions of non-tightness in the realm of collective lower bound estimate of the total volume (i.e. of the thirteen local cells with respect to all possible second layer extensions). In the latter case of the f.c.c. and h.c.p. core packings, one has the respective global extensions of f.c.c. and h.c.p. packings with identical local cells, thus the total volume of such extensions in equal to $52\sqrt{2}$. Hence, the lower bound in the latter case is at most $52\sqrt{2}$. In the former case, the central local cell is a regular circumscribing dodecahedron whose volume is $0.1065\cdots$ smaller than $4\sqrt{2}$. However, the other twelve local cells contain the 60 *build-in* rectilinear slabs coming from the 30 gaps of width $4\sin\frac{1}{2}\tan^{-1}2 - 2 = 0.1029\cdots$. Each of them has a thickness of $(2\sin\frac{1}{2}\tan^{-1}2-1) = 0.05146\cdots$, thus the total volume of the above 60 rectilinear slabs is at least equal to

$$60L(0.05146) > 3.2914778\cdots. \tag{4}$$

Therefore, it is straightforward to apply the volume estimation technique of Chap. 4 to provide the following lower bound of the total volume of the thirteen local cells with respect to any second layer extension, namely

$$\begin{aligned} 13\cdot L_{12} + 60L(0.05146) &= 52\sqrt{2} + 3.2914778 - 13(0.10656325) \\ &= 52\sqrt{2} + 1.9061555\,. \end{aligned} \tag{5}$$

Note that the total amount of edge-excess of the regular spherical icosahedron is equal to $30(\tan^{-1}2 - \pi/3) = 1.798535$, while that of the f.c.c. and h.c.p. configurations is equal to $\pi = 3.14159\cdots$. Although the former has a much smaller total amount of edge-excess than that of the latter, the thirty small edge-excesses of the former produce a much larger volume contribution than the six large edge-excesses of the latter. This rather brief analysis on the two opposite types of extremal distributions of non-tightness already reveals the following geometric intuition which provides a valuable guide to the overall structure of the proof of $(\star)$ for the most critical case of type I core packings:

(i) In the "geography" of type I core packing, small edge-excesses correspond to the gaps of the gully type, while large edge-excesses correspond to gaps of the type of big holes.

(ii) In adding a second layer coating to a given type I core packing, the gullies are *too narrow* for the additional spheres of the second layer to get into any part of them, thus producing the volume contributions of those build-in rectilinear slabs. On the other hand, by putting spheres on the tops of those big holes, such as the six big holes of the f.c.c. (resp. h.c.p.) core packing corresponding to the six squares of the f.c.c. (resp. h.c.p.) type I configuration, it is possible to remove a considerable amount of volume from the big holes, thus making the total volume of the cluster of thirteen local cells smaller than otherwise.

The above geometric intuition not only provides a simple explanation why, on the one hand, the *optimal* locally averaged density *can only be achieved* by a certain specific extensions of the f.c.c. or the h.c.p. core packing (i.e. the only two type I local packings with the maximal number of big holes), and on the other hand, the locally averaged density of any extension of the symmetric type I core packing is always much lower than $\pi/\sqrt{18}$! But it will also serve as a guiding principle for the following organization of the proof of $(\star)$ for type I core packings, namely, those core packings with more big holes are the more critical ones, while those core packings with a considerable amount of gully-width are the less critical cases, for which the proof of $(\star)$ becomes much simpler than otherwise. Based upon the results of Sec. 7.1, it is natural to divide the proof of Case I into the following cases according to their distributions of edge-excesses (i.e. non-tightness):

Case I_1: Core packings of 6□-*type*, namely, the f.c.c. or the h.c.p. core packings.

Case I_2: Core packings with six buckled quadrilaterals, i.e. small deformations of the f.c.c. (cf. Lemma 7.1.5).

Case I_3: Core packings of the 5□-*type* (cf. Sec. 7.1).

Case I_4: Core packings of small deformations of the 5□-*type*.

Case I_5: Core packings of icosahedron type and with at least one small star.

Case I_6: Core packings of icosahedron type without a small star.

We refer to Sec. 7.1 for a rather thorough discussion on the geometry of type I configurations, which will provide the major geometric input for the proof of $(\star)$ in all the above six cases.

Case $\mathbf{I_1}$: Core packings of the f.c.c. or h.c.p. type Let us first consider the f.c.c. core packing whose associated spherical configuration consists of six squares and eight equilateral triangles of side-length $\pi/3$. Moreover, the star-configuration of every vertex has two triangles and two squares arranged in $(\triangle, \square, \triangle, \square)$-pattern. There are exactly three different ways of extending it to a double-layer local packing of hexagonal type so that the twelve local cells of the touching neighbors of S_0 are either of the f.c.c. type or of the h.c.p. type, namely

(i) abaca-stacking consisting of 57 spheres,
(ii) abacb-stacking consisting of 56 spheres,
(iii) cbacb-stacking consisting of 55 spehres.

The volume of each of the thirteen local cells in the above double-layer extensions is uniformly equal to $4\sqrt{2}$, thus having a total volume of $52\sqrt{2}$. The proof for the f.c.c. case amounts to showing that the inequality of $(\star)$ holds for *any other extension* $\mathcal{L}^2$ of the f.c.c. core packing.

Let $\mathcal{L}^2(S_0)$ be a single-layer coating of the f.c.c. local packing $\mathcal{L}(S_0) = \{S_j, 0 \le j \le 12\}$. Then, the local packings of $S_j, 1 \le j \le 12$, in $\mathcal{L}^2(S_0)$ are *local extensions* of $\mathcal{L}(S_j) \cap \mathcal{L}(S_0)$, respectively. However, in this special case of f.c.c. core packing, the twelve *partial ones* are isometric to each other, namely, they all consist of five touching neighbors whose touching points are isometric to the $(\alpha, \beta, \alpha, \beta)$-arrangement. Therefore, it follows from Lemma 7.2.2 that the volume of each local cell of $\{S_j, 1 \le j \le 12\}$ is at least $4\sqrt{2}$ and it is equal to $4\sqrt{2}$ when and only when the local extension $\mathcal{L}(S_j)$ is isometric to either the f.c.c. or the h.c.p. type. Hence, $\sum_{j=0}^{12} \lambda_j(\operatorname{vol} C(S_j, \mathcal{L}^2) - 4\sqrt{2}) \ge 0$ and the equality holds when and only when *all* the local extensions $\{\mathcal{L}(S_j), 1 \leq j \leq 12\}$ are either of the f.c.c. type or of the h.c.p. type. It is not difficult to see that such twelve local extensions can *only* coexist in the above three double-layer local packings of hexagonal close packings.

Next let us consider the case of h.c.p. core packing. Again, there are exactly three different ways of extending it to a double-layer local packing of a hexagonal close packing, namely,

(i) ababa-stacking consisting of 57 spheres,
(ii) acacb-stacking consisting of 56 spheres,
(iii) bcacb-stacking consisting of 55 spheres.

The volume of each of the thirteen local cells in the above extensions is uniformly equal to $4\sqrt{2}$, thus having a total volume of $52\sqrt{2}$. Again, the proof for the h.c.p. case amounts to showing that the *inequality* of $(\star)$ holds for *any other extension* $\mathcal{L}^2$ of the h.c.p. core packing.

Let $\mathcal{L}(S_j)$ be the local packing of S_j in a given extension $\mathcal{L}^2(S_0)$ of the h.c.p. core packing $\mathcal{L}(S_0), 1 \leq j \leq 12$. Then, $\mathcal{L}(S_j)$ is a *local extension* of $\mathcal{L}(S_j) \cap \mathcal{L}(S_0), 1 \leq j \leq 12$, while each given *partial one* again consists of five touching neighbors. However, there are two types of geometric arrangements of the five touching neighbors, namely, six of them are of the $(\alpha, \beta, \alpha, \beta)$-type while the other six of them are of the $(\alpha, \alpha, \beta, \beta)$-type. Therefore, it follows from Lemmas 7.2.2 and 7.2.3 that the volume of $C(S_j, \mathcal{L}^2)$ can be smaller than $4\sqrt{2}$ *only* when $\mathcal{L}(S_j) \cap \mathcal{L}(S_0)$ is of the $(\alpha, \alpha, \beta, \beta)$-arrangement and the local extension $\mathcal{L}(S_j)$ is a small deformation of the 5□-type. As indicated in Figure 8.1, such a local extension can *only* occur around one of the six middle layer spheres. Moreover, the *volume-deficits*, i.e. $4\sqrt{2} - \operatorname{vol} C(S_j)$, if any, are at most 0.01 (cf. Example 7.2.6).

The h.c.p. core packing has six big holes corresponding to the six squares of the h.c.p. configuration. The six positions of fitting a sphere on top of its big holes are indicated in Figure 8.1 by the dotted circles. If the vicinity of one of these six positions is *not*

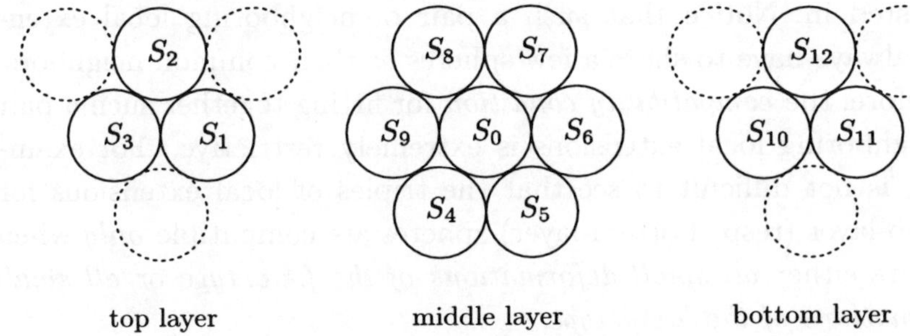

Figure 8.1

occupied by a sphere of the given extension $\mathcal{L}^2$, then it again follows from Lemmas 7.2.2 and 7.2.3 that the local cells of the four corner spheres of that big-hole must be considerably larger than $4\sqrt{2}$. Their volume-excesses, i.e. $\text{vol}\, C(S_j) - 4\sqrt{2}$, are more than enough to compensate whatever those rather small volume-deficits that may occur at other local cells. Therefore, for the proof of the h.c.p. case, it remains to consider those extensions $\mathcal{L}^2(S_0)$ that some of the local extensions, $\mathcal{L}(S_j)$, $4 \leq j \leq 9$, are of the 5□-type and with volumes smaller than $4\sqrt{2}$ and moreover, the vicinity of each of the above six big-hole-positions are all occupied and the other six local extensions $\mathcal{L}(S_j)$, $1 \leq j \leq 3$ and $10 \leq j \leq 12$, are either f.c.c., h.c.p., or their small deformations. For otherwise, the amount of volume-excesses of the latter will be considerably larger than the amount of volume-deficit, if any, of those local extensions of the 5□-type.

Note that the above discussion is strictly *local* in nature which *solely* makes use of the build-in *local arrangement* surrounding a given S_j in $\mathcal{L}(S_0)$ and the condition that the volume of $C(S_j)$ can at most exceed $4\sqrt{2}$ *by only a small amount.* Therefore, the above *geometric restrictions on the possibilities* of the individual local extensions are just the logical consequences of the local arrangement of S_j in the h.c.p. core packing and the natural condition on the *sizes* of local cells. Anyhow, such a collection of highly restrictive local extensions have yet to be fitted together in order to constitute a single-layer coating of the h.c.p. core packing that we are really interested in. Notice that such a pair of neighboring local extensions always have to share a few spheres as their common neighbors. Therefore, the *compatibility condition* for fitting together such a pair of neighboring local extensions is extremely restrictive. For example, it is not difficult to see that the triples of local extensions for the top-layer (resp. bottom-layer) spheres are compatible *only* when they are either *all small deformations of the f.c.c. type* or *all small deformations of the h.c.p. type.*

Finally, let us investigate whether the possibility of having a local extension of the 5□-type (for one of the middle layer spheres) is, at all, compatible with that of the others. Suppose that $\mathcal{L}(S_4)$ is of the 5□-type. Then, $\mathcal{L}(S_4)$ includes two spheres in the vicinity of the pair of "small-holes" surrounded by $\{S_4, S_3, S_9\}$ and $\{S_4, S_9, S_{10}\}$ respectively, namely, in the positions of touching or almost touching the respective triples. However, such a pair of spheres will have a gross *conflict of space* with the local extensions of the top and the bottom triples of spheres which are either all small deformations of the f.c.c. type or all small deformations of the h.c.p. type. Therefore, all local extensions of the middle layer spheres must be (small deformations) of the h.c.p. type. Hence, $\sum_{j=0}^{12} \lambda_j(\operatorname{vol} C(S_j) - 4\sqrt{2}) \geq 0$ and equality holds only when $\mathcal{L}^2(S_0)$ is a double-layer local packing of a hexagonal close packing. This completes the proof for the Case I_1.

Case I_2: Core packings with six buckled quadrilaterals Let $\mathcal{L}(S_0)$ be a type I core packing whose associated configuration $\mathcal{S}(\Sigma)$ has six *slightly buckled* quadrilaterals, which are small deformation of cocircular quadrilaterals. By Lemma 7.1.5, such a type I configuration is necessarily a small deformation of the f.c.c. configuration while the h.c.p. configuration is a rigid, isolated type I configuration. The star-configuration of each vertex in the f.c.c. configuration consists of a pair of $\pi/3$-equilateral triangles separated by a pair of $\pi/3$-squares, (i.e. $\pi/3$-equilateral cocircular quadrilaterals). Under the small deformation, each of the six "squares" buckled into a pair of triangles, while each of the eight triangles has at most second order

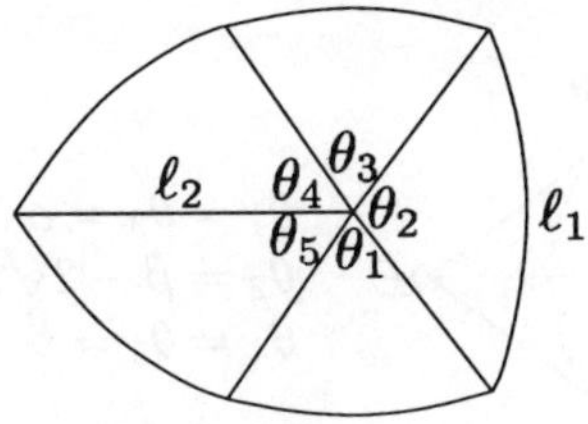

Figure 8.2

change in its side-lengths. Anyhow, the star-configuration of each vertex of such a small deformation consists of five triangles, in which the pair of small triangles are separated by a buckled quadrilateral and one of the halves of another buckled quadrilateral, as indicated in Figure 8.2.

Such a small deformation has six diagonals of the six buckled quadrilaterals whose diagonal lengths $\{\ell_i\}$ are slightly shorter than $\pi/2$. It is convenient to use $\varepsilon = \frac{1}{12}\{\sum_{i=1}^{6}(\frac{\pi}{2} - \ell_i)\}$ as the quantitative measurement of the small deformation. Lemma 7.1.5 proves that the *difference between the diagonal lengths* are at most of the order of ε^2, while the *edge-excesses* of the other twenty-four edges are also at most of the order of ε^2, namely, a general ε-deformation of the f.c.c. configuration (within type I) is always a ε^2-deformation of the specific ε-deformation given by formula (23) of Sec. 7.1, which has its six diagonal lengths uniformly equal to $(\frac{\pi}{2} - 2\varepsilon)$.

The buckling effects of the six buckled quadrilaterals are given by Lemma 7.2.1, namely

$$h_i = \frac{2\cos\ell_i}{1+\cos\ell_i} \equiv 4\varepsilon \ (\mathrm{mod}\ \varepsilon^2)\,, \quad 1 \le i \le 6\,. \tag{6}$$

Therefore, the total amount of the buckling effect of the six buckled quadrilateral is equal to 24ε (mod ε^2). The local arrangement around each touching neighbor S_j of the given core packing $\mathcal{L}(S_0)$ is represented by the star configuration of Figure 8.2, while the *region on* S_j that can be used as the touching point of an additional touching

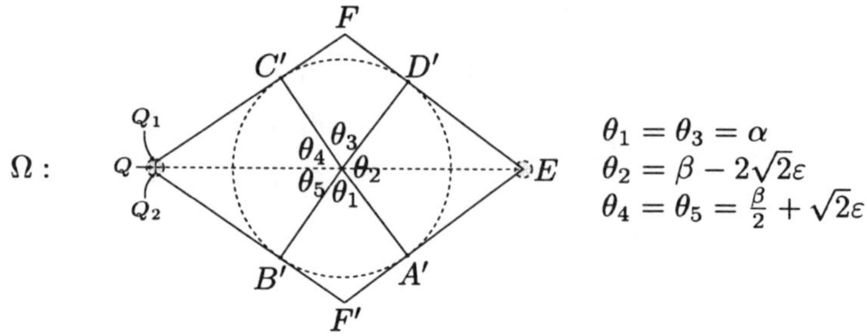

Figure 8.3

neighbor of its local extension is, in general, an ε^2-deformation of the region as represented by the stereographic projection in Figure 8.3.

The two small shaded discs in Figure 8.3 represent those positions of putting spheres on the top of the two adjacent big-holes of S_j. If the local extension of $\mathcal{L}(S_j)$ does *not* include spheres on the top of both big-holes, then $\mathcal{L}(S_j)$ either has at most eleven close neighbors or has a considerable amount of buckling height. Anyhow, the volume of $C(S_j)$ of such a local extension will exceed $4\sqrt{2}$ by a considerable amount. On the other hand, the volume of $C(S_j)$ can be only slightly smaller than $4\sqrt{2}$ even when $\mathcal{L}(S_j)$ *does* put a sphere on top of both big-holes. Therefore, for the proof of this case, it suffices to consider those extensions of $\mathcal{L}(S_j)$ which put a sphere on top of each big-hole, and moreover, each of their local extensions has exactly twelve close neighbors. For otherwise, the considerable volume-excess of one is already large enough to compensate those very small volume-deficits, if any.

Let S' be the sphere of $\mathcal{L}^2(S_0)$ that is on top of the big-hole corresponding to a buckled quadrilateral $\widehat{\square}A_1A_2A_3A_4$. Then, the total sum of the four buckling heights between S' and S_j, $1 \leq j \leq 4$, is at least equal to the *buckling effect* of $\widehat{\square}A_1A_2A_3A_4$ (cf. Sec. 7.2). Therefore, the total volume of those rectilinear slabs between the first layer spheres and the second layer spheres of such an extension $\mathcal{L}^2(S_0)$ is at least equal to $\sum_{i=1}^{6} L(h_i)$, $h_i \equiv 4\varepsilon \pmod{\varepsilon^2}$. Hence, it suffices to consider the case that ε is at most 0.03. For otherwise, the comparatively rather large contribution of $\sum_{i=1}^{6} L(h_i)$ already makes the volume estimates of proving ($\star$) quite simple.

Now, let us first compute the volume of the tangent polyhedron $T(\Sigma_\varepsilon)$ whose associated configuration $\mathcal{S}(\Sigma_\varepsilon)$ has 24 edges of length $a = \cos^{-1}(\frac{1}{2}\cos 2\varepsilon)$ and 6 edges of length $\frac{\pi}{2} - 2\varepsilon$. Set σ_1 (resp. σ_2) to be the spherical triangle whose three side-lengths are given by $\sigma_1 : \{a, a, a\}$ (resp. $\sigma_2 : \{a, a, \frac{\pi}{2} - 2\varepsilon\}$). Then,

$$f(\sigma_1) = \frac{1}{\sqrt{1+\cos 2\varepsilon}} \frac{1 - \frac{1}{2}\cos 2\varepsilon}{1 + \frac{1}{2}\cos 2\varepsilon} \equiv \frac{1}{3\sqrt{2}} \left\{1 + \frac{19}{6}\varepsilon^2\right\} (\mathrm{mod}\ \varepsilon^4)\,,$$

$$f(\sigma_2) = \frac{\sqrt{2}}{3} \left\{ \frac{\frac{5}{2}\cos 2\varepsilon - 1 - \frac{1}{2}\cos^2 2\varepsilon + \sin 2\varepsilon(1 + \frac{1}{2}\cos 2\varepsilon)}{(1 + \frac{1}{2}\cos 2\varepsilon)(1 + \sin 2\varepsilon)^{3/2}} \right\} \tag{7}$$

$$\equiv \frac{2\sqrt{2}}{9} \left(1 - \frac{23}{6}\varepsilon^2\right) (\mathrm{mod}\ \varepsilon^3)\,.$$

Hence,

$$\mathrm{vol}\, T(\Sigma_\varepsilon) = 8f(\sigma_1) + 12f(\sigma_2) = 4\sqrt{2} - 6\sqrt{2}\varepsilon^2\ (\mathrm{mod}\ \varepsilon^3)\,. \tag{8}$$

For example, direct numerical computation shows that

$$\begin{aligned} &\varepsilon = 0.03,\ \mathrm{vol}\, T(\Sigma_\varepsilon) = 8f(\sigma_1) + 12f(\sigma_2) = 4\sqrt{2} - (7.223745)\varepsilon^2\,, \\ &\varepsilon = 0.06,\ \mathrm{vol}\, T(\Sigma_\varepsilon) = 8f(\sigma_1) + 12f(\sigma_2) = 4\sqrt{2} - (6.1316806)\varepsilon^2\,. \end{aligned} \tag{9}$$

On the other hand, the edge-excess of the 24 edges of length a is given by $\cos^{-1}(\frac{1}{2}\cos 2\varepsilon) - \pi/3 \sim \varepsilon^2$. Thus, there are 24 gaps of width given by

$$4\sin\frac{a}{2} - 2 = 4\sqrt{\frac{1 - \frac{1}{2}\cos 2\varepsilon}{2}} - 2 \equiv 2\sqrt{1 + 2\varepsilon^2} - 2 \equiv 2\varepsilon^2 (\mathrm{mod}\ \varepsilon^4)\,. \tag{10}$$

Hence, the total volume of those build-in thin slabs is at least equal to $48\varepsilon^2$.

In the case of general ε-deformations of the f.c.c. configuration (within type I configurations), they are themselves tiny deformations of the above ε-deformation with uniform short (resp. long) edge-lengths, while the maximal change of those short edge-lengths is at most of the order of ε^2. Therefore, the *total amount* from the 24 small edge-excesses remains roughly the same but no longer uniformly distributed.

Next let us study the local extensions at those S_j, $1 \leq j \leq 12$, whose complementary regions are ε^2-deformations of the Ω as indicated in Figure 8.3.

On the one hand, the geometric analysis of such an Ω in the proof of Case 2 of Lemma 7.1.3 has already shown that Ω cannot accomodate a septuple of $\frac{\pi}{3}$-separated points, and on the other hand, the total amount of additional buckling height contained in the twelve local extensions is at least equal to $\sum h(\boxtimes_i)$ which is approximately equal to 24ε (i.e. mod ε^2). As indicatded in Figure 8.3, set E to be the east pole. Then,

$$\cos \overline{D'E} = \cos \overline{A'E} = \frac{\sqrt{3}}{2} \cos\left(\frac{\beta}{2} - \sqrt{2}\varepsilon\right) \equiv \frac{1}{2}(1 + 2\varepsilon) \qquad (11)$$

while the separations between $\{Q, S, A', B', C', D'\}$ are at least equal to $\frac{\pi}{3}$. Therefore, the local extension of adding six touching neighbors with the above six points as their touching points together with another additional neighbor with buckling height of 2ε in the direction of E, provides such a local extension with 2ε as its total buckling height. Moreover, it is not difficult to see that, up to ε^2-deformation, this is the only kind of local extensions which are mutually compatible and also keep their buckling height within 2ε modulo ε^2.

Set $\mathcal{S}(\Sigma'_\varepsilon)$ to be the associated spherical configuration of the above specific local extension. Then, it consists of the following collection of spherical triangles specified by their side-lengths, namely, quadruples of $\sigma(\frac{\pi}{3},\frac{\pi}{3},\frac{\pi}{3})$, $\sigma(\frac{\pi}{3},\frac{\pi}{3},\frac{\pi}{2})$ and $\sigma(\frac{\pi}{3},\frac{\pi}{2},a_2)$, and pairs of $\sigma(\frac{\pi}{3},\frac{\pi}{3},a_1)$, $\sigma(\frac{\pi}{3},a_2,a_2)$, $\sigma(\frac{\pi}{3},\frac{\pi}{3},d)$ and $\sigma(\frac{\pi}{3},a_1,\pi-d)$ where

$$\cos a_2 = \frac{\sqrt{3}}{2}\cos\left(\frac{\beta}{2}-\sqrt{2}\varepsilon\right) \equiv \frac{1}{2}+\varepsilon-\frac{1}{2}\varepsilon^2 \pmod{\varepsilon^3},$$

$$\cos d = \frac{-\cos d'}{1+\cos d'}, \quad \cos d' = \frac{3}{4}\cos(\beta+2\sqrt{2}\varepsilon)+\frac{1}{4}$$

$$\Rightarrow \cos d \equiv 2\varepsilon+3\varepsilon^2 \pmod{\varepsilon^4}, \tag{12}$$

$$\begin{aligned}\cos a_1 &= \frac{\sqrt{3}}{2}\sin d\cos\left(\frac{\beta}{2}+\sqrt{2}\varepsilon\right)-\frac{1}{2}\cos d\\ &\equiv \frac{1}{2}-2\varepsilon-3\varepsilon^2 \pmod{\varepsilon^3}.\end{aligned}$$

Hence, straightforward computations will show that

$$\operatorname{vol} T(\Sigma'_\varepsilon) \equiv 4\sqrt{2}+2\sqrt{2}\varepsilon^2 \pmod{\varepsilon^3}. \tag{13}$$

Therefore, the total volume of the thirteen local cells is approximately $(52\sqrt{2}+24\varepsilon)$ modulo ε^2. This not only proves that $\bar{\rho}(S_0,\mathcal{L}^2)$ are always less than $\pi/\sqrt{18}$ for all single-layer coatings of ε-deformations of the f.c.c. core packing (within type I core-packings), but also proves that $\bar{\rho}(S_0,\mathcal{L}^2)$ are always smaller than $\pi/\sqrt{18}$ *at least by an amount of the order of* ε instead of the order of ε^2.

This completes the proof of $(\star)$ for Case I_2.

Case I_3: Core packing of 5□-type The associated spherical configurations of this type of core packings have five quadrilaterals and two cluster of triangles around the two poles. Such a core packing has *five unbuckled big-holes*, each of them can be fitted with a sphere touching its four corner spheres. There are two different types of local arrangements for the twelve first-layer spheres, namely, the two spheres touching the two poles are surrounded by five touching spheres of the first-layer, while the other ten spheres are

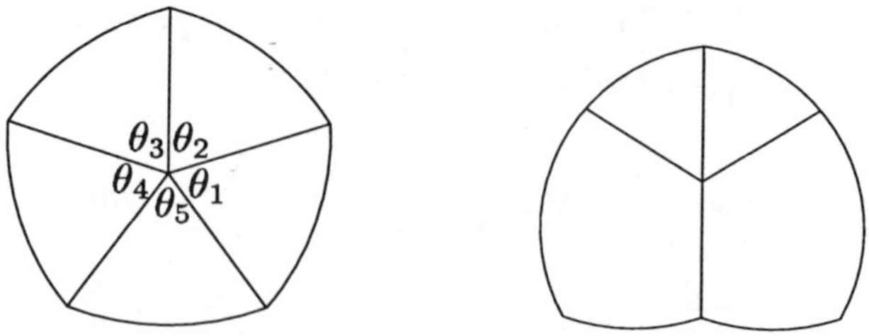

(i) (ii)

Figure 8.4

surrounded by four spheres of the first layer. The star-configurations that represent these two types of local arrangements are as indicated in Figure 8.4.

The radial edges of the pair of 5△-stars are all of length $\pi/3$. However, the ten boundary edges of the pair of 5△-stars must contain a certain amount of small edge-excess, due to the fact that 2π exceeds $5\cos^{-1}\frac{1}{3}$ by $0.128388\cdots$. Therefore, there is always a certain amount of built-in gaps between the ten spheres other than the pair at the poles. Let us first estimate the total amount of the thickness of those built-in rectilinear slabs between the ten spheres as follows. Set the central angles of the 5△-star to be $\{\theta_i,\ 1 \le i \le 5\}$. Then $\sum_{i=1}^{5}\theta_i = 2\pi$, $\theta_i \ge \cos^{-1}\frac{1}{3} = \alpha$ and the corresponding five gaps between the five surrounding spheres (i.e. of the sphere at north pole or south pole) is given by $2\sqrt{3}\sin\frac{\theta_i}{2} - 2$, $1 \le i \le 5$. By the concavity of $y = \sin x$, $0 < x < \pi/2$, their sum is minimal when the angular distribution is the most lopsided, namely

$$\sum_{i=1}^{5}\left(2\sqrt{3}\sin\frac{\theta_i}{2} - 2\right) \ge 2\sqrt{3}\sin\frac{1}{2}(2\pi - 4\alpha) - 2$$

$$= \frac{8\sqrt{2}}{3\sqrt{3}} - 2 = 0.177324\cdots. \tag{14}$$

herefore, the total amount of the thickness of those build-in tilinear slabs between the ten spheres is at least equal to

$0.3546484\cdots$, while the total volume of those slabs is at least equal to $4L(0.0886621) = 0.341295\cdots$.

Set S_1, S_2 to be the pair of spheres at the poles and S_j, $3 \le j \le 12$, to be the other ten spheres of the first layer. Let $C'(S_j)$ be the subset of $C(S_j)$, $3 \le j \le 12$, which *excludes* those *build-in rectilinear* slabs (i.e. between S_j, $3 \le j \le 12$) it contains. We shall call it the *reduced local cell of* S_j. Let us first consider the case that the total amount of buckling effect of $\mathcal{L}^2(S_0)$ is at most 0.1. Then, from Examples 7.2.5 and 7.2.6 and straightforward computation:

$$\sum_{j=3}^{12} \operatorname{vol} C'(S_j) \ge 40\sqrt{2} - 0.1, \quad \operatorname{vol} C(S_0) \ge 4\sqrt{2} - 0.007\,,$$

$$\begin{aligned} \operatorname{vol} C(S_0) + \sum_{j=3}^{12} \operatorname{vol} C(S_j) &\ge 44\sqrt{2} - 0.1 - 0.007 + 0.341295 \\ &= 44\sqrt{2} + 0.234295\,. \end{aligned} \tag{15}$$

On the other hand, by Example 7.2.7, one has

$$\begin{aligned} \operatorname{vol} C(S_1) + \operatorname{vol} C(S_2) &\ge 8\sqrt{2} - 0.14\,, \\ \sum_{j=0}^{12} \operatorname{vol} C(S_j) &\ge 52\sqrt{2} + 0.094295\,. \end{aligned} \tag{16}$$

In the remaining case that the total amount of buckling effect of $\mathcal{L}^2(S_0)$ is more than 0.1, the volume estimates of those additional slabs corresponding to the buckling effects are always about ten times more than the possible decrements in the volume estimates of the core parts (cf. Example 7.2.7). Hence, the estimate of (16) can, in fact, be improved by a bigger margin. This completes the proof of $(\star)$ for Case I_3.

Remarks (i) In the collective volume estimate of this case, the contribution of at least 0.34129... from the built-in thin slabs makes the proof of $(\star)$ considerably easier than that of Cases I_1 and I_2, while the estimate of (16) exceeds $52\sqrt{2}$ by a comfortable margin of 0.094295.

(ii) The existence of those built-in thin slabs geometrically corresponds to those small edge-excesses. Intuitively speaking, the five quadrilaterals of a $5\square$-type configuration already consume an amount of $5(\square_{\pi/3} - 2\triangle_{\pi/3})$ from the total amount of $6(\square_{\pi/3} - 2\triangle_{\pi/3})$ in area excess, while the amount of edge-excess in a $5\square$-type configuration is actually the geometric consequence of the remaining amount of area excess, i.e. $(\square_{\pi/3} - 2\triangle_{\pi/3})$.

Case I_4: Core packings of small deformations of the $5\square$-type Let $\mathcal{L}(S_0)$ be a small deformation of a core packing of the $5\square$-type (within type I core packings) and $\{S_1, S_2\}$ be the pair of (touching) neighbors in the vicinity of the two poles. If S_1 (resp. S_2) and some of its five surrounding neighbors become non-touching under such a small deformation, then it not only *creates new gaps* between them, but it *also increases* the total amount of gap-width between the five surrounding ones. Even in the case that both S_1 and S_2 remain touching all of their surrounding neighbors, some of the five longitudinal "tri-segments" become longer than π, thus creating new edge-excesses in the middle segments. Anyhow, the total volume of the collection of built-in rectilinear slabs contained in such a small deformation is always larger than that of its corresponding undeformed core packing. On the other hand, the deformation also makes some of the five quadrilaterals become buckled which create *buckling effects.* If the total amount of buckling effect is still small, say at most 0.15, then the deformation, itself, must be very small (cf. Lemma 7.2.1). Otherwise, the volume-contribution of buckling effects will make the proof of estimate (15) quite easy. Moreover, the proof of the estimates of $\operatorname{vol} C(S_i)$, $i = 1, 2$, are essentially the same. Therefore, the proof of the estimate of (16) is just the same as that of Case I_3 if the total amount of buckling effect is at most 0.15, while the proof actually becomes much easier as the total amount of buckling effect gets larger. This completes the proof of ($\star$) for Case I_4.

Remarks (i) Notice that a type I configuration either with one to four quadrilateral faces or with one to nine 4-fork vertices is necessarily a small deformation of a 5$\square$-type I configuration (cf. Subsecs. 7.1.2 and 7.1.3). Therefore, the remaining type I core packings are all of the icosahedron type, meaning that their associated configurations are of the combinatorial type of regular icosahedron.

(ii) In the moduli space of congruence classes of type I configurations, the f.c.c. is a *cusp point whose tangent cone is one-dimensional*, represented by the ε-parametrized deformations of $\{\Sigma_\varepsilon,\ 0 \le \varepsilon \le \varepsilon_0\}$ with uniform short (resp. long) edge-lengths. All those configurations in the vicinity of f.c.c., except f.c.c. itself, are of the icosahedron-type. The proof of Case I_2 already covers those type I core packings whose associated configurations are in the 0.1-vicinity of the f.c.c. configuration. Thus, it suffices to consider the proof of ($\star$) for those type I core packings *outside of the* 0.1-*vicinity of the f.c.c. core packing* in the following.

(iii) The set of 5$\square$-type I configurations constitutes a 4-dimensional strata on the boundary of the moduli space of type I configurations. Besides those configurations with some quadrilateral faces and/or with some 4-fork vertices, there are also those type I icosahedrons with five buckled quadrilaterals in the vicinity of the above "5$\square$-strata." Such a type I icosahedron are characterized by having *two small stars* and *five separated long edges* (i.e. edges with lengths close to $\pi/2$). We shall use the *average value* of the *five buckling effects* (i.e. of its five buckled quadrilaterals) as the quantitative measurement of the "distance" between a given type I icosahedron and the above "5$\square$-strata," namely

$$\delta = \frac{1}{5}\sum_{j=1}^{5} h(\boxslash_j)\,.$$

In the range $0.03 < \delta < 0.1$, the volume-contribution of those rectilinear slabs belonging to $\bigcup_{j=3}^{12} C'(S_j)$ with a *total thickness* at least equal to 5δ already makes the lower bound estimate very simple. Therefore, it suffices to consider the proof of $(\star)$ for those type I core packings *outside* of the neighborhood of the $5\square$-strata given by $\delta \leq 0.1$.

(iv) To each vertex A_i of a given type I icosahedron $\mathcal{S}(\Sigma)$, the star configuration $St(A_i)$ consists of five triangular faces with A_i as their common vertex. The total area of $St(A_i)$ is defined to be the *spread of* $\mathcal{S}(\Sigma)$ *at* A_i, and is denoted by $\sigma(A_i)$. The amount of area-excess contained in $St(A_i)$, namely, $\sigma(A_i) - 5\triangle_{\pi/3}$, can be regarded as the amount of local "*non-tightness*" around A_i. Another measurement of the non-tightness around a vertex A_i is the total amount of radial edge-excess of $St(A_i)$. We shall call those stars with at most 0.1 total radial edge-excess *small stars.*

Case I$_5$: Core packings of icosahedron type and with at least one small star Let us first consider the geometric situation of a type I icosahedron containing at least a pair of small stars, say $St(A_1)$ and $St(A_2)$. There are the following three possibilities of relative position between such a pair, namely

(i) separated: $St(A_1) \cap St(A_2) = \phi$,
(ii) adjacent: $St(A_1)$ and $St(A_2)$ sharing a common edge,
(iii) connected: $St(A_1)$ and $St(A_2)$ sharing two triangles.

In the case that $St(A_1)$ and $St(A_2)$ are separated, the six vertices of $St(A_2)$ are all situated inside the complementary region of $St(A_1)$ whose stereographic representation is as indicated in Figure 8.5.

If A_1 and A_2 are antipodal or almost antipodal to each other, and moreover, the radial edges of $St(A_1)$ and $St(A_2)$ are arranged in pairs with almost the same longitude degrees, then $\mathcal{S}(\Sigma)$ is itself a small deformation of a $5\square$-configuration. Thus, the proof of $(\star)$ for such a core packing has already been given in Case I$_4$. Otherwise, there

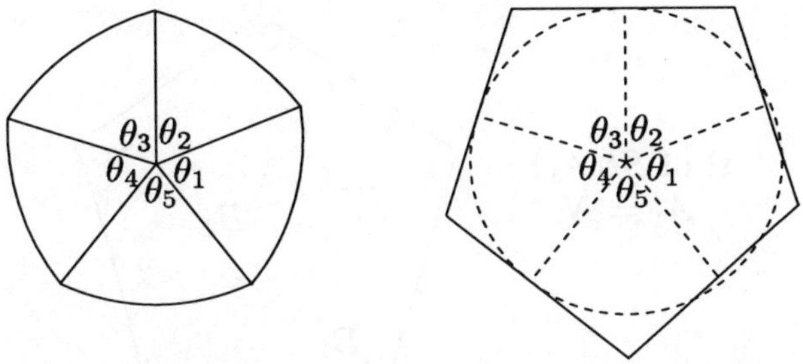

$A_1 = N$, $St(A_1)$ $\qquad$ Ω of $St(A_1)$

Figure 8.5

must be a considerable amount from built-in slabs *in addition* to those built-in slabs corresponding to the *boundary* edge excesses of both $St(A_1)$ and $St(A_2)$. The volume contribution of the latter is at least 0.34 while the additional volume contribution of the former is at least 0.2 and it becomes much larger than 0.2 when the relative position between $St(A_1)$ and $St(A_2)$ deforms further away from that of almost antipodal and with close angular alignment of their radial edges. Moreover, in the case that the volume contribution of the former is at most 0.25, the configuration has five buckled quadrilaterals whose total amount of buckling effect is at least 0.5, cf. Remark (iii) following the proof of Case I_4. Anyhow, the total volume contribution of the above three groups of rectilinear slabs is already large enough to make the task of collective volume estimate of $(\star)$ rather simple.

In the case that $St(A_1)$ and $St(A_2)$ are *adjacent* to each other, as indicated in Figure 8.6, the complementary region Ω of $St(A_1)$ contains four vertices of $St(A_2)$, say $\{A_2, B_1, B_2, B_3\}$, while the remaining two points of Σ are located in the narrow shaded subregion of Ω as indicated in Figure 8.6, which is the complement in Ω of those $\pi/3$-discs centered at B_1, B_2, B_3 and A_2.

In the case that $St(A_1)$ and $St(A_2)$ are *connected* to each other, the complementary region Ω of $St(A_1)$ contains a pair of vertices of $St(A_2)$ which are located in the close vicinity of a boundary arc of Ω,

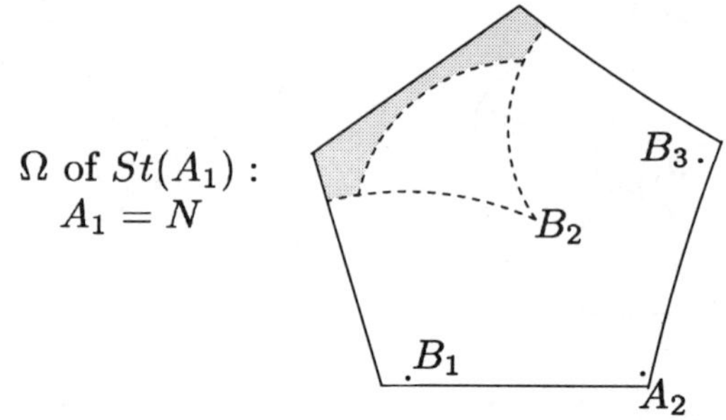

Figure 8.6

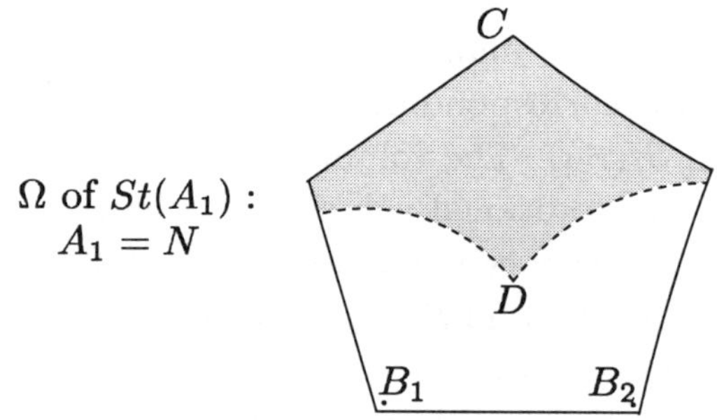

Figure 8.7

while the remaining four vertices of $\mathcal{S}(\Sigma)$ are situated in the shaded subregion of Ω as indicated in Figure 8.7.

The maximal distance between the top and bottom corner points $\{C, D\}$ is at most equal to 1.507. In both cases, it is not difficult to show that there are at most three edges of lengths exceeding 1.4 in $\mathcal{S}(\Sigma)$ and moreover, at most one of them can be close to 1.5 while the shorter ones are quite close to 1.4. Therefore, the *sum* of edge-excesses smaller than $(2\sin^{-1} 0.65 - \frac{\pi}{3}) = 0.36797\cdots$ is at least 1 which will produce a total volume of at least 0.7 from their corresponding build-in slabs. Such a large volume contribution from those built-in slabs makes the volume estimate for all possible first-layer coatings of such a core packing quite simple.

Now, let us proceed to study the remaining case of Case I_5, namely, the associated configuration $\mathcal{S}(\Sigma)$ has exactly one small star $St(A_1)$. By Lemma 7.1.6, *none of* the *other eleven stars* can be a small deformation of the type indicated in Figure (7.33(ii)), lest the "opposite" star of $St(A_1)$ in $\mathcal{S}(\Sigma)$ would also be a small star.

Remarks (i) The same proof as that of the case of two separated small stars also applies to those *not so small* deformations of 5□-ones whose "opposite" star of $St(A_1)$, say $St(A_2)$, is no longer a small star, such as with a total radial edge-excess lying between 0.1 and 0.12. In fact, the additional amount of small edge-excess contained in the edges of $St(A_2)$ simply makes the volume contribution of those thin built-in slabs considerably larger.

(ii) Again, the same proof as that of the case of two adjacent or connected small stars also applies to the ones where $St(A_2)$ contains slightly more radial edge-excess than 0.1, say up to 0.12, in particular, in the case that $St(A_2)$ and $St(A_1)$ are connected to each other and the sum of the edge-excesses of the pair of radial edges of $St(A_2)$ *outside of* $St(A_1)$ is at most 0.1. On the one hand, the additional two edge-excesses simply provides an additional volume contribution of thin slabs, and on the other hand, the shaded subregion of Figure 8.7 actually becomes smaller while there is almost no change of shape.

Therefore, in the following proof of the remaining part of Case I_5, it suffices to consider those core packings whose associated configuration satisfying the following additional conditions, namely

(i) all the other eleven stars contain at least 0.12 total radial edge-excess,

(ii) set $\{A_j, 2 \leq j \leq 6\}$ to be the boundary vertices of $St(A_1)$, the sum of edge-excesses of the pair of radial edges of $St(A_j)$ *outside* of $St(A_1)$ is at least 0.1, $2 \leq j \leq 6$,

(iii) none of the eleven stars $\{St(A_j),\ 2 \leq j \leq 12\}$ is a small deformation of the type of Figure (7.33(ii)).

Set $C'(S_j)$, $2 \le j \le 6$, to be the *reduced* local cells of the five neighboring spheres of S_1 which *exclude* those thin slabs corresponding to edges of $St(A_1)$. It follows from the above conditions (i), (ii) and (iii), one has the following localized individual volume estimates for local extensions with at most 0.1 total buckling effect, namely

$$\begin{aligned} \text{vol } C(S_1) &\ge 4\sqrt{2} - 0.07\,; \\ \text{vol } C'(S_j) &\ge 4\sqrt{2} + 0.01\,, \quad 2 \le j \le 6\,; \\ \text{vol } C(S_j) &\ge 4\sqrt{2} + 0.01\,, \quad 7 \le j \le 12\,. \end{aligned} \tag{17}$$

On the other hand, the total volume of those rectilinear slabs between $\{S_j, 2 \le j \le 6\}$ is at least 0.1706, which is *not* included in the estimates of $\text{vol}\, C'(S_j)$, $2 \le j \le 6$. Therefore, the collective estimate of $(\star)$ for such a subcase holds with a margin of at least 0.1.

This completes the proof of Case I_5.

Case I_6: Core packings of icosahedron type without small star In the case that the total radial edge-excess of $St(A_j)$ is at most 0.47, either the local extension at S_j must have a considerable amount of buckling effect or a large portion of its radial and boundary edge-excesses are of the short type, or both. For others, the volume of $C(S_j)$ can be slightly smaller than $4\sqrt{2}$ only when $St(A_j)$ is a small deformation of the type of Figure (7.33(ii)) and the local extension is of the type described in Example 7.2.4. As indicated in Figure 8.8, suppose that $St(A_1)$ is such a star configuration contained in $\mathcal{S}(\Sigma)$. If the buckling effect of $\overset{\frown}{\square}\, A_1A_2A_3A_4$ is at least 0.1 or if $St(A_2)$ and $St(A_4)$ are *not both* small deformations of the type of Figure (7.33(ii)), then at least one of the three local cells must be considerably larger than $4\sqrt{2}$ thus providing the following collective estimate, namely

$$\text{vol}\, C(S_1) + \text{vol}\, C(S_2) + \text{vol}\, C(S_4) \ge 12\sqrt{2} + 0.035\,. \tag{18}$$

Otherwise, $\{St(A_j),\ j = 1, 2, 4\}$ as indicated in Figure 8.8 are all small deformations of the type of Figure (7.33(ii)) and hence, $\mathcal{S}(\Sigma)$

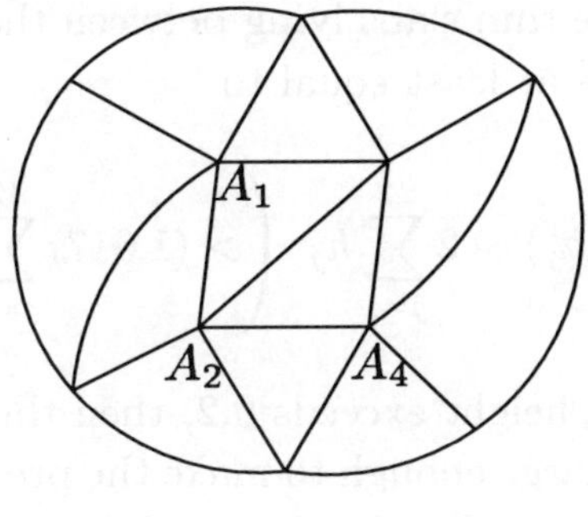

Figure 8.8

must be a small deformation of a 5□-type I configuration, for which the estimate of ($\star$) has already been proved for such a core packing in Case I_4. Therefore, one has the collective estimate of

$$\sum_{j=1}^{12} \lambda_j(\text{vol } C(S_j) - 4\sqrt{2}) \geq 0.14 \tag{18'}$$

and hence, the collective estimate of ($\star$) holds with a margin of at least 0.03.

This completes the proof of Case I_6 and hence also the proof of the whole case of type I core packings.

8.2.2 *Case* II: *Non-type* I *core packings with twelve close neighbors*

In this subsection, we shall prove the collective estimate for the case of core packings with twelve close neighbors but not all of them are touching neighbors. Set $\{h_j,\ 1 \leq j \leq 12\}$ to be their *buckling heights* which record their individual non-touchingness, $0 \leq h_j \leq 0.09$, while the *total buckling height*, Σh_j, provides a quantitative measurement of the total amount of non-touchingness of the given core packing. From the viewpoint of collective volume estimate, the existence of each non-zero h_j produces a pair of rectilinear slabs of thickness h_j lying between S_j and the central sphere S_0. Therefore, the *volume contribution* of those thin slabs lying between the central sphere and its close neighbors is at least equal to

$$2\sum_{j=1}^{12} L(h_j) \sim 2\sum_{j=1}^{12} h_j \left(> (1.917) \sum_{j=1}^{12} h_j \right) . \tag{19}$$

If the total buckling height exceeds 0.2, then the above volume contribution is already large enough to make the proof of this case rather simple. Therefore, we shall only discuss the proof of this type of core packing with at most 0.2 total buckling height in the following. Set $\mathcal{S}(\Sigma)$ to be the associated configuration of such a core packing $\hat{\mathcal{L}}(S_0)$. The only difference between such a configuration $\mathcal{S}(\Sigma)$ and a type I configuration is that some of its edges may become slightly shorter than $\pi/3$. For example, corresponding to a neighboring pair $\{S_i, S_j\}$ with buckling heights $\{h_i, h_j\}$, the edge length of $\overset{\frown}{A_i A_j}$ is bounded below by $\ell(h_i, h_j)$, which is approximately $\frac{1}{\sqrt{3}}(h_i + h_j)$ smaller than $\frac{\pi}{3}$, while the equality holds when and only when S_i and S_j are touching each other. Therefore, they are, in general, small deformations of type I configurations with at most a small amount of edge-shortening, while the edge-excess of $\overset{\frown}{A_i A_j}$ is *redefined* to be $\ell(\overset{\frown}{A_i A_j}) - \ell(h_i, h_j)$. Moreover, in the case that $\{S_1, S_2, S_3\}$ is a triple of close neighbors which are mutually touching each other, the three side-lengths of $\triangle A_1 A_2 A_3$ are given by $\ell(h_1, h_2)$, $\ell(h_2, h_3)$ and $\ell(h_3, h_1)$. We shall denote the area of such a triangle simply by $\triangle(h_1, h_2, h_3)$. In general, the area-excess of a triangular face σ in $\mathcal{S}(\Sigma)$ is *redefined* to be $A(\sigma) - \triangle(h_1, h_2, h_3)$. We refer to Sec. 5.1 for a discussion of some simple properties of the above function $\triangle(h_1, h_2, h_3)$, e.g.

$$\begin{gathered} \triangle(3\bar{h}, 0, 0) \leq \triangle(h_1, h_2, h_3) \leq \triangle(\bar{h}, \bar{h}, \bar{h}), \quad \bar{h} = \frac{1}{3}\sum_{i=1}^{3} h_i , \\ \triangle(h_1, h_2, h_3) \geq \triangle_{\pi/3} - (0.4757122) \sum_{i=1}^{3} h_i . \end{gathered} \tag{20}$$

Suppose that $\widehat{\triangle} A_1A_2A_3$ and $\widehat{\triangle} A_1A_3A_4$ form a buckled quadrilateral in $\mathcal{S}(\Sigma)$ and h_i is the buckling height of S_{A_i}, $1 \leq i \leq 4$. Then, it is not difficult to show that the *actual buckling effect* is approximately equal to the amount given by Formula (9) of Sec. 7.2 plus $(h_1+h_3-h_2-h_4)$.

Basically, the proof of this case is essentially the same as that of the type I case, especially when the total buckling height is very small, such as in the range of at most 0.05, while such a $\mathcal{S}(\Sigma)$ is geometrically a tiny deformation of a type I configuration. On the other hand, in the case that the total buckling height exceeds 0.05, the additional volume contribution of at least $2\Sigma L(h_j)$ plus the corresponding *increments* in *edge-excess* and *area-excess* already provides a considerable amount of *additional* volume estimate which, of course, will make the task of proving the collective estimate easier than that of the type I case. Therefore, we shall mainly present here a comparison analysis between the proof of this case and the proof of the type I case in order to clarify the following two aspects.

(i) Which part of the proof of type I case can be directly applied to this case, and what is the geometric "common ground" that makes such a direct generalization from the type I case to the case with buckling heights still applicable?
(ii) For the remaining part that the proof of type I case is *not* directly applicable, provide the necessary modifications which enable us to prove the collective estimate also for such a case.

Remark In the whole proof of the type I case that we present in Subsec. 8.2.1, we pay special attention to the choice of methods and techniques so as to maximize the portion which is directly generalizable to the case with bucklings.

The following are the major points of such a comparison analysis:

On the division of cases Core packings of this case can be regarded as small deformations of type I core packings. For example, the ones with six or five *cocircular* quadrilaterals in their associated configurations are, themselves, small deformations of 6□- or 5□-type I packings. Therefore, it is more natural to consider them just as some special cases of the ones with six or five *buckled quadrilaterals.* Anyhow, for the overall organization of the proof of this case, it is natural to subdivide into the following four cases:

Case II_1: Small deformation of 6□-type I core packings.

Case II_2: Small deformation of 5□-type I core packings.

Case II_3: Icosahedron type with small stars.

Case II_4: Icosahedron type without small star.

Case $\mathbf{II}_1$: Small deformation of 6□-type I core packings (i.e. f.c.c. and h.c.p.) The h.c.p. core packing is a rigid, isolated one within type I core packings. However, it is no longer rigid and isolated in the realm of core packings with twelve close neighbors. Let $\hat{\mathcal{L}}(S_0)$ be a small deformation of either the f.c.c. or the h.c.p. core packing. Then, the build-in partial surroundings of all its close neighbors S_j, $1 \leq j \leq 12$, are also small deformations of the ones in either the f.c.c. or h.c.p. core packing. Therefore, by Lemmas 7.2.2 and 7.2.3, the volumes of $C(S_j)$, $1 \leq j \leq 12$, can only be slightly smaller than $4\sqrt{2}$ in the least. Hence, for the purpose of proving the collective estimate, it suffices to consider the case that *none* of the volume of $C(S_j)$ is considerably larger than $4\sqrt{2}$. Thus, again by Lemmas 7.2.2 and 7.2.3, one may assume that all the local extensions $\mathcal{L}(S_j)$, $1 \leq j \leq 12$, are small deformations of either the f.c.c., the h.c.p., or the 5□-type local packings. Moreover, the same proof as in the Case I_1 will also show that the above types of local extensions can be compatibly fitted together only when the double-layer extension $\mathcal{L}^2(S_0)$ is itself a *small deformation of* a hexagonal double layer local packing (cf. Case I_1 in Subsec. 8.2.1). Thus, all the local packings of S_j, $0 \leq j \leq 12$, in $\mathcal{L}^2(S_0)$ are, themselves, small deformations of either the f.c.c. or the h.c.p. local packing. Hence, the volumes of the core parts of $C(S_j)$, $0 \leq j \leq 12$, can only be *minutely* smaller than $4\sqrt{2}$ in the least (i.e. of the ε^2-order with respect to the magnitude of deformation of $\mathcal{L}^2(S_0)$.)

Let $\mathcal{S}(\Sigma)$ be the associated configuration of a non-type I small deformation, $\hat{\mathcal{L}}(S_0)$, of either the f.c.c. or the h.c.p. core packing. If $\mathcal{S}(\Sigma)$ is still a type I configuration, then $\mathcal{S}(\Sigma)$ must be a small deformation of the f.c.c. configuration and $\hat{\mathcal{L}}(S_0)$ is, itself, a simple deformation of another *type*-I deformation of the f.c.c. core packing which just adds a certain amount of buckling to some of its (touching) neighbors *without* any change in their directions. The proof of Case I_2 clearly also applies to such a core packing. Therefore, it suffices to discuss here the proof of those core packings in Case II_1 whose associated configurations are no longer type I (i.e. with some edges slightly shorter than $\pi/3$).

Let us first compare some of the volume-influential geometric features between the type I-configurations and the non-type I-configurations as follows.

(i) In the case of a *type* I small deformations of the f.c.c. configuration, the diagonal lengths $\{\ell_i\}$ of its six buckled quadrilaterals are almost of the same length and it is natural to use $\varepsilon = \frac{1}{12}\{3\pi - \sum_{i=1}^{6} \ell_i\}$ as the quantitative measurement of the "magnitude" of deformation. Then, Lemma 7.1.5 proves that such a configuration has the following crucial geometric features, namely

(a) modulo possible differences of ε^2-order, the six diagonal lengths are equal to $\pi/2 - 2\varepsilon$ while the average length of the twenty-four short edges is given by $\cos^{-1}(\frac{1}{2}\cos 2\varepsilon)$,
(b) the buckling effects of the six quadrilaterals is equal to 4ε (modulo ε^2), while the areas of the six quadrilaterals are smaller than $\square_{\pi/3}$ only by amounts of ε^2-order.

Note that the proof of Lemma 7.1.5 is based upon the following geometric facts, namely, $4\pi = 6\,\square_{\pi/3} + 8\triangle_{\pi/3}$ and the shearing ε-deformations of cocircular quadrilaterals can have area decrements of at most ε^2-order.

(ii) In the case of *non-type* I small deformations of either the f.c.c. or the h.c.p. configurations, the areas (resp. diagonal lengths) of some of its quadrilaterals (buckled or cocircular) may be exceeding $\square_{\pi/3}$ (resp. $\pi/2$). However, the occurrence of such a quadrilateral is necessarily accompanied by a certain amount of buckling whose amount can be estimated in terms of the sum of such area-excesses (resp. diagonal length-excesses). For the purpose of collective volume estimation, it is advantageous to use the following two types of geometric invariants as another kind of quantitative measurement of the deformation of the given core packing (i.e. from either the f.c.c. or the h.c.p. core packing), namely

(a) the total sum of the actual buckling effects of its six quadrilaterals as a measurement of the amount of the "*horizontal*" *component of deformation*,

(b) the total sum of gully-widths plus twice of the total buckling heights as a measurement of the amount of "*vertical*" *component of deformation.*

The volume contribution of those rectilinear slabs in $\mathcal{L}^2(S_0)$ is roughly equal to the sum of the above two kinds of invariant, while the volumes of each core parts of $C(S_j)$, $0 \leq j \leq 12$, can only be ε^2-order smaller than $4\sqrt{2}$ in the least. Therefore, the total volume of the thirteen local cells will certainly exceed $52\sqrt{2}$ by a magnitude of the linear order of the deformations.

This completes the proof of $(\star)$ for Case II_1.

Case II_2: Small deformation of $5\square$-type I core packings In this case, the associated configurations $\mathcal{S}(\Sigma)$ of such a core packing $\hat{\mathcal{L}}(S_0)$ always consists of two small $5\triangle$-stars separated by five buckled or unbuckled quadrilaterals. In the proofs of Cases I_3 and I_4, the lower bound estimate of 0.34 for the volume contribution of those build-in slabs corresponding to the edge-excesses contained in the two small stars plays a major role in the proof of a collective estimate for

$St(A_1)$ or $St(A_2)$:
$(r = \sin^{-1}(\frac{1}{2\sin\pi/5}))$

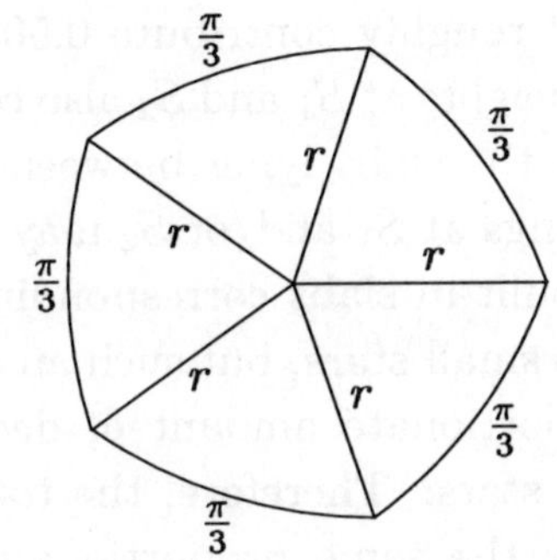

Figure 8.9

these two kinds of type I core packings. In the more general case of non-type I core packings of this kind, if both of the two spheres with small stars are still touching neighbors of S_0, then it is easy to show that the total amount of the volume contribution of those build-in slabs is again at least 0.34. In fact, the bucklings of some of their surrounding spheres always make the total volume of those build-in slabs corresponding to the boundary edge-excesses of $St(A_1)$ and $St(A_2)$ larger than that of the case of no bucklings. However, in the case that h_1 and/or h_2 are non-zero, the total volume of those build-in slabs corresponding to the boundary edge-excesses of $St(A_1)$ and $St(A_2)$ may become smaller than 0.34. For example, in the special case of $h_1 = h_2 = 2\cos\sin^{-1}\frac{1}{2\sin\pi/5} - 1 = 0.05146\cdots$, one may surround both S_1 and S_2 by quintuples of spheres touching $\{S_1, S_0\}$ and $\{S_2, S_0\}$ respectively such that their star configurations are as indicated in Figure 8.9.

In such a special case, both $St(A_1)$ and $St(A_2)$ contains no edge-excess at all and hence the volume contribution of those built-in slabs actually become zero. However, the edge-excesses contained in the five longitudinal tri-segments of such a configuration at least have a total amount of

$$5\left(\frac{2\pi}{3} - 2\sin^{-1}\frac{1}{2\sin\pi/5}\right) = 0.2997558\cdots, \tag{21}$$

which will roughly contribute 0.50 in build-in slabs. Moreover, the buckling heights at S_1 and S_2 also contribute at least $2(h_1+h_2)$ in the volume of the slabs lying between $\{S_0, S_1\}$ and $\{S_0, S_2\}$. Anyhow, the bucklings at S_1 and/or S_2 may decrease the volume contribution of those built-in slabs corresponding to the boundary edge-excesses of the two small stars, but such an decrement is always accompanied by a proportionate amount of decrement in the total area of the two small stars. Therefore, the total area of the five quadrilaterals must have the same proportionate amount of area increment and hence, it follows from the isoperimetric type of estimate that *total edge-excesses* of the *side-lengths* of the five quadrilateral will, in fact, have a larger amount of increment than that of the above decrement. Therefore, the same proof as that of Cases I_3 and I_4 in Subsec. 8.2.1 also applies to this type of core packing while the above discussion on the volume contribution of those build-in slabs is the only needed modification.

Case $\mathbf{II_3}$: Icosahedron type with small stars Again, the same modification on the build-in slabs of the small stars already suffices to make the same proof of Case I_5 in Subsec. 8.2.1 also applicable to this type of core packings. For example, in the most critical case that $\mathcal{S}(\Sigma)$ has a single small star $St(A_1)$ and it is as indicated in Figure 8.9, one has the following modified individual volume estimates for local extensions with at most 0.1 total buckling effects (cf. (17)), namely

$$\begin{aligned} &\text{vol } C(S_0) \geq 4\sqrt{2} - 0.0929 + L(0.05416) \geq 4\sqrt{2} - 0.0381 \\ &\text{vol } C'(S_1) \geq 4\sqrt{2} - 0.055 + L(0.05416) \geq 4\sqrt{2} - 0.0002 \qquad (17') \\ &\text{vol } C(S_j) \geq 4\sqrt{2} + 0.01\,, \quad 2 \leq j \leq 12 \end{aligned}$$

Therefore, the collective estimates of $(\star)$ for such a subcase still holds with a margin of at least 0.07.

Case $\mathbf{II_4}$: Icosahedron type without small star The proof of the Case I_6 in Subsec. 8.2.1 is so chosen that no modification is needed in order to generalize it to this more general type of core packings.

8.2.3 *Case III: Core packings with at most eleven close neighbors*

We shall divide the proof of Case III into two cases, namely

Case III$_1$: Core packings with eleven close neighbors.

Case III$_2$: Core packings with at most ten close neighbors.

Case III$_1$: Core packings with eleven close neighbors Let us first discuss the case that the eleven close neighbors are all touching or almost touching. Set $\mathcal{S}'(\Sigma)$ to be the triangulated associated configuration of such a core packing. It has eleven vertices, twenty seven edges and eighteen triangles and we shall simply ignore the amount of buckling height in the accountings of area excesses and edge-excesses because it is assumed to be very small anyway. The total amount of area-excess is given by

$$\sum_{j=1}^{18}(A(\sigma_j) - \triangle_{\pi/3}) = 4\pi - 18\triangle_{\pi/3} = 2.64323\,, \tag{22}$$

which exceeds the total area-excess of type I configurations by $2\triangle_{\pi/3} = 4\tan^{-1}\frac{\sqrt{2}}{5}$. Therefore, $\mathcal{S}'(\Sigma)$ must have a much larger amount of edge-excess than that of type I configurations. However, it is again necessary to study the possible distributions of edge-excesses into that of short edges or that of long edges in order to estimate their corresponding volume contributions. For example, if a considerable amount of edge-excess is of the size of at most $0.368 = (2\sin^{-1}0.65 - \frac{\pi}{3})$, then the volume contribution of those corresponding build-in slabs already suffices to make the collective estimate rather simple. Therefore, we shall study here the amount from long edges and their possible arrangement as the buckling diagonals of different types of "big-holes."

As indicated in Figure 8.10, the diagonal of a buckled (or unbuckled) quadrilateral with short sides is an "*isolated*" *long edge*, the pair of diagonals of a buckled pentagon with short sides consists of a *pair*

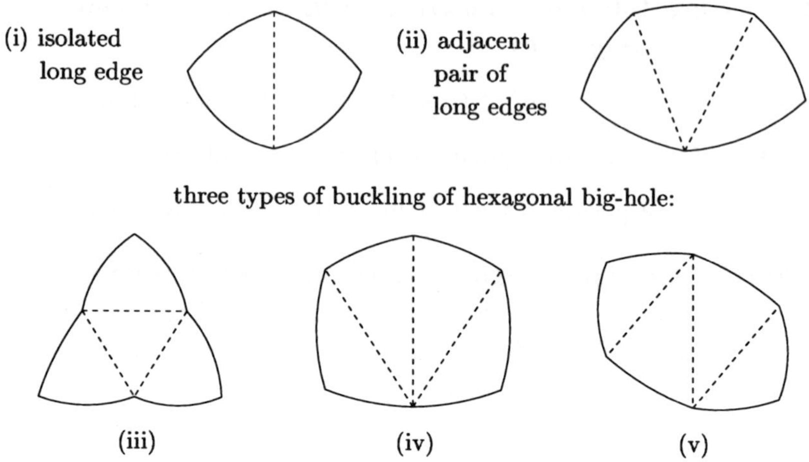

Figure 8.10

of adjacent long edges, while the triple of buckling diagonals of a hexagon consists of a *triple* of *adjacent long edges.*

Roughly speaking, edges longer than 1.52 are regarded to be long, edges shorter than 1.4 are regarded to be short while edge-lengths in between are regarded to be medium. We shall call the above types of big-hole as indicated in Figure 8.10(i) (resp. (ii), (iii) or (iv) or (v)) quadrilateral (resp. pentagonal, hexagonal) big-holes. The following is a comparison analysis of some of the volume-influential geometric invariants such as area-excesses, circumradii and buckling effects of the three types of big-holes. Let us begin with some examples.

Examples of big-holes with $\pi/3$-boundary edges If one sets the edge-lengths of all the boundary edges, represented by solid segments in Figure 8.10, to be $\pi/3$ and all the diagonal edges, represented by dotted segments in Figure 8.10, to be $\pi/2$, then their areas and area-excesses are given as follows:

(i) Quadrilateral:

$$A = \square_{\pi/3} = 1.3593476 = 2\triangle_{\pi/3} + 0.2567764\,.$$

(ii) Pentagon:

$$A = \pi/3 + \square_{\pi/3} = 2.406545 = 3\triangle_{\pi/3} + 0.752688\,.$$

(iii) Hexagon with triangular buckling:

$$A = \frac{\pi}{2} + \frac{3}{2}\square_{\pi/3} = 3.6098118 = 4\triangle_{\pi/3} + 1.4046694\,.$$

(iv) Hexagon of type-(iv) or (v) buckling:

$$A = \frac{2\pi}{3} + \square_{\pi/3} = 3.4537427 = 4\triangle_{\pi/3} + 1.248003\,.$$

Remarks (i) The buckled hexagons of (iv) and (v) in Figure 8.10 can also be regarded as a pair of quadrilaterals sharing a common long edge. For example, if the common long edge is equal to $\pi/2$ while the pair of quadrilaterals is cocircular and with triples of $\pi/3$-sides, then straightforward computation shows that the diagonal lengths (resp. area) of such a quadrilateral is given by

$$\ell = \cos^{-1}\left(\frac{1}{2}(1-\sqrt{2})\right) = 1.779413\cdots$$

$$\left(\text{resp. } A = 2\left\{\cos^{-1}\frac{1}{\sqrt{3}}(1-\sqrt{2}) + \cos^{-1}\frac{1}{3}(1-2\sqrt{2}) - \pi\right\}\right.$$
$$\left. = 1.79377138\right).$$

Hence, the area-excess of such a buckled hexagon is equal to 1.3824.

(ii) The area-excess of a big-hole is a natural, basic geometric measurement of the amount of non-tightness contained in the given big-hole, while the short edge-excesses, the long edge-excesses and the buckling effects are three different kinds of volume-influential manifestations of the given big-hole. Thus, big-holes with uniform $\pi/3$-boundary edges are the kind of big-holes without short boundary edge-excesses, namely, all the edge-excesses are concentrated in their long diagonals.

(iii) Set $\square_a$ to be the area of a-square. Then, it is easy to show that

$$\square_a = 4\cos^{-1}\frac{\cos a - 1}{1+\cos a} - 2\pi\,.$$

Therefore, a quadrilateral big-hole with its area A exceeding $4\tan^{-1}\frac{1}{\sqrt{8}}$ contains a total amount of boundary edge-excesses at least equal to

$$4\left(\cos^{-1}\frac{1-\sin\frac{A}{4}}{1+\sin\frac{A}{4}} - \frac{\pi}{3}\right).$$

(iv) Let $\pentagon_{\frac{\pi}{3}}$ be the area of the cocircular pentagon with uniform $\frac{\pi}{3}$-side-lengths. Then, $\pentagon_{\frac{\pi}{3}} = 3\triangle_{\frac{\pi}{3}} + 0.98069\cdots$. Therefore, a pentagonal big-hole with its area smaller than $\pentagon_{\frac{\pi}{3}}$ is necessarily buckled. It is not difficult to see that the *buckling effect* of such a pentagonal big-hole with a given area is bounded below by that of the buckled pentagon as indicated in Figure 8.12.

Circumradii of big-holes First of all, the spherical polygon of a big-hole, in general, does *not* have a circumcircle of itself. However, it is natural to regard the *maximum* of the circumradii of its triangles as the circumradius of the given big-hole. Geometrically, area is of course a natural measurement of the size of a big-hole, while the above circumradius is another measurement of its size. In the case that the corresponding corner spheres of the face with maximal circumradius are touching neighbors of S_0, the minimal buckling height of a sphere positioning on the top of such a big-hole is given by

$$h = 2\cos r - 1\,, \tag{23}$$

where r is the circumradius of the given big-hole. Otherwise, the minimal buckling height h can be a tiny bit smaller than $(2\cos r - 1)$ (cf. Sublemma 4.1.3).

Examples (i) The circumradius of the buckled hexagon with triangular buckling diagonals of lengths $\pi/2$ is equal to that of the equilateral triangle of side-length $\pi/2$, namely

$$r_1 = \cos^{-1}\sqrt{\frac{1}{3}} = 0.9553\cdots\,, \quad 2\cos r_1 - 1 = 0.1547\,. \tag{24}$$

(ii) The circumradius of the buckled hexagon which actually consists of a pair of cocircular quadrilaterals with side-lengths $\{\pi/2, \pi/3, \pi/3, \pi/3\}$ is given by

$$\begin{gathered} r_2 = \tan^{-1}\frac{1}{\sqrt{3}}\sec\frac{1}{2}\cos^{-1}\frac{1}{3}(1-2\sqrt{2}) = 0.91753\cdots\,, \\ 2\cos r_2 - 1 = 0.2155625\cdots\,. \end{gathered} \tag{25}$$

(iii) The circumradius of the cocircular pentagon with uniform $\pi/3$-side-lengths is given by

$$\begin{gathered} r_3 = \sin^{-1}\frac{1}{2\sin\pi/5} = 1.017222\,, \\ 2\cos r_3 - 1 = 0.05146224,\ A = 2.634547 = 3\triangle_{\pi/3} + 0.98069\,. \end{gathered} \tag{26}$$

(iv) If A_1 is a vertex in a small deformation of a type I configuration with $St(A_1)$ being a small deformation of the $(\triangle, \triangle, \square, \square)$-star (resp. $(\triangle, \square, \triangle, \square)$-star), then the removal of A_1 creates a big-hole of circumradius close to $\pi/3$ which is a small deformation of the ones as indicated in Figure 8.11.

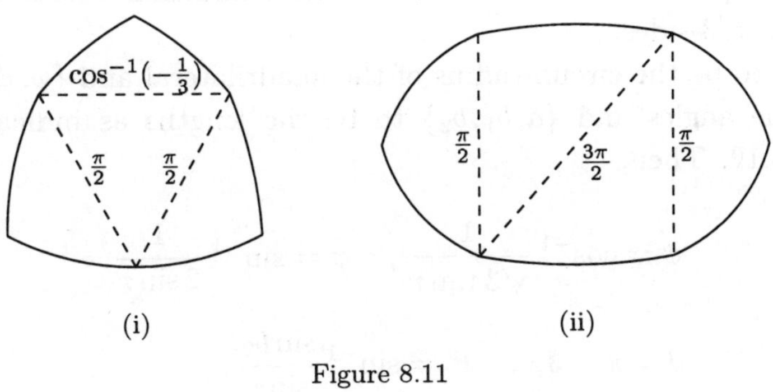

Figure 8.11

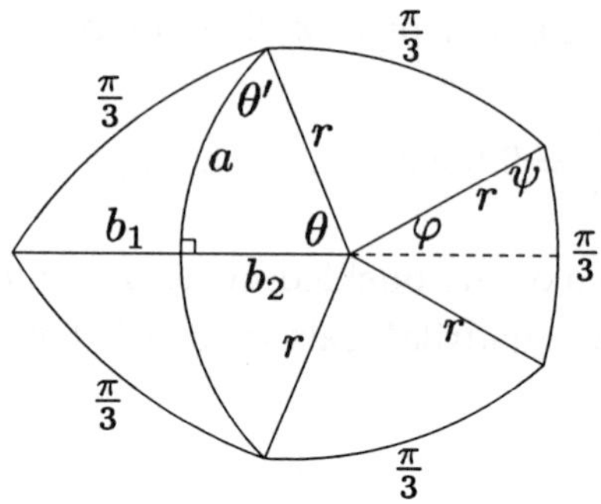

Figure 8.12

The areas of the above two buckled hexagons, themselves, are equal to $2\cos^{-1}(-\frac{1}{3}) = 4\triangle_{\pi/3} + 1.616124\cdots$. Those small deformations, shortening the longest diagonal, will make the above *area-excess* of 1.6161 smaller, roughly varying between 1.6 to 1.4.

(v) Set r to be the circumradius of an equilateral cocircular quadrilateral (i.e. a spherical "square"). Then, its area (resp. side-length) is given by

$$A(\square(r)) = 8\tan^{-1}\frac{1-\cos r}{1+\cos r}, \quad (\text{resp. } \ell = \cos^{-1}\cos^2 r)\,. \tag{27}$$

(vi) As indicated in Figure 8.12, the buckled pentagon consists of a cocircular quadrilateral with three $\pi/3$-sides and an isosceles triangle with two $\pi/3$-sides.

Set r to be the circumradius of the quadrilateral and $\{\varphi, \psi, \theta, \theta'\}$ to be the angles and $\{a, b_1, b_2\}$ to be the lengths as indicated in Figure 8.12. Then,

$$\psi = \cos^{-1}\frac{1}{\sqrt{3}\tan r}, \quad \varphi = \sin^{-1}\frac{1}{2\sin r},$$

$$\theta = \pi - 3\varphi\,, \quad \theta' = \sin^{-1}\frac{\sin b_2}{\sin r}\,,$$

$$a = \sin^{-1} \sin\theta \sin r\,, \quad b_2 = \tan^{-1} \cos\theta \tan r\,, \quad b_1 = \cos^{-1} \frac{1}{2\cos a}\,,$$

$$A(\pentagon) = (6\psi + 2\theta' - 2\pi) + 2\cos^{-1}\frac{2}{3}(\cos a + \cos b_1)\,. \tag{28}$$

In the special case of $r = 0.96$, straightforward computation shows that

$$\begin{aligned} &\psi = 1.15486,\ \varphi = 0.656512\,,\ \theta = 1.172056,\ \theta' = 0.633574\,, \\ &a = 0.855542,\ b_1 = 0.703756,\ b_2 = 0.506346\,, \\ &A(\pentagon) = 2.576558 = 3\Delta_{\pi/3} + 0.9227\,. \end{aligned} \tag{29}$$

Buckling effects of big-holes (i) A big-hole of hexagonal type with triangular buckling (cf. Figure 8.10(iii)) and with circumradius of at most 1 is always accompanied by a rather large amount of buckling effect. For example, in the special case of uniform $\pi/3$-boundary edges and $\pi/2$-diagonal length, it has three buckling effects, each of them is equal to

$$\left[\frac{4}{3} + 1 - \frac{4}{\sqrt{3}}\cos\left(\frac{\pi}{4} + \cos^{-1}\sqrt{\frac{2}{3}}\right)\right]^{\frac{1}{2}} - 1 = 0.65041\,. \tag{30}$$

Therefore, the occurrence of a hexagonal big-hole of this type already brings with it such a large amount of buckling effect, which makes the estimate of $(\star)$ quite simple.

However, a hexagonal big-hole consists of a pair of *cocircular* quadrilaterals sharing a long edge actually has no buckling effect, while in the case of a pair of buckled quadrilaterals sharing a long edge, its buckling effect is, in general, equal to the sum of the buckling effects of the individual quadrilaterals.

(ii) The circumradius of the *cocircular* pentagon with uniform $\pi/3$-boundary edges is equal to $r_3 = 1.017222$ whose corresponding buckling height is equal to $2\cos r_3 - 1 = 0.05146\cdots$. Therefore, a pentagonal big-hole with circumradius of at most $\cos^{-1} 0.545 = 0.9944\cdots$

is always accompanied by a certain amount of buckling effect. For example, the buckling effect of a pentagonal big-hole is at least 0.12 (resp. 0.227) if its circumradius is at most $\cos^{-1} 0.55$ (resp. 0.96).

(iii) In the case that the circumradius r of a big-hole in $\mathcal{S}'(\Sigma)$ exceeds $\cos^{-1} 0.545$, the placement of a *non-close neighbor* S' on top of such a big-hole always produces buckling effects between S' and its corner spheres. The average value of such a set of buckling effects is bounded below by $(1.1881-2.18\cos r)^{\frac{1}{2}}-1$. For example, in the specific case of cocircular pentagon of uniform side-lengths of $\frac{\pi}{3}$, the average value of the quintuple of buckling effects of placing a sphere S' with buckling height of 0.09 is bounded below by 0.02078, thus producing a total amount of at least 0.104 in buckling effects. In another example of a buckled hexagonal big-hole which is a small deformation of the ones as indicated in Figure 11, the placement of a sphere S' with buckling height of 0.09 will also produce a quadruple of such buckling effects. Moreover, the volume of the 0.09-slab is considerably larger than $L(0.09)$.

Remark It is impossible to have a pair of hexagonal big-hole or one hexagonal big-hole together with another pentagonal big-hole in a given configuration $\mathcal{S}'(\Sigma)$. The total amount of area-excess in $\mathcal{S}'(\Sigma)$ is always equal to 2.64323. The occurrence of such a pair of super-big-holes implies that the remaining ten or eleven faces of $\mathcal{S}'(\Sigma)$ must all have rather small area-excesses and hence they *must be all quite close to be $\pi/3$-equilateral.* It is easy to show the *non-existence* of such a spherical configuration.

Now, let us proceed to combine the above results on the geometric invariants of big-holes and the techniques of volume estimation to provide a proof of the collective volume estimate of $(\star)$ for Case III_1 as follows.

Case 1: $\mathcal{S}'(\Sigma)$ has a hexagonal big-hole Set q to be the number of quadrilateral big-holes in $\mathcal{S}'(\Sigma)$. If $q \geq 4$, then q must be equal to 4 and the four quadrilateral big-holes and the hexagonal big-holes, again, use up the total amount of 2.64323 in area-excess *almost* completely. Therefore, the remaining six faces of $\mathcal{S}'(\Sigma)$ must be all quite close to be $\pi/3$-equilateral. Moreover, the hexagonal big-hole must be a small deformation of either one of the two buckled hexagons of Figure 8.11. Set A_{12} to be the circumcenter of the largest face contained in it and $\Sigma^* = \Sigma \cup \{A_{12}\}$. Then, it is not difficult to show that $\mathcal{S}(\Sigma^*)$ is actually a small deformation of either the f.c.c. or the h.c.p. configuration. Hence, the volume of $C(S_0)$ is at least $4\sqrt{2} + 0.08$, while the volumes of $C'(S_j)$, $1 \leq j \leq 11$, are at least $4\sqrt{2} - 0.005$. Therefore, the estimate of $(\star)$ already holds without counting the contributions of buckling effects and built-in slabs. Next let us consider the case that $q = 3$. Then, either $\mathcal{S}(\Sigma^*)$ is still a small deformation of the f.c.c. or the h.c.p. configuration (except that the buckling diagonal of one quadrilateral is not so long), or $\mathcal{S}(\Sigma^*)$ is a small deformation of a 5□-type I configuration. In either case, the amount of volume contribution of the built-in slabs and the buckling effects is already large enough that makes the estimate of $(\star)$ for such a core packing quite straightforward. In the remaining case of $q \leq 2$, the volume contributions of the built-in slabs and the buckling effects is considerably larger than that of the previous case of $q = 3$. Hence, the estimate of $(\star)$ actually becomes simpler than that of the case $q \geq 3$.

Case 2: $\mathcal{S}'(\Sigma)$ has two pentagonal big-holes One of their circumradii must be less than $\cos^{-1} 0.545$. Moreover, such a configuration cannot have any other big-holes. Therefore, $\mathcal{S}'(\Sigma)$ must contain a considerable amount of short edge excess and buckling effect, say at least 0.5, and hence such a core packing always contains a large enough amount from built-in slabs, which makes the estimate of $(\star)$ quite simple.

Case 3: $\mathcal{S}'(\Sigma)$ has one pentagonal big-holes Set q to be the number of quadrilateral big-holes in $\mathcal{S}'(\Sigma)$. Then, it is easy to show that $q \leq 5$ and $q = 5$ only when $\mathcal{S}(\Sigma^*)$ is a small deformation of a 5□-type I configuration. The rest of the proof of this case is essentially the same as that of Case 1. (Set $\Sigma^* = \Sigma \cup \{\text{pt}\}$ on top of the pentagonal big-hole, then $\mathcal{S}(\Sigma^*)$ can be regarded as a small deforamtion of a type I configuration with a small star.)

Case 4: $\mathcal{S}'(\Sigma)$ has neither pentagonal nor hexagonal big-holes The configurations of this case always have a rather large amount of short or medium edge-excess and hence, core packings of this case always have a rather large volume contribution of built-in slabs. Thus, the estimate of $(\star)$ for this case is again quite simple.

Finally, it is not difficult to check that the above proof for the case of eleven almost touching close neighbors also works without modification if the total buckling height, $\sum_{j=1}^{11} h_j$, is at most 0.1. On the other hand, the additional volume contribution of $2\sum_{j=1}^{11} L(h_j)$ in the case of at least 0.1 total buckling height already makes the estimate of $(\star)$ quite straightforward.

This completes the proof of Case III_1.

Case III_2: Core packings with at most ten close neighbors Let $\mathcal{L}(S_0)$ be a given single-layer local packing, $\mathcal{L}'(S_0)$ (resp. $\hat{\mathcal{L}}(S_0)$) be its reduced local packing (resp. core packing) and $\mathcal{L}^2(S_0)$ be one of its *single-layer coating*, namely

$$\hat{\mathcal{L}}(S_0) \subset \mathcal{L}'(S_0) \subset \mathcal{L}(S_0) \subset \mathcal{L}^2(S_0)\,,$$

where $\mathcal{L}^2(S_0)$ is a double-layer local packing of S_0.

Set m (resp. n) to be the number of close (resp. reduced) neighbors in $\hat{\mathcal{L}}(S_0)$ (resp. $\mathcal{L}'(S_0)$) and we shall assume that $m \leq 10$ in the following discussion. Set $\mathcal{S}(\Sigma)$ (resp. $\mathcal{S}(\Sigma^*)$) to be the associated configuration of $\hat{\mathcal{L}}(S_0)$ (resp. $\mathcal{L}'(S_0)$), and $\mathcal{S}'(\Sigma)$ (resp. $\mathcal{S}'(\Sigma^*)$) to be a triangulation of $\mathcal{S}(\Sigma)$ (resp. $\mathcal{S}(\Sigma^*)$) by diagonal cuttings of its non-triangular faces, if any. Moreover, we shall label the reduced neighbors by S_j, $1 \leq j \leq n$, so that their buckling heights are in increasing order, namely

$$0 \leq h_1 \leq \cdots \leq h_m \leq 0.09 < h_{m+1} \leq \cdots \leq h_n \leq 0.3\,. \tag{31}$$

Note that there is *no* intrinsic reason for the choice of 0.09 as the upper limit of the buckling heights of *close neighbors.* Geometrically, there is *no* essential difference between a neighbor with its buckling height equal to 0.09 and another neighbor with its buckling height slightly larger than 0.09. For example, the replacement of a former one by a latter one in a local packing will make very little difference in its geometric structure. However, due to the *definition of the locally averaged density*, such a small difference in buckling height *acrossing* 0.09 will actually make quite a difference in their contribution toward the estimation of $(\star)$. For example, a close neighbor of buckling height 0.09 contributes a pair of 0.09-slab-volumes towards the estimate of $(\star)$ while a non-close neighbor of buckling height $0.09 + \varepsilon$ only contribute a single $(0.09 + \varepsilon)$-slab-volume towards the estimate of $(\star)$, thus only making half as much contributions. On the other hand, such a change will decrease the number of close neighbors by one, which makes the weight factor λ_0 changing from $\frac{13}{m+1}$ to $\frac{13}{m}$. Thus, almost the same amount of *volume-excess of the central cell*, i.e. $\operatorname{vol} C(S_0) - 4\sqrt{2}$, will contribute slightly more towards the estimate of $(\star)$.

Remark In the case that $m \leq 5$, it is easy to check that

$$\begin{aligned} \operatorname{vol} C(S_0) &> L_n + (n-m)L(0.09) \\ &\geq L_{15} + 10L(0.09) = 4\sqrt{2} + 0.42459 \end{aligned} \tag{32}$$

and hence, it suffices to use the lower bound estimate of L_{12} for the volumes of the other local cells, namely

$$\begin{aligned} \sum_{j=0}^{m} \lambda_j (\operatorname{vol} C(S_j) - 4\sqrt{2}) &\geq \frac{13}{6}(0.42459) + m(L_{12} - 4\sqrt{2}) \\ &\geq 0.919945 - 5(0.106563) = 0.38713\,. \end{aligned} \tag{33}$$

Therefore, it suffices to discuss the proof of the remaining cases of $6 \leq m \leq 10$ in the following.

Again, both the total amount and the distribution of edge-excesses in such an $\mathcal{S}'(\Sigma^*)$ will play a major role in the following proof of Case III$_2$. Recall that the *area-excess* (resp. *edge-excess*) of a face $\sigma \in \mathcal{S}'(\Sigma^*)$ is defined to be

$$\begin{aligned} A_\varepsilon(\sigma) &= \triangle(\sigma) - \triangle(h_i, h_j, h_k)\,, \\ \ell_\varepsilon(\sigma) &= \ell(\sigma) - \ell(h_i, h_j) - \ell(h_j, h_k) - \ell(h_k, h_i)\,, \end{aligned} \tag{34}$$

where $\{h_i, h_j, h_k\}$ are the buckling heights of the triple of reduced neighbors corresponding to the triple of vertices of σ. Moreover, one has (cf. (33) of Sec. 5.1)

$$\triangle(h_i, h_j, h_k) \sim \triangle_{\frac{\pi}{3}} - \frac{\sqrt{2}}{3}(h_i + h_j + h_k) \tag{35}$$

and hence, the *total area-excess* of $\mathcal{S}'(\Sigma^*)$ is approximately given as follows, namely

$$\begin{aligned} \sum_\sigma A_\varepsilon(\sigma) &\sim \sum_\sigma \left(\triangle(\sigma) - \triangle_{\frac{\pi}{3}} + \frac{\sqrt{2}}{3}(h_i + h_j + h_k)\right) \\ &= 4\pi - (2n-4)\triangle_{\frac{\pi}{3}} + \frac{\sqrt{2}}{3}\sum_{j=1}^{n} k_j h_j\,, \end{aligned} \tag{36}$$

where k_j is the number of triangles in $St(A_j)$ and it follows from Euler's formula that $\sum_{j=1}^{n} k_j = 6n - 12$.

On the other hand, one has the following approximation of $\ell_\varepsilon(\sigma)$, namely

$$\ell_\varepsilon(\sigma) \sim \ell(\sigma) - \pi + \frac{2}{\sqrt{3}}(h_i + h_j + h_k) \tag{37}$$

and Corollary 2.1.2 provides a simple lower bound estimate of $\ell(\sigma)$ in terms of $\triangle(\sigma)$. In particular, in the special case that $\triangle(\sigma) = \triangle_{\frac{\pi}{3}}$, one has

$$\begin{aligned} A_\varepsilon(\sigma) &\sim \frac{\sqrt{2}}{3}(h_i + h_j + h_k)\,, \\ \ell_\varepsilon(\sigma) &\sim \ell(\sigma) - \pi + \frac{2}{\sqrt{3}}(h_i + h_j + h_k), \quad \ell(\sigma) - \pi \geq 0\,. \end{aligned} \tag{38}$$

Therefore, $\sqrt{6}A_\varepsilon(\sigma)$ is an approximate lower bound for $\ell_\varepsilon(\sigma)$ for such a special case. In general, set $\triangle(\sigma) = \triangle_{\frac{\pi}{3}} + 3\delta$. Then, the edge length of the equilateral triangle with $\triangle_{\frac{\pi}{3}} + 3\delta$ as its area is given by

$$a(\delta) = \cos^{-1}\frac{\cos(\alpha+\delta)}{1-\cos(\alpha+\delta)}, \quad \alpha = \cos^{-1}\frac{1}{3} \tag{39}$$

and it follows from the Corollary 2.1.2 that

$$\begin{aligned} \ell(\sigma) \geq 3a(\delta) &= 3\cos^{-1}\frac{\cos(\alpha+\delta)}{1-\cos(\alpha+\delta)} \\ &\equiv \pi + 3\sqrt{6}\delta - \frac{27\sqrt{3}}{4}\delta^2 \pmod{\delta^3}\,. \end{aligned} \tag{40}$$

Therefore, even in the most critical case of equilateral triangles with their areas close to $\triangle_{\frac{\pi}{3}}$,

$$(\ell(\sigma) - \pi) \sim \left\{\sqrt{6} - \frac{3\sqrt{3}}{4}(\triangle(\sigma) - \triangle_{\frac{\pi}{3}})\right\}(\triangle(\sigma) - \triangle_{\frac{\pi}{3}}) \tag{41}$$

while in the general case that σ is not very close to be equilateral, but $\triangle(\sigma)$ is still quite close to $\triangle_{\frac{\pi}{3}}$, it is quite simple to use (65), (66) and (67) in Sec. 2.1 to show that

$$\ell(\sigma) - \pi \geq \sqrt{6}(\triangle(\sigma) - \triangle_{\frac{\pi}{3}}) \tag{42}$$

actually holds for such a triangle.

In the application of the above geometric fact to the estimation of *total edge-excess* of an $\mathcal{S}'(\Sigma^*)$, $\#(\Sigma^*) = n \geq 13$, it is not difficult to show that at most a few of its faces can be close to equilateral while there are always quite a few of them with rather lopsided distributions of their side-lengths, i.e. far from being equilateral. Therefore, only at most a few of $\ell_\varepsilon(\sigma)$ can be slightly smaller than $\sqrt{6}$-times of $A_\varepsilon(\sigma)$ while quite a few of $\ell_\varepsilon(\sigma)$ are always considerably larger than $\sqrt{6}$-times of $A_\varepsilon(\sigma)$. Hence, one always has the following *collective* lower bound estimate of the *total edge-excess* of such an $\mathcal{S}'(\Sigma^*)$, $\#(\Sigma^*) \geq 13$, namely

$$\frac{1}{2}\sum_\sigma \ell_\varepsilon(\sigma) \geq \sqrt{\frac{3}{2}}\sum_\sigma A_\varepsilon(\sigma) \tag{43}$$

while the right-side is approximately given by

$$\sqrt{\frac{3}{2}}\left\{4\pi - (2n-4)\triangle_{\frac{\pi}{3}} + \frac{\sqrt{2}}{3}\sum_{j=1}^{n} k_j h_j\right\}, \quad 4 \leq k_j \leq 6,$$

$$\sum_{j=1}^{n} k_j = 6n - 12 > 5n\,. \tag{43$'$}$$

Remark For the sake of simplicity in presentation, we shall use

$$\sqrt{\frac{3}{2}}\left\{4\pi - (2n-4)\triangle_{\frac{\pi}{3}} + \frac{5}{3}\sqrt{2}\sum_{j=1}^{n} h_j\right\} \tag{43$''$}$$

as an *approximate lower bound estimate* of the *total edge-excess* of $\mathcal{S}'(\Sigma^*)$.

Geometrically, the gap-width between a pair of neighboring spheres $S_i, S_j \in \mathcal{L}'(S_0)$ is roughly equal to $\sqrt{3}$-times of the edge-excess of $\overline{A_iA_j}$. Therefore, corresponding to each edge-excess of small or medium size, there are a pair of rectilinear slabs situated between S_i and S_j whose thickness is roughly equal to $\frac{\sqrt{3}}{2}$-times of the edge-excess. On the other hand, an edge $\overline{A_iA_j}$ with a large edge-excess is actually a diagonal of a big-hole whose contribution to the volume of its corner local cells makes their volumes at least very close to $4\sqrt{2}$.

Before we proceed to the actual proof of Case III_2, it is important to clarify here the difference between the following three types of gaps in terms of their contributions towards the estimation of $(\star)$, namely

(i) the gap between a pair of close neighbors,
(ii) the gap between a close neighbor and a non-close reduced neighbor,
(iii) the gap between a pair of non-close reduced neighbors.

A gap of width $2h$ between a pair of close neighbors $\{S_i, S_j\}$ will contribute the volume estimate of an h-slab towards that of both $C(S_i)$ and $C(S_j)$, thus providing a pair of such estimates towards that of $(\star)$. A gap of width $2h$ between a close neighbor S_i and a non-close reduced neighbor S_j only contributes the volume estimate of a single h-slab towards that of $C(S_i)$, while its contribution in the S_j-part is restricted to that of an h-strip in addition to the estimate of $L(h_j)$ for the h_j-slab belonging to $C(S_0)$. Finally, a gap of width $2h$ between a pair of non-close reduced neighbors $\{S_i, S_j\}$ only contributes that of an h-strip in addition to the estimates of both $L(h_i)$ and $L(h_j)$ for the h_i-slab and h_j-slab (belonging to $C(S_0)$), respectively. Anyhow, a sufficiently large amount of total edge-excess of $\mathcal{S}'(\Sigma^*)$ will always contribute a considerable amount from the above two kinds of volume estimates towards the collective estimate of $(\star)$.

Now, let us proceed to the proof of the estimate of $(\star)$ for Case III_2. We shall divide the proof into two cases according to the number of close neighbors, namely, Case 1: $m = 10$ and Case 2: $m \leq 9$.

Case 1: $m = 10$. We shall always assume, without loss of generality for the proof of $(\star)$, that the circumradii of all faces of $\mathcal{S}(\Sigma^*)$ are less than $\cos^{-1} 0.65$. For otherwise, the proof of $(\star)$ can be reduced to the subcase of $n+1$ reduced neighbors, instead of n reduced neighbors. Moreover, the proof of Case III_1 is also directly applicable to the subcase of $n \leq 11$. Therefore, we shall only discuss the proof of this case (i.e. $m = 10$) for $n \geq 12$ as follows.

Subcase (i): $n = 12$. By the above assumption on the circumradii of faces of $\mathcal{S}(\Sigma^*)$, all the big-holes of $\mathcal{S}(\Sigma^*)$, if any, are of the quadrilateral type. Set q to be the number of buckled (or cocircular) quadrilaterals of $\mathcal{S}(\Sigma^*)$. There are at most six such quadrilaterals in $\mathcal{S}(\Sigma^*)$. For otherwise, there would be at least four vertices whose stars contain triples of such quadrilaterals. On the other hand, the total buckling height, $\sum_{j=1}^{12} h_j$, is at least 0.18, and hence the total amount of area-excess of $\mathcal{S}(\Sigma^*)$ exceeds $6(\square_{\frac{\pi}{3}} - 2\triangle_{\frac{\pi}{3}})$ by at least 0.4 (cf. (36)). Therefore, $\mathcal{S}(\Sigma^*)$ always contains a considerable amount of short or medium edge-excess even when q is equal to 6, while the amount becomes much larger in the case of smaller q. Anyhow, the volume contribution of those rectilinear slabs corresponding to such a considerable amount of short or medium edge-excess easily makes

$$\sum_{j=1}^{10} \lambda_j(\operatorname{vol} C(S_j) - 4\sqrt{2}) > 0 \tag{44}$$

while the first term of $(\star)$ is at least 0.078. For examples, the volumes of the *core parts* of $\{C(S_j), 0 \leq j \leq 10\}$ are all at least $(4\sqrt{2} - 0.01)$ in the case of $q = 6$, while in the case of $q = 5$, nine of them are at least $(4\sqrt{2} - 0.01)$.

Subcase (ii): $n = 13$. The associated configuration $\mathcal{S}(\Sigma^*)$ contains at least two $6\triangle$-stars, while the set of buckling heights $\{h_j, 1 \leq j \leq 13\}$ contains a triple exceeding 0.09. If $\sum_{j=1}^{10} h_j$ is at least 0.05, then the contribution of the slab volumes of at least $2\sum_{j=1}^{10} L(h_j) + 3L(0.09)$ already makes the proof of $(\star)$ for such a geometric situation quite simple. Therefore, we shall only discuss the situation that $\sum_{j=1}^{10} h_j$ is *less than 0.05*. On the one hand, it is straightforward to check that the same estimates of Lemma 5.3.1 will again prove that $\sum_{j=1}^{13} h_j$ is at least 0.3, and on the other hand, it follows from the occurrence of quite a few triangular faces in $\mathcal{S}'(\Sigma^*)$ with rather uneven side-lengths that the volume of the *core part* of

$C(S_0)$ always exceeds the lower bound estimate of L_{13} by at least 0.015. Hence, one has the following estimate of the first term of $(\star)$, namely

$$\begin{aligned}&\operatorname{vol} C(S_0) > L_{13} + 0.015 + 2L(0.09) + L(0.12)\,,\\&\lambda_0(\operatorname{vol} C(S_0) - 4\sqrt{2}) > 0.064\,.\end{aligned} \tag{45}$$

If *none* of $St(A_j)$, $1 \le j \le 10$, is a small $5\triangle$-star, then the volume of each $C(S_j)$, $1 \le j \le 10$, is at least $(4\sqrt{2} - 0.01)$ and it can be smaller than $4\sqrt{2}$ only when $St(A_j)$ is a small deformation of that of Figure 7.33(ii). However, $\mathcal{S}'(\Sigma^*)$ can contain at most five such $5\triangle$-stars. Therefore, the estimate of $(\star)$ holds with a margin of at least 0.014 without counting other contributions such as the buckling effects of buckled quadrilaterals, etc.

Next let us consider the only remaining possibility that one of $St(A_j)$, $1 \le j \le 10$, is a small $5\triangle$-star. Suppose that $St(A_1)$ is a small $5\triangle$-star. Then *none* of the quintuple of stars of its boundary vertices can be of $6\triangle$-type, unless the total buckling height, $\sum_{j=1}^{13} h_j$, is at least 0.6 which will make the estimate of $(\star)$ quite simple. Note that the matching of a small $5\triangle$-star and a $6\triangle$-star with a pair of small triangles in common will place the remaining triple of vertices of the $6\triangle$-star in the vicinities of the middle portions of three boundary arcs of the complementary region of $St(A_1)$, as indicated in Figure 8.13.

$\Omega(St(A_1))$:

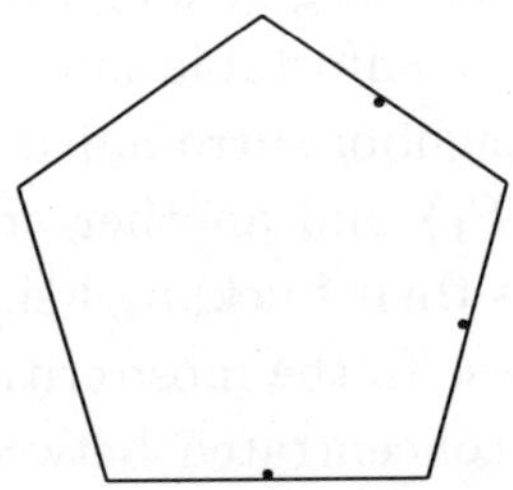

Figure 8.13

Ω :

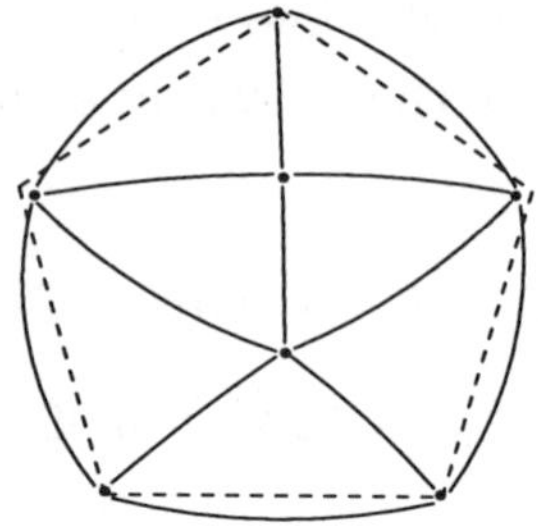

Figure 8.14

On the other hand, straightforward combinatorial analysis of configurations with thirteen vertices will show that the one indicated in Figure 5.6(i) is, in fact, the *only* one which contains some 5$\triangle$-star having *no* overlapping triangles with any of the 6$\triangle$-stars in $\mathcal{S}'(\Sigma^*)$, unless the total buckling height is, again, larger than 0.6. Let $\widehat{\mathcal{S}}$ be the subconfiguration of such an $\mathcal{S}'(\Sigma^*)$ spanned by the seven vertices outside of such a small 5$\triangle$-star. Then, it is not difficult to see that $\widehat{\mathcal{S}}$ must have five of its vertices situated in the vicinities of the five corners of the $\Omega(St(A_1))$, respectively, while the remaining pair of vertices are situated in the central region of $\Omega(St(A_1))$ (cf. Figure 8.14).

Hence, it is quite straightforward to check that such a configuration $\mathcal{S}'(\Sigma^*)$ always contains a rather large amount of short or medium edge-excesses. The volume contribution of their corresponding rectilinear slabs is already large enough to make the estimate of $(\star)$ very simple and with a comfortable margin. For example, suppose that S_1 is a touching neighbor surrounded by a triple of common touching neighbors of $\{S_0, S_1\}$ and another adjacent pair of touching neighbors with 0.09^+ as their buckling heights. Then, $St(A_1)$ contains no radial edge-excesses. In the most critical case that all of its *boundary* edge-excesses are concentrated between the pair of non-close neighbors, the geometry of such a $St(A_1)$ and its complementary region is as indicated in Figure 8.15, where

$St(A_1)$:

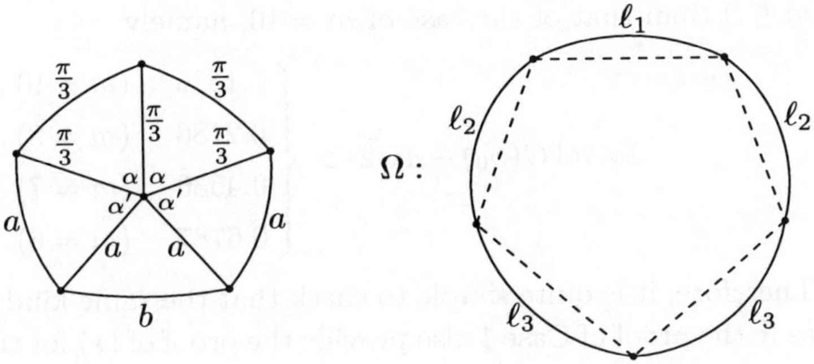

Figure 8.15

$$\begin{gathered} a = \cos^{-1} 0.545\,, \quad b = 1.17843682\,, \\ \ell_1 = 1.1511829\,, \quad \ell_2 = 1.2227587\,, \quad \ell_3 = 1.2052773\,. \end{gathered} \tag{46}$$

The volume contribution from those edge-excesses of $\{\ell_1, \ell_2, \ell_3\}$ is at least equal to

$$2L(0.08866) + 4L(0.14799) + 4L(0.1336) = 1.09\,, \tag{47}$$

which is already more than enough for the proof of the estimate of $(\star)$.

Subcase (iii): $n \geq 14$. In this remaining subcase of Case 1, it is easy to check that the same kind of estimates will provide a slightly larger estimate for the first term of $(\star)$, and the same estimate for the remaining part of $(\star)$. Thus, the same kind of estimates as that of the previous subcase of $n = 13$ also works for $n \geq 14$.

This completes the proof of $(\star)$ for Case 1 of III_2.

Case 2: $m \leq 9$. The proof for the case of $m \leq 9$ is essentially the same as that of the previous case of $m = 10$. Moreover, the estimate of the first term of $(\star)$ becomes considerably larger for the case of

$m \leq 9$ than that of the case of $m = 10$, namely

$$\lambda_0(\operatorname{vol} C(S_0) - 4\sqrt{2}) > \begin{cases} 0.1516 & (m = 9)\,, \\ 0.2786 & (m = 8)\,, \\ 0.4536 & (m = 7)\,, \\ 0.6787 & (m = 6)\,. \end{cases} \tag{48}$$

Therefore, it is quite simple to check that the same kind of estimates as in the proof of Case 1 also provide the proof of ($\star$) for the remaining case of $m \leq 9$.

This completes the proof of ($\star$) for Case III_2 and also the whole case of at most eleven close neighbors.

Chapter 9
Retrospects and Prospects

After the long and laborious journey of the mathematical proof of the *least action principle of crysal formation of the dense packing type* and *Kepler's Conjecture on sphere packing*, it is natural to have a *retrospective overview of the whole journey*, in order to reflect and to analyze where the crucial ideas, the key steps, the basic techniques as well as the major geometric insights are that constitute those important components of achieving the *optimal estimates* of both the *local density* and the *locally averaged density.*

Of course, it is also a natural juncture to look forward to the prospects of further developments in the realm of sphere packings in specific and classical geometry in general.

9.1 Retrospects

9.1.1 *Main results and their significances*

Kepler's Conjecture on sphere packing is undoubtedly one of the greatest outstanding problems in solid geometry. However, the establishment of a *least action principle of crystal formation of the dense packing type* is much more important for the *understanding of Nature*. The former is a neat statement on the *optimal density* of *infinite sphere packings* which has challenged geometers since 1611, while the latter seeks to provide a kind of mathematical understanding of the underlying mechanism of crystal formation, namely, whether the remarkable regularities and precision of crystals are indeed the mathematical consequence of some suitable kinds of optimization. Moreover, the mathematical task of establishing such a least action principle is much more difficult than the proof of Kepler's Conjecture; one needs to search for a *suitable concept of density* for a *finite cluster of spheres* (cf. Chaps. 1 and 3) and then to prove a strong uniqueness theorem that a finite cluster of spheres with *optimal density* is necessarily isometric to a hexagonal cluster, just the same as the regular arrangement of atoms in a crystal of the dense packing type. Anyhow, the most important result of this book is the *proof* of such a *strong uniqueness theorem on finite* clusters of spheres with the *optimal global density of the second kind*, i.e. $\pi/\sqrt{18}$ (cf. Chaps. 1 and 3). Both the proof of such a strong uniqueness theorem and the proof of Kepler's Conjecture can be logically reduced to the proof of Main Theorem I of this book, which not only proves the *optimality of* $\pi/\sqrt{18}$ for the *locally averaged density*, but also *determines* the geometric structures of those double layer local packings with *optimal locally averaged density*. It was exactly the philosophical belief that those least action principles of Nature are always *highly localizable* that led to the *discovery* of the crucial local invariant of locally averaged density. In retrospect, the *search* for a *suitable local invariant*, which enables us to reduce the kind of least action principle to the *optimization* of the chosen local invariant *everywhere*, is always one of the most decisive steps toward the establishment of such a *highly localizable least action principle*.

9.1.2 *Localizations and reductions*

The reduction of the proof of both the least action principle of crystal formation and Kepler's Conjecture to the proof of Main Theorem I (cf. Chap. 3) is, of course, one of the most important steps of localization which, in just one stroke, reduces both of the global optimization problems to a local optimization problem. However, the local invariant of locally averaged density, itself, depends on the geometric structure of a *double-layer local packing* $\mathcal{L}^2(S)$ whose geometry is still very, very complicated. Therefore, the task of achieving its *optimal estimation* together with the determination of those double layer local packings with optimal locally averaged density is still unmanageably difficult *unless* one can find a systematic way of stepwise localization and reduction which enables one to *further reduce* it to a large collection of much simpler geometric problems and estimates. In retrospect, the whole structure of the proof of Main Theorem I that we present in this book is essentially an elaborate system of stepwise localizations and reductions outlined in [Hsi-2]. Of course, one still needs to prove all kinds of lemmas, sublemmas and estimates in order to actually carry out such an elaborate scheme of localizations and reductions, all the way to producing a complete proof of Main Theorem I. However, the most difficult part of the whole proof is, in fact, the finding of such a workable scheme. First of all, there is no *a priori* reason that the *optimal estimation* of such a complicated local invariant can, at all, be systematically reduced to the *combination of some much more localized estimates.* It, of course, all depends on the geometric reality of the totality of all possible double-layer local packings together with how such geometric structures interacting with the specific local invariants. Therefore, the very existence of such a workable scheme is, in itself, a *remarkable reality of Nature* which, on the one hand, makes such a proof of Main Theorem I feasible, and on the other hand, may also play a role in the actual formation of crystals of the dense packing type. (Suppose, contrary to the geometric reality, that 4π is *not* equal to $6\Box_{\frac{\pi}{3}} + 8\triangle_{\frac{\pi}{3}}$, then the optimal global density would be difficult to conjecture and crystals would be difficult to form.)

9.1.3 *Subdivision of a local cell and basic techniques of volume estimation of local cells*

First of all, the *local cell* associated to a given local packing is the most basic geometric object in the study of sphere packings. All kinds of densities of sphere packings such as the local density, the locally averaged density, global densities of the first or the second kind are naturally defined in terms of the volumes of local cells. Therefore, the techniques of volume estimation of local cells are clearly of primary importance in order to achieve various kind of estimates on densities of sphere packings. Geometrically, the first and also the decisive step for developing a set of volume estimation techniques, specifically tailored for that of local cells, is the choice of a *suitable subdivision of a local cell.* Here, a subtle difference of subdivisions will make a huge difference for the task of achieving the optimal estimates of both the local density and the locally averaged density. In retrospect, it is how well such a corresponding set of volume estimation techniques works in the above task that, in the end, justifies the choice of the subdivision. Anyhow, the subdivision we choose (cf. Chap. 4) works quite well. Moreover, the subdivision we choose is also geometrically rather natural and simple, namely, we first separate a local cell into its *peripheral part* and its *core part* and then further decompose the former into rectilinear slabs, while the decomposition of the latter for the critical, generic case that all the faces of $\mathcal{S}(\Sigma)$ are triangular and containing their own circumcenters actually *corresponds* to the restriction to the core part of the Delaunay decomposition with respect to the *reduced local packing* $\mathcal{L}'(S)$ (instead of the local packing $\mathcal{L}(S)$).

The volume estimation lemmas of Chap. 4 not only provide a set of clean-cut lower bound volume estimates for the individual pieces of the above subdivision, but also help us to understand what those *volume-influential geometric invariants* of a local packing are. Among them, the number of close (resp. reduced) neighbors and the total buckling height of the core (resp. reduced) packing $\widehat{\mathcal{L}}(S)$ (resp. $\mathcal{L}'(S)$) are the most volume-influential ones. Such a geometric insight clearly provides valuable guidance for the search of the geometric understanding of Chaps. 5 and 7 which provides the crucial geometric inputs needed in the proofs of Chaps. 6 and 8.

9.1.4 *Vector algebra and spherical geometry*

Vector algebra and spherical geometry are well-known basic mathematical tools for the study of solid geometry, while Kepler's Conjecture and the least action principle of crystal formation are some outstanding problems on the optimal arrangement of identical spheres. Therefore, vector algebra and spherical geometry should naturally be an important part of the basic mathematical tools in the study of these two problems. In retrospect, one finds that they are the *only basic mathematical tools* that are needed for the proof of both outstanding problems, namely, there is *no* need to create a new theory; one just needs to have a better understanding of vector algebra and spherical geometry in order to achieve the proof of both problems. Thus, on the one hand, the proof of both problems that we give in this book is entirely elementary in terms of mathematical tools, and on the other hand, one achieves a much better understanding on the *usage* of vector algebra and spherical geometry through the experience of such an elementary proof of some challenging natural problems. Such a better understanding on their usage will, naturally, also lead to the solutions of many other unsolved problems in solid geometry.

9.2 Prospects

In the concluding section of this book, we shall briefly discuss some open problems in the realm of sphere packings in specific and classical geometry in general.

9.2.1 *Problems in solid geometry*

Solid geometry studies the geometry of the *physical space* in which we and everything else of the universe are situated. Thus, it is forever the most natural and most important kind of geometry, both for the practical applications of daily life and for the theoretical understanding of Nature. From antiquity up to the present day, the study of solid geometry has always been a central part of the civilization of the rational mind, and there were many major advancements such as Greek geometry, the invention of analytical geometry, the development of vector analysis and Grassmann algebra, etc. However, there are still many challenging natural problems in solid geometry that remain unsolved. The study of them will always be a canonical way to improve our understanding of solid geometry specifically and the physical universe in general. For example, the study of the sphere packing problem naturally leads to the least action principle of crystal formation. Anyhow, the following are just a few examples of such problems.

9.2.1.1 *Isoperimetric problem for polyhedra with a given number of faces (cf. Sec. 2.2)*

Although the isoperimetric problem for general regions has already been solved (cf. [Tal]), the isoperimetric problem *restricted* to polyhedra with a given number of faces remains unsolved for most cases (cf. [Gol]). The theorem of Lindelöf and Minkowski reduces the above problem to the following problem of optimal distribution of a given number of points on the unit sphere, namely

Problem 1 *Among all circumscribing polyhedra of the unit sphere with k faces, which one has the minimal volume? In other words, what is the optimal arrangement of k points, Σ, on the unit sphere that makes the volume of $T(\Sigma)$ minimal?*

(The volume formula of $T(\Sigma)$ (cf. $(17')$ of Sec. 4.2), the gradient formula of $\operatorname{vol} T(\Sigma)$ together with the volume estimation lemmas of

Sec. 4.2 naturally constitute a useful system of tools for the study of the above problem.)

9.2.1.2 *Problems on the optimal distribution of points on* $S^2(1)$

Of course, the above Problem 1 is just one example of the family of problems on *various kinds of optimal distributions* of a given number of points on the unit sphere. For example:

Problem 2 *Among all inscribing polyhedra of the unit sphere with* k *vertices, which one has the maximal volume?*

Problem 3 *Among all arrangements of* k *unit charges on the unit sphere, which one has the minimal Coulomb potential?*

Problem 4 (spherical codes) *Among all arrangements of* k *points on the unit sphere, which one has the maximal separation? (The separation of a set* $\Sigma \subset S^2(1)$ *is defined to be the minimal pairwise spherical distances among points of the given set* Σ*.)*

Remarks (i) The problem of spherical codes is a natural generalization and refinement of the problem of thirteen spheres in classical geometry. Up to now, this problem has been solved only for $k \leq 12$ and $k = 24$ (see [CFG]). The area estimates and combinatorial analysis, developed in Chaps. 2 and 5 for the purpose of estimates on total buckling heights, naturally also constitute a set of useful techniques in the study of spherical codes. We believe that such an approach should be able to provide the solutions to the special cases of $k = 13$, 14, 16 and 36 to begin with.

(ii) In general, the associated *spherical configuration* $\mathcal{S}(\Sigma)$ provides a canonical efficient organization of the spherical geometry of a given set of points, thus facilitating the systematic analysis of the geometric correlation between lengths, angles and areas. Anyhow, the geometry of spherical configurations, such as the ones developed in Chaps. 4, 5 and 7, will always be a natural and efficient way of studying this family of problems.

9.2.1.3 *Generalized Dodecahedron Conjecture*

In Chaps. 1 and 3, we discussed the concept of global density for finite clusters of spheres (without a container), and introduced two kinds of intrinsic densities, namely, the first kind and the second kind. The least action principle of crystal formation proves that $\frac{\pi}{\sqrt{18}}$ is *always* the optimal global density of the second kind (i.e. for clusters of arbitrary cardinality) and $\overline{\rho}(\mathcal{C}) = \frac{\pi}{\sqrt{18}}$ when and only when $\mathcal{C}$ has a double-layer extension $\mathcal{C}^{**}$ which is an hexagonal cluster. However, the situation for the global density of the first kind is quite different. First of all, the optimal value, namely

$$\rho_N = l.u.b.\{\rho(\mathcal{C});\ \#(\mathcal{C}) = N\} \tag{1}$$

will certainly depend on the cardinality N. Secondly, both the exact value of ρ_N (for $N \geq 2$) and the optimal geometric arrangements of those clusters $\mathcal{C}$ with $\rho(\mathcal{C}) = \rho_N$ are yet to be determined, namely

Problem 5 (cf. Sec. 3.2) *For a given cordinality N of at least 2, what is the least upper bound of the global density of the first kind of N-clusters, and what are the arrangements of N-clusters with optimal global density of the first kind?*

We mention here the following generalization of the Dodecahedron Conjecture:

Generalized Dodecahedron Conjecture

$$\rho_N = 0.75082643 \cdot \left[1 - \frac{1}{N}(0.007487489)\right]^{-1} \tag{2}$$

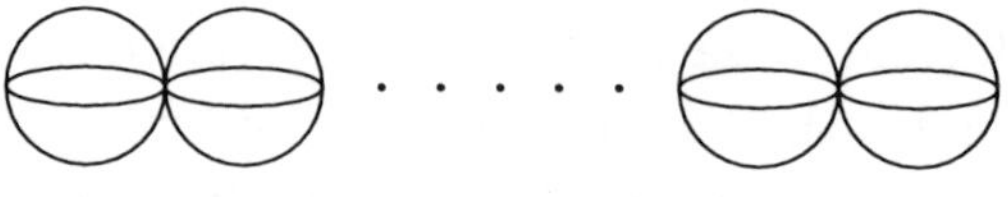

Figure 9.1

and $\rho(\mathcal{C}) = \rho_N$ when and only when $\mathcal{C}$ consists of a linear string of N-spheres as indicated in Figure 9.1.

9.2.1.4 *Another characterization of hexagonal double-layer local packings and a conjecture of Fejes Tóth*

The *locally averaged density,* $\overline{\rho}(\mathcal{L}^2)$, is defined in terms of the geometry of a *double-layer local packing* $\mathcal{L}^2(S)$, (cf. Chap. 1). The main theorem on sphere packing proves that those six types of *hexagonal double-layer local packings* (cf. the last paragraph of Chap. 3) are the only ones that achieve the optimal locally averaged density of $\frac{\pi}{\sqrt{18}}$. There should be another geometrically simpler characterization of the six hexagonal double-layer local packings (i.e. than the above one), namely

Condition T *The central sphere has twelve touching neighbors and each of them also has twelve touching neighbors.*

The proof of such a characterization of those hexagonal double-layer local packings will also deduce the following conjecture of L. Fejes Tóth, namely

Conjecture *Those hexagonal close packings are the only (infinite) packings that each sphere has twelve touching neighbors.*

Remark The characterization of those hexagonal double-layer local packings by Condition T should follow readily from a suitable combination of the results and the techniques of Chap. 7. However, it is interesting to provide a simpler direct proof of it rather than use the full strength of Chap. 7, such as Lemmas 7.1.3 and 7.1.4.

In concluding this rather brief discussion on problems in solid geometry, let us also mention here several important areas of applications that solid geometry will certainly play an important role, namely

(i) robotics (vector algebra and spherical geometry),

(ii) configurations of proteins,
(iii) the least action principles of other types of monoatomic crystals (i.e. other than the type of hexagonal close packings).

9.2.2 *Problems in higher dimensional Euclidean and spherical geometry*

Although physical space is 3-dimensional, the geometry of higher dimensional Euclidean spaces (i.e. of dimensions exceeding 3) is also very useful and clearly *indispensible*. For example, the configuration space of k particles is, in fact, a $3k$-dimensional Euclidean space. From the modern viewpoint, the entire geometric structure of $\mathbf{E}^n$ is *encoded in its isometry group*, usually called the *nth Euclidean group*, which is the semi-direct product of the n-dimensional *translation subgroup*, $\mathbf{R}^n$, and the nth *orthogonal group*, $O(n)$. To be more precise, the whole geometry of $\mathbf{E}^n$ can be intrinsically formulated as the *invariant theory* of the above transformation group. Moreover, the basic invariants of the nth Euclidean group can be organized into two major parts, namely

(i) the orthogonal invariants (or covariants) of the translation (normal) subgroup, which can be concisely organized into the setting of Grassmann algebra,
(ii) the spherical geometry of $S^{n-1}(1)$, which represents the *non-linear* part of orthogonal invariants.

Thus, the basic features of the higher dimensional Euclidean geometry are essentially the same as that of the solid geometry, except that the higher dimensional spherical geometry becomes more difficult than that of $S^2(1)$. One of the major difficulties of higher dimensional spherical geometry lies in the transcendental functions that occur in the volume formulas of higher dimensional spherical simplices.

Anyhow, one of the canonical ways to develop higher dimensional Euclidean geometry is, again, the study of some challenging natural problems. The following are just a few of such examples:

9.2.2.1 *Sphere packings in $S^n(1)$ and the problem of kissing numbers*

Note that the spherical spaces are *no longer* homothetically invariant. Thus, the local or global densities of sphere packings in $S^n(1)$ will certainly depend on the given size of the spherical balls. Moreover, there are several specific sizes of spherical balls whose packing problems are geometrically more meaningful than that of the other sizes. For example, the problem of kissing numbers in $\mathbf{E}^{n+1}$ is geometrically equivalent to the problem of optimal global density of packing $\frac{\pi}{6}$-balls in $S^n(1)$, while the problem of kissing number of $\frac{\pi}{6}$-balls in $S^{n+1}(1)$ is geometrically equivalent to the problem of optimal global density of packing r-balls in $S^n(1)$, $r = \frac{1}{2}\cos^{-1}\frac{1}{3}$. Therefore, it is geometrically preferable to study ball packings in $S^n(1)$ with those specific radii, namely, $r = \frac{1}{2}\cos^{-1}\frac{1}{m}$, $m = 2, 3, 4, \ldots$ etc.

Problem 6 *What are the optimal local densities of r-ball-packings in $S^n(1)$ with $r = \frac{1}{2}\cos^{-1}\frac{1}{m}$, $m = 2, 3, 4, \ldots$ etc.?*

Problem 7 *What are the optimal global densities of r-ball-packings in $S^n(1)$ with $r = \frac{1}{2}\cos^{-1}\frac{1}{m}$, $m = 2, 3, 4, \ldots$ etc.?*

Remarks (i) The case of $r = \frac{\pi}{6}$ in Problem 7 is geometrically equaivalent to the problem of kissing number in $\mathbf{E}^{n+1}$.

(ii) In the case that optimal local density for a given pair of n and $r = \frac{1}{2}\cos^{-1}\frac{1}{m}$ *cannot* be realized as the optimal global density of r-ball-packings in $S^n(1)$, it is natural to define a kind of locally averaged density (just the same way as we did for sphere packings in $\mathbf{E}^3$) and proceed to study the corresponding problem of optimal locally averaged density so defined.

(iii) In the case of $n = 3$ and $r = \frac{\pi}{6}$, the optimal local density and the optimal global density are *unequal.* In a recent paper by the author [Hsi-4], it has been proved that the kissing number in $\mathbf{E}^4$ is equal to 24, and moreover, the D_4-local packing is the only possible arrangement with 24 touching neighbors. The basic idea of such a proof is exactly the optimal estimation of a suitably defined locally averaged density for $\frac{\pi}{6}$-ball-packings in $S^3(1)$.

Conjectures on the kissing numbers in $\mathbf{E}^5$ (resp. $\mathbf{E}^6$ and $\mathbf{E}^7$) *The kissing number in* $\mathbf{E}^5$ (*resp.* $\mathbf{E}^6$ *and* $\mathbf{E}^7$) *should be equal to* 40 (*resp.* 72 *and* 126), *and moreover, the only possible arrangement in* $\mathbf{E}^5$ (*resp.* $\mathbf{E}^6$ *and* $\mathbf{E}^7$) *with* 40 (*resp.* 72 *and* 126) *touching neighbors is that of the local packings in the* D_5- (*resp.* E_6- *and* E_7-) *lattice packing.*

Remarks (i) A better understanding of the volume formula of higher dimensional spherical simplices naturally becomes an *indispensible* tool for the study of densities of various kinds of ball-packings in spheres.

(ii) The geometric ideas and techniques of localization, symmetrization and comparison will, again, play important roles in the study of such problems of spherical geometry.

(iii) If the above conjectures on the kissing number in $\mathbf{E}^5$ (resp. $\mathbf{E}^6$ and $\mathbf{E}^7$) happens to be true, then it is intuitively a strong indication that such a unique local packing with a maximal number of touching neighbors will most likely be the unique local packing with the maximal local density.

9.2.2.2 *Sphere packings in* $\mathbf{E}^n$, $n \geq 4$

Problem 8 *What is the optimal local density of sphere packings in* $\mathbf{E}^n, n \geq 4$, *and what are those local packings in* $\mathbf{E}^n$ *that achieve the optimal local density?*

Let $\mathcal{C}$ be a finite cluster of spheres in $\mathbf{E}^n$. Then, we can, again, define the *global density of the first kind* of $\mathcal{C}$ just the same way as we did in Sec. 3.2.

Problem 9 *For a given cardinality* N *of at least* 2, *what is the optimal global density,* ρ_N, *of the first kind for all* N*-clusters of spheres* *in* $\mathbf{E}^n$, *and what are those arrangements which achieve the optimal global density, namely, with* $\rho(\mathcal{C}) = \rho_N$?

Remarks (i) In the beginning dimension of $n = 2$, the optimal local density is equal to $\frac{\pi}{\sqrt{12}}$ and the local cell of minimal area is a regular circumscribing hexagon, which happens to be a tiling domain of $\mathbf{E}^2$ (cf. Sec. 3.1). Therefore, the optimal local density can be globalized to the optimal global density of the first kind thus settling the whole sphere packing problem completely and neatly.

(ii) In the next dimension of $n = 3$, the proof of the Dodecahedron Conjecture determines that the optimal local density is equal to

$$\frac{4\pi}{3} : 20\left(1 - 2\cos\frac{2\pi}{5}\right)\tan\frac{\pi}{5}$$

and the local cell of minimal volume is a regular circumscribing dodecahedron. However, it is *not* a tiling domain of $\mathbf{E}^3$ and the optimal local density cannot be realized as the optimal global density (either of the first kind or of the second kind). In fact, this is exactly the reason that, on the one hand, the Generalized Dodecahedron Conjecture is a result yet to be proved, and on the other hand, it becomes necessary to introduce the locally averaged density and the global density of the second kind in order to settle the sphere packing problem in $\mathbf{E}^3$.

(iii) In general, if the optimal local density in $\mathbf{E}^n$ happens to be realized by a unique kind of tiling domain as the tightest local cell, then such a solution for Problem 8 in $\mathbf{E}^n$ automatically also provides the solution of Problem 9 in $\mathbf{E}^n$ in an obvious way. Otherwise, on the one hand, Problem 9 in $\mathbf{E}^n$ becomes much more difficult than Problem 8, and on the other hand, it becomes necessary to introduce other kinds of local (resp. global) invariant, such as in the case of $n = 3$, in order to settle the sphere packing problem in $\mathbf{E}^n$ satisfactorily.

Let us conclude our brief discussion on this topic by the following two conjectures for the cases of $n = 4$ and $n = 8$.

Conjecture 1 *The optimal local density of sphere packings in* $\mathbf{E}^4$ *is equal to*

$$\frac{\pi^2}{16} = 0.61685\cdots$$

and $\rho(\mathcal{L}) = \frac{\pi^2}{16}$ *when and only when the local packing* $\mathcal{L}(S)$ *is the same as that of the* D_4*-lattice packing.*

Conjecture 2 *The optimal local density of sphere packings in* $\mathbf{E}^8$ *is equal to*

$$\frac{\pi^4}{384} = 0.25366\cdots$$

and $\rho(\mathcal{L}) = \frac{\pi^4}{384}$ *when and only when the local packing* $\mathcal{L}(S)$ *is the same as that of the* E_8*-lattice packing.*

References

[Bar] Barlow, *Nature*, 1883.

[B-H] J. D. Berman and K. Hanes, *Volumes of polyhedra inscribed in the unit sphere in E^3*, Math. Ann. **188** (1970) 78–84.

[B-S] E. Bannai and N. J. A. Sloane, *Uniqueness of certain spherical codes*, Canad. J. Math. **33** (1981) 437–449.

[BöH] J. Böhm and E. Hertel, *Polyedergeometrie in n-Dimensionalen Räumen konstanter Krümmung*, Birkhäuser Verlag, Basel, Boston, Stuttgart, 1981.

[BHW] U. Betke, M. Henk and J. M. Wills, *Finite and infinite packings*, J. reine angew. Math. **453** (1994) 165–191.

[BKM] A. Bezdek, W. Kuperberg and E. Makai Jr., *Maximum density space packing with parallel strings of balls*, Discrete & Comput. Geom. **6** (1991) 277–283.

[Bli] H. F. Blichfeldt, *The minimum value of quadratic forms and the closest packing of spheres*, Math. Ann. **101** (1929) 605–608.

[CFG] H. T. Croft, H. J. Falconer and R. K. Guy, *Unsolved Problems in Geometry*, Springer-Verlag, New York, 1997.

[C-S] J. H. Conway and N. J. A. Sloane, *Sphere Packings, Lattices and Groups*, Springer-Verlag, 1988.

[Cox-1] H. S. M. Coxeter, *The function of Schläfli and Lobatschefsky*, Quart. J. Math. Oxford **6** (1935) 13–29.

[Cox-2] H. S. M. Coxeter, *On Schläfli's generalization of Napiers Pentagramma Mirificum*, Bull. Calcutta Math. Soc. **28** (1936) 125–144.

[Fej-1] L. Fejes Tóth, *Über die dichteste Kreislagerung und dünnste Kreisüberdeckung*, Comment. Math. Helv. **23** (1949) 342–349.

[Fej-2] L. Fejes Tóth, *Lagerungen in der Ebene, auf der Kugel und im Raum*, Springer-Verlag, Berlin, 1971.

[Fej2-K] G. Fejes Tóth and W. Kuperberg, *Packing and covering with convex sets*, Handbook of Convex Geometry (eds P. M. Gruber and J. M. Wills), North-Holland, Amsterdam (1993) 799-860.

[Gau] C. F. Gauss, *Besprechung des Buchs von L. A. Seeber: Untersuchungen über die Eigenschaften der positiven ternären quadratischen Formen USW*, Gottingsche Gelehrte Anzeigen (1831) = Werke, II (1876) 188–196.

[Gol] M. Goldberg, *The isoperimetric problem for polyhedra*, Tôhoku Math. J. **40** (1935) 226–236.

[Hla] E. Hlawka, *Ausfullung und Überdeckung konvexer Körper durch konvexe Körper*, Monatsh. Math. **53** (1949) 81–131.

[Hil] D. Hilbert, *Mathematische probleme*, Archiv. Math. Phys **1** (1901) 44–63. English translation in Bull. AMS **8** (1902) 437–479.

[Hsi-1] W. Y. Hsiang, *On infinitesimal symmetrization and volume formula for spherical or hyperbolic tetrahedrons*, Quart. J. Math. Oxford **39** (2) (1988) 463–468.

[Hsi-2] W. Y. Hsiang, *On the sphere packing problem and the proof of Kepler's conjecture*, International J. of Math. **4** (1993) 739–831.

[Hsi-3] W. Y. Hsiang, *Addendum to "On the sphere packing problem and the proof of Kepler's conjecture"*, Preprint, U. C. Berkeley PAM-658 (1995).

[Hsi-4] W. Y. Hsiang, *On the kissing number of sphere packings in E^4 and a strong uniqueness theorem*, Preprint, Department of Mathematics, HKUST, Hong Kong.

[Kep] J. Kepler, *On the Six-Cornered Snowflakes*, Oxford Univ. Press, 1966. (English translation of latin booklet of 1611)

[Lee] J. Leech, *The problem of the thirteen spheres*, Math. Gazette **40** (1956) 22–23.

[Lin1] L. Lindelöf, *Propiétés générales des polyèdres*, St. Petersburg Bull. Ac. Sc. XIV (1869), 258–269; *Recherches sur les polyèdres maxima*, Helsingfors Acta Soc. Sc. Fenn. XXIV (8) (1899).

[Lin2] J. H. Lindsey, *Sphere packings in R^3*, Mathematika **33** (1986) 137–147.

[Mar] J. Martinet, *Les Réseaux Parfaits des Espaces Euclidiens*, Masson, Paris, 1996.

[Mil] J. Milnor, *Hilbert's Problem 18: On crystallographic groups, fundamental domains and on sphere packings*, Proc. of Symp. in Pure Math. AMS **28** (1976) 491–506.

[Min] H. Minkowski, *Allgemeine Lehrsätze über die konvexen Polyeder*, Nachr. Ges. Wiss. Göttingen, Math.-Phys. KL. (1897) 198–219.

[Mud] D. J. Muder, *A new bound on the local density of sphere packings*, Discrete & Comput. Geom. **10** (1993) 351–375.

[O-S] A. M. Odlyzko and N. J. A. Sloane, *New bounds on the unit spheres that can touch a unit sphere in n-dimensions*, J. Combinat. Theory A **26** (1979) 210–214.

[Ran] R. A. Rankin, *On the closest packing of spheres in n dimensions*, Ann. Math. **48** (1947) 1062–1081.

[Rog-1] C. A. Rogers, *The packing of equal spheres*, Proc. LMS **8** (1958) 609–620.

[Rog-2] C. A. Rogers, *Packing and Covering*, Cambridge Univ. Press, Cambridge, 1964.

[See] L. A. Seeber, *Untersuchung über die Eigenschaften der Positiven Ternären Quadratischen Farmen*, Freiberg, 1831.

[Sch] L. Schlafli, *Gesammelte Math. Abh 1 and 2*, Birkhäuser, Basel, 1953.

[SW-1] K. Schütte and B. L. van der Waerden, *Auf welcher Kugel haben 5, 6, 7, 8 oder 9 Punkte mit Mindesabstand Eins Platz?*, Math. Ann. **123** (1951) 96–124.

[SW-2] K. Schütte and B. L. van der Waerden, *Das Problem der dreizehn Kugeln*, Math. Ann. **125** (1953) 325–334.

[Tal] G. Talenti, *The standard isoperimetric theorem*, Handbook of Convex Geometry (eds P. M. Gruber and J. M. Wills), (1993) 73–123, North-Holland, Amsterdam.

[Thu] A. Thue, *Über die dichteste zuzammenstellung von kongruenten Kreisen in der Ebene*, Norske vid Selsk. Skr. **1** (1910) 1–9.

[Wat-1] G. L. Watson, *On the minimum of a positive quadratic form in $n(\leq 8)$ variables, verification of Blichfeldt's calculations*, Proc. Cambridge Phil. Soc. **62** (1966) 719.

[Wat-2] G. L. Watson, *The number of minimum points of positive quadratic forms*, Dissertationes Math. **84** (1971) 1–43.

[Zon] C. Zong, *Sphere Packings*, Springer-Verlag, New York, 1999.

Index